INTRODUCTION TO
PROBABILITY

INTRODUCTION TO
PROBABILITY

HAROLD J. LARSON
Naval Postgraduate School

ADDISON-WESLEY
ADVANCED SERIES
IN STATISTICS

ADDISON-WESLEY PUBLISHING COMPANY
Reading, Massachusetts • Menlo Park, California • New York
Don Mills, Ontario • Wokingham, England • Amsterdam • Bonn
Sydney • Singapore • Toyko • Madrid • San Juan • Milan • Paris

Sponsoring Editor: Julia Berrisford
Production Supervisor: Mona Zeftel
Text and Cover Designer: Darci Mehall
Copy Editor: Susan Middleton
Illustrator: Scientific Illustrators
Compositor: ICPC
Senior Manufacturing Manager: Roy E. Logan

Library of Congress Cataloging-in-Publication Data
Larson, Harold J., 1934–
    Introduction to probability / Harold J. Larson.
    p.   cm.
    Includes index.
    ISBN 0-201-51286-6
    1. Probabilities.   I. Title.
QA273.L35197   1994
519.2—dc20   93-48344
                                            CIP

Copyright © 1995 by Addison-Wesley Publishing Company, Inc. All rights reserved. No part of this publication may be reproduced, stored in a retrieval system, or transmitted, in any form or by any means, electronic, mechanical, photocopying, recording, or otherwise, without the prior written permission of the publisher. Printed in the United States of America.

1 2 3 4 5 6 7 8 9 10 MA 9897969594

PREFACE

Randomness perhaps represents one of the strangest realms whose depths are successfully explored by mathematics; from games of chance to quantum mechanics, probability theory has proved useful (and necessary) in providing satisfactory descriptions of observable phenomena. This text presents a one-semester introduction to probability theory, requiring a modest background in calculus. Examples and illustrations are presented from both the physical and social sciences, in an attempt to indicate the possible breadth of application of this subject.

The first chapter presents the basic ideas of probability, showing the interrelationships between probabilities of unions and intersections of events. A relative frequency interpretation for probability is stressed as an aid to understanding. Conditional probability is introduced here, as is the important concept of independent events and experiments consisting of independent trials. Brief introductions to both set theoretic notation and elementary counting techniques are also included.

Random variables are introduced in Chapter 2, for both the continuous and discrete cases; the cumulative distribution function is stressed as a unifying way of describing both types of random variables. Means, variances, and standard deviations are introduced as useful tools for summarizing the probability distribution of a random variable. The Chebychev inequality is derived and used to illustrate the "naturalness" of the standard deviation as a measure of variability of a distribution. Quantiles are defined for continuous distributions, and their descriptive use is also discussed. Both the discrete and continuous uniform distributions are used for illustrative purposes.

Chapter 3 derives the basic discrete distributions stemming from sequences of independent Bernoulli trials, with careful descriptions of how they occur; many of their attributes are derived, as well as relationships between them. The Poisson process is discussed, and the Poisson probability distribution is derived as the limiting case of the binomial; the Poisson approximation to the binomial is presented and illustrated. While perhaps redundant in today's age of cheap and readily available computing power, tables of both the binomial and Poisson distribution functions are given in an appendix. Also given there is a discussion of finite and infinite geometric series, and related results, for those who may appreciate a review of these topics. Sampling without replacement is discussed in the final section, introducing the hypergeometric distribution, and its similarity to the binomial.

Many standard continuous distributions are presented in Chapter 4; the Poisson process, described in the previous chapter, is used to introduce the exponential and Erlang distributions; the generalization to the gamma is discussed. The normal distribution is presented and its use in approximating the binomial and Poisson distributions is illustrated; a table of the standard normal cumulative distribution is provided in the appendix. The beta probability distribution is discussed next; the probability laws for the order statistics for uniform random variables are shown to be beta distributions. The basics of the distributions for transformed random variables is given, for both the continuous and discrete cases; the probability integral transform (both continuous and discrete) is discussed. Several additional continuous distributions (lognormal, Weibull, Pareto) are introduced here through appropriate transformations.

Joint probability distributions are introduced in Chapter 5, with simple illustrations of both the discrete and continuous cases; most discussion and derivation is presented explicitly for two random variables, but the obvious extensions to an arbitrary number of random variables are also noted. Marginal distributions are discussed, as is the important concept of conditional distributions and the simplifications to occur with independence. The multinomial distribution is discussed, as is the bivariate normal; the multihypergeometric is introduced through examples and exercises.

Expectation for the multivariate case is discussed in Chapter 6; means and variances of linear functions are derived, with the latter leading to the natural introduction of the covariance of two random variables, as well as their correlation. The moment generating function is introduced here, as are the cumulant, factorial, and probability generating functions, and the relationships between them. Conditional expectation, the final section of this chapter, is used to illustrate the computation of the covariance of two random variables. The mean and variance of the conditional expectation are derived; the usefulness of these concepts is illustrated with several examples.

The final chapter of the text discusses some distributions of two or more random variables, including the convolution formula; conditional expectation is employed in many of these discussions. Several familiar sampling distributions, including the χ^2, Student's t, and Snedecor's F distributions are derived here. Limiting distributions are discussed, and the weak law of large numbers and the Central Limit Theorem are derived. The use of the Central Limit Theorem in approximating other distributions is discussed; it is also used to find Stirling's formula for $n!$. A brief discussion of the Cornish–Fisher approximations for distributions is given.

The answers to all even-numbered exercises, except those of the "show that" type, are presented in the appendix. A solutions manual for all the exercises is available from Addison-Wesley.

The author would like to thank Michael Payne, Mona Zeftel, and Julia Berrisford, of Addison-Wesley, for their encouragement and help in seeing this project to its completion, and to thank the many reviewers of the manuscript for their comments and corrections. Donald Knuth's TEX program and Michael Wichura's PICTEX macros were used to create the manuscript; these provide very impressive tools for such undertakings.

Pacific Grove, California   Harold J. Larson

CONTENTS

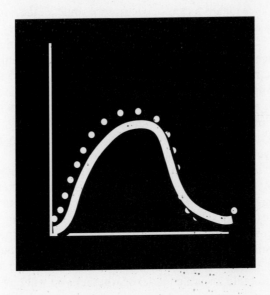

CHAPTER 1

PROBABILITY

We have all, at one time or another, had experiences with *chance*, which, as a noun, modern dictionaries define as (apparent lack of cause or design.) If 3 million people purchase tickets for a lottery, the winner or winners are determined by chance. During an influenza outbreak, it is commonly assumed that the particular individuals who contract the flu are determined by chance. When a tornado scribes a path through a populated area, completely demolishing one house and not touching the one next door, chance is assumed to operate in making the choices. If a number of seeds are planted in a field, some of which germinate and some of which do not, again chance has played a role. Many successful persons in business frequently list chance as a major contributor to their prominence.

This lack of cause or design embodied in the word *chance* is, in fact, so pervasive in human experience that several other common English words are used for the same concept. *Random* is defined as (being without aim or purpose;) *luck* is the seemingly (chance happenings) (either good or bad) that affect a person; and somewhat more technically, one definition of the adjective *stochastic* is "of, pertaining to, or arising from chance." The verb *happen* means to occur or befall by chance or without plan, while the noun *happenstance* means a chance occurrence. Indeed, the word *chance* itself can be used as a noun, a verb (either transitive or intransitive), an adjective, or an adverb.

Chance or randomness has been studied for many centuries, resulting in a rich literature concerning this rather vague but important and influential concept. Probability, the subject of this text, provides a way to measure chance, employing numerical values to indicate the likelihood of occurrence of chance or random events. As will be seen, this measure is derived from an assumed model of the situation where chance is operating. The probability of an event occurring then is meant to reflect the relative frequency of occurrence of the event.

To illustrate this discussion, suppose you daily walk from point *A* to point *B*, along a route that takes you through an intersection containing a stoplight. If the stoplight is green in your direction of travel, you can (with some degree of safety) cross the intersection without stopping for passing automobile traffic; if the stoplight is red in your direction of travel, you should wait for it to turn green before proceeding through the intersection. Which of these two situations will occur, on a given day, is subject to chance. We can build a simple model of whether you will have to stop at the intersection on any given day by considering the possibilities for each day (each time you take this walk). Only two different *outcomes* are possible: Either you find the stoplight is green when you arrive, or you find it is red. If, in your experience, you must wait at the intersection (i.e., the light is red) 70% of the time (and thus you don't have to wait 30% of the time), the model would set the probability that you must wait at the intersection equal to .7 and the probability that you don't have to wait equal to .3. These two numbers, called *probabilities of the respective events* (*wait* and *don't wait*), thus represent the relative frequencies of the two different results.

In a formal sense, this model consists of a list of possible outcomes for the situation (whether you encounter a red or a green stoplight) and an assignment of numbers to each outcome (called *their probabilities of occurrence*). The word *experiment* will be used as the generic name for the situation modeled; for a given day, then, this experiment consists of observing whether the stoplight is red or green when you arrive. The list of possible outcomes for the experiment will be formally treated as a set; the set of outcomes for this experiment is {red, green}, a set with two elements. The number .7 is taken as the probability of finding the stoplight red, which will be written $P(\{red\}) = .7$; similarly, $P(\{green\}) = .3$. The probability model for this experiment consists of the set of outcomes and the specification of their probabilities of occurrence.

As you might expect, the earliest known historical uses of probability were concerned with games and gambling, a natural setting where chance is explicitly recognized. Most of the early games that stimulated interest in measuring chance are no longer popular, having been replaced over time with other games. One of these early games proceeds as follows: You as the player pay $1; you then are allowed to roll a fair pair of dice at most 24 times. If you roll a double-6 (in which both dice show a 6 on their uppermost faces) within these rolls, you win $1 (meaning you get back what you paid plus one more dollar); if you do not get a double-6 within these 24 rolls, you lose your dollar. The question of interest then is whether this is a game you should choose to play. There are, of course, many different facets to this kind of choice; probability and expected values provide rational ways of approaching questions of this type.

While probability seems somewhat natural to use in gaming situations, its more important applications lie elsewhere. Many of the tools and results to be discussed are now routinely employed in making decisions in the face of uncertainty, in many different applied areas. Some simple examples follow. A commercial bakery sells freshly baked bread each day; since it is not known how many loaves of bread will be demanded by customers on a given day, how many loaves should be baked? A large chain of franchise stores is interested in opening a new store in a given geographic area; which particular site in the area would seem the best location? A doctor hears the symptoms expressed by a patient; what medication (if any) would seem appropriate to prescribe? A meteorological station has gathered weather-related data from nearby locations over the previous 48 hours; what should the station forecast as the weather to come in its area? These provide a small sample of specific applications in which probability theory has proved useful.

The theory of probability is concerned with modeling "random" phenomena. The word *experiment* is used for any operation whose outcome is not certain; what may in fact be observed defines a collection, or set, of possibilities. The probability for each possible outcome to occur then should represent the proportion of the time it would be observed. The probability model for a random phenomenon consists of an assumed specification of all the possible results that could occur, together with a rule or procedure for evaluating the probability of occurrence for every possible event. Many different probability models have been studied in the literature and have found practical uses in many different areas; some of the simpler, frequently occurring models are introduced in this text.

Some knowledge of set theory is useful in any study of probability; Section 1.1 gives a short synopsis of the pertinent concepts. This is followed by a discussion in Section 1.2 of the basic idea of an event, and of the special meaning of the phrase "occurrence of an event" for the definition of a probability measure. The simple axioms of Kolmogorov will be discussed, along with some of the immediately apparent consequences of these axioms.

Section 1.3 describes the special simplifications available to you if you are willing to assume that all the outcomes are equally likely, and Section 1.4 gives some rudiments of counting procedures to enumerate the numbers of elements belonging to certain sets. We shall then explore the important and interesting realm of conditional probability (Section 1.5), along with the fundamental concept of independent events and their role in common applications of probability theory (Section 1.6).

Most people have an intuitive feeling about probability and to a large extent they generally find that the formal structure pretty much bears out those intuitive feelings. It is also true that most people find some of the results that will be presented to be counterintuitive; that can at first be uncomfortable, but in the long run it should be an enjoyable, as well as challenging, experience.

1.1 | Set Theory Concepts

A *set* is a collection of objects called *elements*; the elements themselves can be essentially any type of entity, similar or not to any other element of the same set. In a sense, these elements share a bond in belonging to the same set. Sets will be denoted by uppercase letters A, B, C, ... while elements of sets will be denoted by lowercase letters $x, y, \ldots$. The fact that x belongs to the set A is written $x \in A$, while $y \notin A$ indicates that the element y does not belong to A.

Sets can be defined in two ways; one method we will call the *roster method*. A *roster* is a list of all the elements in a set. Simple sets with short rosters can easily be defined or displayed using printer's braces ({ }, also called set builders) with the roster listed between them; thus

$$A = \{1, 2, 4, 8, 16\}$$

means that the set A has five elements: the integers 1, 2, 4, 8, and 16. Even infinite sets can be indicated in this way; the specification

$$B = \{1, 2, 3, \ldots\}$$

means that B is the set of all positive integers. The three lower dots are read "and so on" and mean that the listing of the integers is to continue with no end. The statement $C = \{1, 2, 3, \ldots, n\}$, where n is a given positive integer, defines the set whose elements are the first n integers. The succeeding integers $n + 1, n + 2, \ldots$ are *not* elements of C.

It is also of use to define a set using the *rule method*; again we employ the set builders, this time including a generic symbol for an element occurring immediately after the opening brace. The symbol is followed by a colon (read "such that") and an expression of the rule that defines the elements. Thus the same set C defined in the previous paragraph can be specified by writing $C = \{x : x = 1, 2, \ldots, n\}$. The symbol used ($x$ in this case) to define the rule is a dummy variable; the symbol itself has no relevance to the definition of the set. The same set can be specified or defined by $C = \{y : y = 1, 2, \ldots, n\}$ or $C = \{c : c = 1, 2, \ldots, n\}$.

Equality of two sets is defined as follows.

DEFINITION 1.1

Two sets A and B are *equal* (written $A = B$) if they contain exactly the same elements. If some element belongs to only one of them, they are not equal; written $A \neq B$.

This definition implies that the order of listing of the elements is not important; thus

$$\{1,2,4\} = \{2,1,4\} = \{4,1,2\}.$$

Example 1.1

Definition 1.1 also allows the possibility that the same element might be listed more than one time in the roster. Thus, if $A = \{5,6,7,6,7\}$ and $B = \{7,6,5,5\}$, it would be true that $A = B$, since the two 6's and two 7's listed in A as well as the two 5's listed in B represent the same element. We have no use for such redundant lists, so each distinct element will be listed only once in the roster.

It could be that each element of A is also an element of B, but not necessarily vice versa. In such a case, A is called a subset of B.

DEFINITION 1.2

A is a *subset* of B, written $A \subset B$, if every element that belongs to A also belongs to B; that is, $A \subset B$ if $x \in A$ implies $x \in B$.

Example 1.2

Set equality can be defined in terms of subsets. $A = B$ if and only if $A \subset B$ and $B \subset A$.

It is necessary to recognize the difference between elements of a set and subsets of a set; if $C = \{a,b,c,d\}$, $D = \{b,d\}$ then it is true that $a \in C$, $a \notin D$, $D \subset C$ but $D \notin C$ while $d \in D$, and $d \not\subset D$. The set of all married persons is a subset of the set of all persons, but it is not an element of this set; an individual married person belongs to the set of all married persons, as well as the set of all persons, but it is not a subset of either.

The null set (also called the empty set), $\emptyset$, is the set with no elements; thus $\emptyset = \{\}$. It plays a role rather similar to 0 in the real number system. Since $\emptyset$ has no elements, it is a subset of any other set A; there is no $x \in \emptyset$ to make the statement $\emptyset \subset A$ false. It is important to draw the distinction between $\emptyset$ and $\{\emptyset\}$; the first set is empty, but the second is not since it has one element ($\emptyset$). If you think of the set builders ($\{\ \}$) as representing a paper bag and a set as representing a paper bag plus its contents, then $\emptyset$ would be just the paper bag while $\{\emptyset\}$ would be one paper bag inside another.

Venn diagrams prove very useful in picturing set relationships and operations with sets. Sets are represented by geometric shapes; shading is employed to illustrate relations and operations. If two equal sets were to be pictured, they would completely cover each other and be identical in all details; the set A would be a subset of B if and only if the figure representing A were completely contained in the figure representing B, as illustrated in Fig. 1.1.

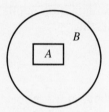

Figure 1.1 $A \subset B$

The empty set $\emptyset$ cannot be pictured in a Venn diagram as any nonempty region without running into logical difficulties.

Unions and intersections of sets are useful for many purposes; these are defined next.

DEFINITION 1.3 ────────────────────────

The *union* of two sets, written $A \cup B$, is the set of all elements that belong to A, or B, or both. The *intersection* of two sets, written $A \cap B$, is the set of all elements that belong to both A and B.

If $A = \{1, 2, \ldots, 10\}$ and $B = \{7, 8, \ldots, 16\}$, then $A \cap B = \{7, 8, 9, 10\}$ and $A \cup B = \{1, 2, \ldots, 16\}$. Figures 1.2 and 1.3 present Venn diagrams of two sets A and B; the shading in Fig. 1.2 indicates their intersection, while in Fig. 1.3 it shows their union. It is always true that $A \cap B \subset A \subset A \cup B$ and $A = A \cap A = A \cup A$, while $\emptyset \cap A = \emptyset$, and $\emptyset \cup A = A$. Both unions and intersections are commutative: $A \cap B \equiv B \cap A$ and $A \cup B \equiv B \cup A$. If $A \cap B = \emptyset$, so A and B have no elements in common, then A and B are *disjoint sets*.

Example 1.3 ─────────────

Suppose $A = \{(x, y) : x + y = 2\}$ and $B = \{(x, y) : y - x = 4\}$. Then A and B are each sets of points on straight lines, in the Cartesian plane; $A \cup B$ is the collection of all points on the two lines, and $A \cap B = \{(-1, 3)\}$, the single point that lies on both lines, the solution of the simultaneous equations

$$x + y = 2,$$
$$y - x = 4.$$

The union of three or more sets is the set of elements that belong to any one or more of them, while their intersection is the set of elements that

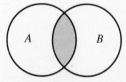

Figure 1.2 $A \cap B$

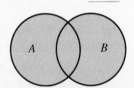

Figure 1.3 $A \cup B$

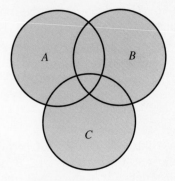

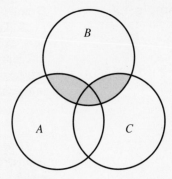

Figure 1.4 $A \cup B \cup C$

Figure 1.5 $(A \cap B) \cup (B \cap C)$

belong to all of them. $A \cup B \cup C$ is shaded in Fig. 1.4 while $(A \cap B) \cup (B \cap C)$ is shaded in Fig. 1.5. This latter set is identical to $(A \cup C) \cap B$; that is, $(A \cap B) \cup (B \cap C) = (A \cup C) \cap B$, illustrating one of two distributive laws linking unions and intersections. Use a similar Venn diagram to convince yourself that $B \cup (A \cap C) = (A \cup B) \cap (B \cup C)$ as well.

In applications of set theory to probability problems, the sample space S (which we will formally define in Section 1.2) plays the role of the universal set; all sets employed in a given problem are subsets of S. The complement of a set $A \subset S$ is the set of elements that do not belong to A (but do belong to S), as in the following definition.

DEFINITION 1.4

Given any set $A \subset S$, the *complement* of A is

$$\overline{A} = \{x : x \notin A, x \in S\}.$$

Thus $\overline{A}$ contains all the elements in S that are not in A.

Figure 1.6 pictures $\overline{A}$, S, and A. It follows then that $A \cap \overline{A} = \emptyset$ and $A \cup \overline{A} = S$. If one takes a complement twice, the original set returns; that is, $\overline{\overline{A}} = A$.

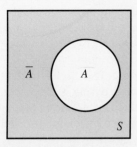

Figure 1.6 $\overline{A}$, complement of A

Example 1.4

Granted $S = \{1, 2, 3, \ldots\}$, the set of positive integers, define $A_i = \{i\}$ for $i \in S$. Then A_i and A_j are disjoint for $i \neq j$; $\overline{A_i}$ is the set of all positive integers, excluding i; $\overline{A_i} \cup \overline{A_j} = S$ for all $i \neq j$; and $\overline{A_i} \cap \overline{A_j}$ is the set of positive integers excluding i and j. $E_j = A_2 \cup A_4 \cup A_6 \cup \cdots \cup A_{2j}$ is the set of even integers no larger than $2j$, where $j = 1, 2, 3 \ldots$, and E_∞ is the set of even integers.

DeMorgan's laws illustrate equality of certain sets employing complements. Consider the set $\overline{A \cup B}$. If $x \in \overline{A \cup B}$, then necessarily $x \notin A \cup B$, which in turn implies that $x \notin A$ and $x \notin B$, that is, $x \in \overline{A}$ and $x \in \overline{B}$; thus $x \in \overline{A} \cap \overline{B}$. This shows that $\overline{A \cup B} \subset \overline{A} \cap \overline{B}$. This argument also can be reversed to show that $\overline{A} \cap \overline{B} \subset \overline{A \cup B}$, which establishes the following theorem.

THEOREM 1.1 If $A \subset S$ and $B \subset S$, then

$$\overline{A \cup B} = \overline{A} \cap \overline{B}.$$

Example 1.5

What is wrong with the following reasoning? Let A be the set of married males while B is the set of married females. Then $\overline{A \cup B} = \overline{A} \cap \overline{B}$ is the intersection of the set of unmarried males with the set of unmarried females. This latter intersection is, of course, empty; thus all males and females are married.

A second law of DeMorgan (really a restatement of Theorem 1.1) is gotten by replacing A and B with $\overline{A}$ and $\overline{B}$, and then taking complements of the resulting equation. This gives

$$\overline{A \cap B} = \overline{A} \cup \overline{B}.$$

As mentioned earlier, the order of listing of the elements in the roster of a set is immaterial. In many probability applications it is important to be able to let order count, that is, to recognize not only the objects themselves, but also the order in which they occur. The standard way of listing coordinates in a Cartesian plane makes this distinction; that is, in the (x, y)-plane, the pair $(1, 2)$ refers to the point with $x = 1, y = 2$, while the pair $(2, 1)$ refers to the point with $x = 2, y = 1$. Since these are two different points in the plane, it is useful to distinguish between them by saying $(1, 2) \neq (2, 1)$, thus the ordering is of importance. Exactly the same distinction occurs in the use of vector notation. We shall call an ordered collection of n objects an n-tuple. (When the elements are numeric entries, the n-tuple is also an n-dimensional vector.)

DEFINITION 1.5 ───────────────────────────

An ordered collection of n elements, written

$$(x_1, x_2, \ldots, x_n),$$

is called an *n-tuple*. Two *n*-tuples are equal if and only if they contain the same elements, written in the same order.

n-tuples occur as the elements of Cartesian products of sets, defined next.

DEFINITION 1.6 ───────────────────────────

The Cartesian product of n sets $A_1, A_2, \ldots, A_n$ is the set of *n*-tuples

$$A_1 \times A_2 \times \cdots \times A_n = \{(a_1, a_2, \ldots, a_n) : a_i \in A_i, i = 1, 2, \ldots, n\}.$$

Thus a Cartesian product of two sets has 2-tuples as elements, the Cartesian product of three sets has 3-tuples as elements, and so forth. If A_1 has m elements and A_2 has n elements, then $A_1 \times A_2$ has mn 2-tuples as elements. Note that this set operation involves a change in dimensionality; if the elements of A_1 mark positions on an x-axis (a one-dimensional space), and the elements of A_2 mark positions on a y-axis (another one-dimensional space), then the elements of $A_1 \times A_2$ give the locations of points in a two-dimensional (x, y)-space.

Example 1.6

Take $A = \{1, 2, 3\}$ and $B = \{4, 5\}$. Then

$$A \times B = \{(1, 4), (1, 5), (2, 4), (2, 5), (3, 4), (3, 5)\}.$$

If we wanted to represent things graphically, we could picture A as locating three points on the x-axis and B as locating two points on the y-axis. The Cartesian product $A \times B$ then consists of the six points in the (x, y)-plane whose coordinates are listed above, while the Cartesian product

$$B \times A = \{(4, 1), (5, 1), (4, 2), (5, 2), (4, 3), (5, 3)\}$$

is a different collection of six points. Thus $A \times B \neq B \times A$. Both of these Cartesian products are pictured in Fig. 1.7. In similar manner, the elements of the Cartesian product of three sets can be thought of as points in a three-dimensional space. For $n \geq 4$, the n-tuples represent points in higher dimensional spaces, which one can not visualize in the same sense.

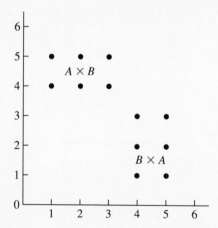

Figure 1.7 Cartesian products $A \times B$ and $B \times A$

EXERCISES 1.1

1. Suppose $A = \{1,2,3,4,5\}$, and $B = \{3,1,2,4\}$. Are the following statements true or false?
   a. $B \subset A$ b. $A = B$
2. If $A \subset B \subset C$ and $C \subset B \subset A$, does it follow that $A = C$?
3. Give an example to show that $E \subset F$ and $D \subset F$ does not imply that $E \subset D$ or $D \subset E$.
4. Let $F = \{y : 0 < y < 1\}$, $G = \{\frac{1}{4}, \frac{1}{2}, \frac{3}{4}\}$, $H = \{x : 0 \le x \le 1\}$, and indicate whether each of the following is true or false.
   a. $G \subset F$ b. $G \subset H$ c. $F = H$ d. $\frac{1}{2} \in F$
   e. $F \subset H$ f. $0 \in F$ g. $1 \in H$
5. Define $A = \{(x,y) : x + y = 2, x - y = 1\}$, $B = \{(x,y) : x = \frac{3}{2}, y = \frac{1}{2}\}$.
   a. Is it true that $A \subset B$?   b. How many elements does A have?
6. Define $B = \{x : 0 \le x \le 2\}$ and $C = \{y : 1 \le y \le 3\}$.
   a. What is $B \cup C$?   b. What is $B \cap C$?
7. Sketch a picture of the region of points in the (x,y)-plane defined by $A = \{(x,y) : 0 < x < 1, 0 < y < x\}$. What is the relation between A and $B = \{(x,y) : y < x < 1, 0 < y < 1\}$?
8. Is the set of all persons living in the United States at a given time equal to the set of all U.S. citizens at the same time?
9. Intersections and complements can be used to decompose a union of two sets into disjoint sets. Draw a Venn diagram of $A \cup B$, and show that it equals $(A \cap \overline{B}) \cup (A \cap B) \cup (\overline{A} \cap B)$.
10. Draw Venn diagrams to convince yourself of the truth of the two distributive laws

$$A \cup (B \cap C) = (A \cup B) \cap (A \cup C), \qquad A \cap (B \cup C) = (A \cap B) \cup (A \cap C).$$

11. Prepare simple Venn diagrams that illustrate DeMorgan's laws

$$\overline{A \cup B} = \overline{A} \cap \overline{B}, \qquad \overline{A \cap B} = \overline{A} \cup \overline{B}.$$

12. Show that if $A \subset B$ then $\overline{B} \subset \overline{A}$.

13. Mark each of the following statements as true (if always true), or false.
    a. $\overline{A} \cup B = A \cap \overline{B}$ b. $B \cap C \subset (A \cup B) \cap (A \cup C)$
    c. $A \subset (A \cap D) \cup (A \cup B)$
14. Draw a Venn diagram including a sample space S and two sets A and B; identify $A \cap \overline{B}$, $\overline{A} \cap B$, and $\overline{A} \cap \overline{B}$.
15. What must be true if $A \times B = B \times A$?

1.2 | Experiments, Sample Spaces, and Probability

Probability theory was developed to model chance, random phenomena. We shall use the word *experiment* as the generic name for any operation that may result in any of several different possible outcomes. Games fall into this category: Which face will be uppermost when a die stops rolling, or which card will be drawn from a deck? Sporting events do as well: Which of the horses in a race will cross the finish line first (or second or third)? Which of several teams in a tournament will be the winner? Indeed, we intend the inclusion of *any* sort of situation whose outcome is not known in advance: What will be the rate of inflation in the United States next year? Which of several possible crops would produce the best yield, on average, if planted in a given area? What will your grade-point average be, overall, at the end of the current term? Each of these is an example of an experiment.

Granted that an experiment has many different possible outcomes, any of which could be observed, we will not know the result to occur in advance; indeed if the experiment were repeated two or more times, we might anticipate getting different results on the various repetitions. Probability theory is used to build a model of the resulting possibilities; to the extent that the model correctly represents reality, the model can be used to investigate many questions of typical interest. Such a probability model consists of listing all the possible outcomes that could be observed, together with a rule that allows us to compute the probability that any possible outcome might occur. This list of all possible outcomes is called the sample space, a concept we first encountered in Section 1.1 and which we now define formally.

DEFINITION 1.7

The *sample space* S for an experiment is the set of all possible outcomes that could be observed.

It is tacitly assumed that *all* possible outcomes for the experiment have been listed in the sample space, that no possibilities have been overlooked. In any realistic, complex situation this step in itself may be far from trivial.

Generally, sample spaces for experiments are not unique; different persons might very well adopt different sets for S, depending possibly on which facets of the experiment are deemed of interest. If one sample space appears easier to use than another for a given purpose, then the easier one would naturally be appealing.

Example 1.7

A coin is flipped three times. One possible sample space for this experiment is the set that has as elements the integers $0, 1, 2,$ and 3; the elements here represent the number of heads that will occur when the experiment (flipping the coin three times) is performed. Thus

$$S_1 = \{0, 1, 2, 3\}$$

is a candidate sample space for this experiment. A more complex-looking sample space, which has certain advantages when solving probability problems, is given by

$$S_2 = T \times T \times T.$$

Here $T = \{h, t\}$, where h indicates the occurrence of a head and t indicates the occurrence of a tail. This sample space has eight elements:

$$S_2 = \{(h, h, h), (h, h, t), (h, t, h), (h, t, t), (t, h, h), (t, h, t), (t, t, h), (t, t, t)\}.$$

The positions in the 3-tuple represent the coin flips. In many senses, S_2 captures all the data to be generated by the experiment, including the possible orders in which the three faces can occur. In contrast, S_1, with integers as elements, does not capture all this information. Generally, if several pieces of information are to be observed (three in this case), there are advantages to adopting a sample space that reflects all these individual pieces.

The word *event* is defined to mean a happening or occurrence in common English usage. This same word is used in probability theory, with essentially the same meaning. We have discussed the idea of the sample space listing all possible outcomes for an experiment; when the experiment is in fact performed, we assume that exactly one of the elements of S (the sample space) will be observed. The individual element observed, though, belongs to many different events. It is a common misinterpretation to equate the concept *event* with the phrase "outcome of the experiment" (meaning the element of S that occurs). In probability theory the word *event* represents anything that may be observed when the experiment is performed.

Consider the example of flipping a coin three times, just discussed in Example 1.7. The event {two heads occur} may be observed when this experiment is performed (or, of course, it may not be observed). This event can be expressed as a subset of either of the two sample spaces mentioned, S_1 or S_2. If we employ S_1, then, when the experiment is performed, two heads would occur if and only if the outcome observed is 2. The set $A_1 = \{2\}$ is a subset of S_1, that is, event A_1 is a subset of the sample space, a subset that contains only one element of S_1. If we use S_2, then this same event, {two heads occur}, would again be a subset of the sample space. This subset has three elements,

$$A_2 = \{(h,h,t),(h,t,h),(t,h,h)\},$$

since we would observe (exactly) two heads if *any* one of these outcomes were to happen when the experiment is performed. Any event that depends on the order in which specific faces occur defines a subset of S_2 but not of S_1. For example, the result (event) "two heads followed by a tail" (the set $\{(h,h,t)\}$) and which is a subset of S_2 but not of S_1.

There are two important things to notice here:

1. S lists all the individual outcomes that could occur; when the experiment is performed, exactly one element of S will occur.
2. Events are subsets of S, and may contain more than one element of S; each $x \in S$ is an element of many different events (subsets).

Events are subsets of the sample space employed; the choice of the sample space S determines the list of possible events.

DEFINITION 1.8

Events are subsets of the sample space S; an event is said to occur if any one of its elements is the outcome observed when the experiment is performed. If A and B are events such that $A \cap B = \emptyset$, A and B are called *mutually exclusive events*.

Since $S \subset S$, the sample space S itself is an event, called the *certain event*, since it will always occur.

Example 1.8

A coin is flipped three times; we shall use the sample space S_2 listed in Example 1.7. Let A_i be the event in which i heads occur when the coins are flipped, for $i = 0, 1, 2, 3$. Then

$$A_0 = \{(t,t,t)\},$$
$$A_1 = \{(h,t,t),(t,h,t),(t,t,h)\},$$
$$A_2 = \{(h,h,t),(t,h,h),(h,t,h)\},$$
$$A_3 = \{(h,h,h)\}.$$

Any two of these events, A_i and A_j with $i \neq j$, are mutually exclusive.

Example 1.9

A supermarket milk case contains 40 identical-looking cartons of milk, of which 10 are spoiled. A customer selects two of these cartons to purchase. Suppose we wanted to model the number of milk cartons she selects that are not spoiled. Since she will select two cartons, there are two pieces of information to be observed, so a sample space with 2-tuples as elements would seem useful. One possible sample space would be $S_1 = T \times T$, where $T = \{g,b\}$; g represents the selection of an unspoiled carton, while b represents the selection of a spoiled carton. The order in the 2-tuple represents the order of selection.

A second, more basic sample space of 2-tuples is given by assuming the cartons have been numbered from 1 to 40, with those bearing numbers 1 through 10 identifying the spoiled cartons. Then

$$S_2 = \{(x_1, x_2) : x_i = 1, 2, \ldots, 40, \ x_1 \neq x_2\}$$

could also be used as the sample space. Note that this collection of 2-tuples is not the Cartesian product of two sets, since it does not contain any of the 2-tuples with equal coordinates; this condition is necessary since she would not purchase the same carton twice.

Let A be the event that she gets (exactly) one spoiled carton; using S_1 as a sample space, then,

$$A = \{(g, b), (b, g)\}.$$

With S_2 as the sample space,

$$A = \{(x_1, x_2) : x_i = 1, 2, \ldots, 10, \ only \text{ for } i = 1 \text{ or } i = 2 \text{ (but not both)}\}.$$

The probability model for an experiment consists of a sample space S for the experiment, together with a rule that specifies the probability of occurrence for every event $A \subset S$. The *probability* of event A occurring is denoted $P(A)$. Axiomatically, relative frequencies possess three simple properties called the Kolmogorov axioms.

KOLMOGOROV AXIOMS

1. For any event $A \subset S$, $P(A) \geq 0$
2. $P(S) = 1$.
3. If $A_1, A_2, \ldots$ are mutually exclusive events, then

$$P(A_1 \cup A_2 \cup \cdots) = P(A_1) + P(A_2) + \cdots .$$

The first two of these are quite apparent; the relative frequency of any event cannot be negative, so we will not allow negative values for any probability. The sample space S lists every possible outcome that could occur; since S is the certain event, one of its elements will always be observed when the experiment is performed, so its relative frequency (probability) must equal 1.

The third axiom may require a little discussion. In the milk carton purchase described in Example 1.9, let A_1 be the event "no spoiled cartons are selected," while A_2 is the event "exactly one spoiled carton is selected"; then A_1 and A_2 are mutually exclusive and cannot happen together. Axiom 3 insists that the relative frequency of A_1 or A_2 occurring is the sum of the relative frequencies of the two individual events. Thus the proportion of the time that either zero or one spoiled carton is selected must be the sum of the proportion for zero and the proportion for one.

A number of consequences stem immediately from these simple axioms. Recall that $\emptyset$, the empty set, is a subset of any set; in particular then, $\emptyset \subset S$, so $\emptyset$ is an event. What must the probability of this event be? Its probability of occurrence is necessarily 0, for any case: $S \cup \emptyset = S$, so $P(S \cup \emptyset) = P(S) = 1$; $S \cap \emptyset = \emptyset$, so from axiom 3, $P(S \cup \emptyset) = P(S) + P(\emptyset) = 1 + P(\emptyset)$. That is, $1 = 1 + P(\emptyset)$, which gives $P(\emptyset) = 0$.

THEOREM 1.2 $P(\emptyset) = 0$ for any experiment.

If A is any event, and $\overline{A}$ is its complement, then $A \cup \overline{A} = S$ and $A \cap \overline{A} = \emptyset$. Thus $P(A \cup \overline{A}) = P(S) = 1$ (from axiom 2), and $P(A \cup \overline{A}) = P(A) + P(\overline{A})$ (from axiom 3); together these results give $P(\overline{A}) = 1 - P(A)$. The probability of the complement of any event must be 1 minus the probability of the event itself. This establishes the following result.

THEOREM 1.3 For any event $A \subset S$, $P(\overline{A}) = 1 - P(A)$.

We shall see several cases for which it is useful to remember this result; if we desire $P(A)$ and find it easy to evaluate $P(\overline{A})$, then it is a simple matter also to evaluate $P(A) = 1 - P(\overline{A})$.

By referring to Fig. 1.8 we can easily see that an event B can be decomposed into the elements it has in common with A and those it has in common with $\overline{A}$. That is, $B = (B \cap A) \cup (B \cap \overline{A})$, and these two parts are mutually exclusive (since any element in A does not belong to $\overline{A}$); A and $\overline{A}$ are said to *partition* any event B. Thus $P(B) = P(B \cap A) + P(B \cap \overline{A})$, again using axiom 3, from which it follows that $P(B \cap \overline{A}) = P(B) - P(B \cap A)$.

We can equally well decompose the union of A and B in the same way; that is, $A \cup B = A \cup (\overline{A} \cap B)$ (and these two pieces are necessarily mutually exclusive), so $P(A \cup B) = P(A) + P(\overline{A} \cap B) = P(A) + P(B) - P(A \cap B)$, using the above result. This gives the relationship between $P(A)$, $P(B)$, $P(A \cap B)$, and $P(A \cup B)$.

THEOREM 1.4 If $A \subset S$ $B \subset S$ are any two events, then

$$P(A \cup B) = P(A) + P(B) - P(A \cap B).$$

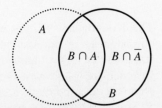

Figure 1.8 $B = (B \cap \overline{A}) \cup (B \cap A)$

Notice that this result also says that $P(A) + P(B) = P(A \cup B) + P(A \cap B)$, for any two events.

Example 1.10

Five coins are flipped; define the five events

$$A_i = \{\text{exactly } i \text{ heads occur}\} \qquad i = 1, 2, \ldots, 5.$$

These events are mutually exclusive; no two (or more) of them could possibly occur at the same time; let their union be B, the event that at least one head occurs. From axiom 3, $P(B) = \sum_{i=1}^{5} P(A_i)$. If A_0 is the event that "no heads occur on the five coins," then $\overline{B} = A_0$, so $P(B) = 1 - P(A_0)$.

Example 1.11

For two events $A \subset S$, $B \subset S$, it is given that $P(A) = .2$, $P(A \cap \overline{B}) = .1$, and $P(B) = .4$. Then it follows that

$$P(A \cap B) = P(A) - P(A \cap \overline{B}) = .2 - .1 = .1$$

and that

$$P(A \cup B) = P(A) + P(B) - P(A \cap B) = .2 + .4 - .1 = .5,$$

while

$$P(\overline{A} \cap \overline{B}) = 1 - P(A \cup B) = .5.$$

There are two further immediate consequences to the axioms that can be easily established. If $A \subset B$, then $A \cap B = A$ and

$$P(B) = P(A \cap B) + P(\overline{A} \cap B) = P(A) + P(\overline{A} \cap B).$$

Since $P(\overline{A} \cap B) \geq 0$, it follows that $P(B) \geq P(A)$; this is called the *monotonic property* of a probability measure. The "larger" set (the one containing the other as a subset) must have probability at least as large as any of its subsets. This is stated formally as the following theorem.

THEOREM 1.5   If $A \subset B$, then $P(A) \leq P(B)$.

As an immediate corollary to this result, we know that every event A is a subset of the sample space S; thus $P(A) \leq P(S) = 1$; that is, the maximum probability for any event is 1.

Example 1.12

The Kolmogorov axioms and their consequences can be used to spot inconsistencies in the assignment of probabilities to events. For example, the statements $P(A) = .4$ and $P(A \cup B) = .3$ are contradictory since $A \subset A \cup B$. In much the same way, the group of statements $P(A) = .4$, $P(B) = .2$, and $P(A \cap B) = .4$ contains a contradiction, while the group $P(A) = .4$, $P(B) = .2$, and $P(A \cup B) = .4$ does not.

Theorem 1.4 states the general relationship between the probability of the union of two sets and the probabilities of the individual events and their intersection. This is easily extended to three sets by realizing that $A \cup B \cup C = A \cup (B \cup C)$, so immediately it is true that

$$P(A \cup B \cup C) = P(A) + P(B \cup C) - P(A \cap (B \cup C)). \tag{1}$$

Note as well that $A \cap (B \cup C) = (A \cap B) \cup (A \cap C)$, so the last term on the right-hand side of Eq. (1) can be written as

$$P(A \cap B) + P(A \cap C) - P((A \cap B) \cap (A \cap C)).$$

Realizing that $(A \cap B) \cap (A \cap C) = A \cap B \cap C$ and expanding $P(B \cup C)$ then gives

$$\begin{aligned} P(A \cup B \cup C) = {} & P(A) + P(B) + P(C) \\ & - P(A \cap B) - P(A \cap C) - P(B \cap C) \\ & + P(A \cap B \cap C), \end{aligned}$$

the final result. In general, the probability of the union of k events is equal to the sum of the probabilities of each of the events, minus the sum of the probabilities of the intersections of every possible pair, plus the sum of the probabilities of the intersections of all possible triples, minus the sum of the probabilities of the intersections of all possible collections of four of the events, and so on.

EXERCISES 1.2

1. A hat contains 10 slips of paper, two of which are yellow and the rest of which are white. The yellow slips bear the numbers 1 and 2, and the white slips bear the numbers $3, 4, \ldots, 10$. Consider the experiment of drawing one slip from this hat; define a reasonable sample space S. Let A be the event that a white slip is drawn, and express A as a subset of S.

2. Using the same hat defined in Exercise 1, now assume that two slips of paper are drawn from the hat, without replacement. Specify a reasonable sample space for the experiment. Let B be the event that both slips drawn are yellow, and express B as a subset of S.

3. A pair of dice, one red and one green, are rolled (one time). Define a reasonable sample space for this experiment. Let A be the event that the sum of the two numbers rolled is 4, and express A as a subset of S.

4. A California municipal court district calls panels of persons registered to vote in the local precincts to serve as possible jurors for jury trials. A typical panel consists of 45 persons. When a jury is to be selected for a specific trial, the panel members are assembled in the court room; 18 (of the 45) are selected by lot, first to be questioned by the judge and the attorneys for both sides of the case and then to possibly be selected as jury members. Describe a reasonable sample space for this selection of 18 persons initially selected for questions. (A word description is adequate; it is not necessary to employ exact set or -tuple notation, as long as the sample space is accurately portrayed.)

5. A homeowner wants to plant two rose bushes in a corner of his property.

On visiting his local nursery, he finds 23 different varieties of roses available. Describe a reasonable sample space of the possibilities available to him in purchasing two rose bushes.

6. A television game show allows the game winner to select one of three closed doors; the participant is told that a new automobile is behind one of the three doors, while there is nothing behind either of the other two. If the participant selects the door with the auto, she then wins it; otherwise she wins nothing. What is a reasonable sample space for the experiment that has as its outcome the door (number) selected by the participant?

7. At a dinner party for four couples, the host wants to award two prizes by lot; to do this, he will place four slips of paper in a bowl, each bearing one of the numbers 1 through 4, and then will draw out two of these slips without looking. The two numbers drawn represent the two couples who will win the prizes.
   a. Look at this drawing as an experiment, and give a reasonable sample space for the possible outcomes.
   b. Let A be the event that couple number 1 wins a prize, and express A as a subset of S.

8. Show that the probability of the intersection of two events can be written
$$P(A \cap B) = P(A) + P(B) - P(A \cup B).$$

9. A small town contains three grocery stores (label them $\mathcal{A}, \mathcal{B}, \mathcal{C}$). Four people living in this town each pick one of these stores to shop in, on the same morning. Give a sample space for all the possible selections of stores by these people, and define the following events (subsets):
A : All the people choose $\mathcal{A}$.
B : Two choose $\mathcal{B}$ and two choose $\mathcal{C}$.
C : All of the stores are chosen (by at least one person).

10. Eight horses are entered in a race (bearing numbers 1 through 8). Monetary prizes are given for win, place, and show (finishing first, second, or third, respectively). Define a reasonable sample space for this experiment, and define the following events:
A : Horse number 1 wins.
B : Horse number 1 places.
C : Horse number 1 does not win any prize.

11. The Lotto $\frac{6}{51}$ game played in California and other states determines winners of prizes by randomly selecting six integers from the sequence $1, 2, \ldots, 51$ without replacement. What is a reasonable sample space for the collection of numbers that might occur for one play of this game? Define the events
C : The selected numbers are $1, 2, \ldots, 6$.
D : One of the numbers selected is 33.

12. It was mentioned in the text that $P(A \cup B) + P(A \cap B) = P(A) + P(B)$; is it also true that $P(A \cup B \cup C) + P(A \cap B \cap C) = P(A) + P(B) + P(C)$?

13. A memo, addressed to two people, is placed into the mail system of a large corporation. It may or may not eventually get to the addressees. With the sample space $S = \{0, 1, 2\}$, where $\omega \in S$ indicates the number of addressees to eventually receive the memo, assume that the probability

it will be received by at most one of the addressees is .5, while the probability it will be received by at least one of the addressees is .9. Evaluate the probability that the number of addressees to receive the memo is (*a*) 0; (*b*) 1; or (*c*) 2.

14. A person modeling the outcome for an experiment, with events A and B already defined, used a probability measure for which $P(A \cap \overline{B}) = .3$, $P(\overline{A} \cap B) = .2$, and $P(\overline{A} \cap B) = .8$. Evaluate the probabilities of occurrence for $P(A)$, $P(B)$, $P(A \cup B)$, and $P(A \cap B)$.

15. Many board games employ spinners to control various aspects of the play. Assume such a spinner has five sectors, spanning equal angles, and that the sector in which the arrow ends up is proportional to the angle spanned. What is the probability of getting an odd number on any given spin?

16. Suppose you must design a spinner with five sectors, like the one mentioned in Exercise 15. Let A_i, where $i = 1, 2, 3, 4, 5$, represent the event that the spinner points at integer i, and assume that we want $P(A_i)$ to be proportional to i; that is, the number 2 should be pointed at with probability $2P(A_1)$, while the number 5 should be pointed at with probability $5P(A_1)$. Describe the implied probability measure.

1.3 | Finite Sample Spaces

In order for probabilities to behave like relative frequencies, the probability function for any experiment must satisfy the Kolmogorov axioms. This section discusses some special simplifications in defining a probability function for finite sample spaces.

A *finite sample space* is a sample space whose number of elements is some positive integer k, no matter how large; a sample space S is finite if the roster of its elements has an end, a last element. No matter what the actual elements may represent, the set $S = \{1, 2, \ldots, k\}$ can be used as the sample space for such an experiment.

If a sample space S has k elements, there are exactly k mutually exclusive distinct subsets of S, each of which has a single element of S belonging to it; these are called *single-element events*. The sample space $S = \{1, 2, 3\}$ has three single-element events,

$$A_1 = \{1\}, \qquad A_2 = \{2\}, \qquad A_3 = \{3\},$$

while the sample space $S = \{(1,1), (1,2), (2,1), (2,2)\}$ has four single-element events:

$$A_1 = \{(1,1)\}, \qquad A_2 = \{(1,2)\}, \qquad A_3 = \{(2,1)\}, \qquad A_4 = \{(2,2)\}.$$

Let $A_i = \{i\}$, where $i = 1, 2, \ldots, k$, represent the k single-elements events for an experiment whose sample space is $S = \{1, 2, \ldots, k\}$; immediately, then,

$$S = A_1 \cup A_2 \cup \cdots \cup A_k,$$

and since all single-element events are mutually exclusive, it must be true that

$$P(S) = 1 = \sum_{i=1}^{k} P(A_i);$$

that is, the sum of all the single-element event probabilities must total to 1.

In addition, any event $B \subset S$ can be written as a union of these mutually exclusive, single-element events. The probability of B, using the third Kolgomorov axiom, is then given by summing together the appropriate single-element event probabilities (i.e., summing over those elements that belong to B). For example, consider again $S = \{1, 2, \ldots, k\}$, and $A_i = \{i\}$. If $B = \{1, 2, 3\} = A_1 \cup A_2 \cup A_3$, then $P(B) = P(A_1) + P(A_2) + P(A_3)$, while if $C = \{2, 5, 8, 10\} = A_2 \cup A_5 \cup A_8 \cup A_{10}$, then $P(C) = P(A_2) + P(A_5) + P(A_8) + P(A_{10})$. Thus the specification of the probability function for an experiment with a finite sample space S with k outcomes can be accomplished simply by assigning values to the k single-element events; the probability for any other event is derived by appropriate sums of these values.

Example 1.13

A die (singular of *dice*) is a six-sided cube with one to six spots on its faces. Consider rolling this die one time, and assume that nothing untoward occurs to the die. That is, it will not disappear into a deep hole never to be seen again, nor will it be grabbed by a passing bird in flight, nor will it shatter into pieces when it hits the playing surface. In short, when the die stops rolling, assume that one of the six possible faces will be uppermost; the set $S = \{1, 2, 3, 4, 5, 6\}$ will serve well as a sample space, with the integer element of S indicating the number of spots on the top face. Building a probability model for this experiment, then, means specifying the probabilities for the six single-element events. If the die is *fair*, these single-element events will all have the same probability, and the common value for this probability must then be $\frac{1}{6}$.

Rather than being fair, assume the die has been weighted in a way that makes the probabilities of all even-numbered faces equal and the probabilities of all odd-numbered faces equal, but the probability of occurrence for an even-numbered face is twice the value for an odd-numbered face. That is, we let $A_1, A_2, \ldots, A_6$ represent the six single-element events and $P(A_1)$, $P(A_2), \ldots, P(A_6)$ their respective probabilities. The information given states that

$$P(A_1) = P(A_3) = P(A_5), \qquad P(A_2) = P(A_4) = P(A_6) = 2P(A_1).$$

Then the fact that the six single-element event probabilities must sum to $P(S) = 1$ gives the equation

$$
\begin{aligned}
1 &= P(A_1) + P(A_2) + \cdots + P(A_6) \\
&= P(A_1)(1 + 2 + 1 + 2 + 1 + 2) \\
&= 9P(A_1).
\end{aligned}
$$

From this conclusion it follows that $P(A_1) = \frac{1}{9}$ and $P(A_2) = \frac{2}{9}$. If B is the event that an odd number occurs, $P(B) = P(A_1) + P(A_3) + P(A_5) = \frac{1}{9} + \frac{1}{9} + \frac{1}{9} = \frac{3}{9}$, and the probability that the number of spots on the upper face is at least 4 then is $P(A_4) + P(A_5) + P(A_6) = \frac{2}{9} + \frac{1}{9} + \frac{2}{9} = \frac{5}{9}$.

An important and frequently occurring class of problems is typified by fair gambling games; the word *fair* means an equal probability of occurrence for the single-element events, as used in Example 1.13. If an experiment with a finite sample space S has k elements and they are "equally likely" to occur, then the probability of occurrence for each of the single-element events is simply $P(A_i) = \frac{1}{k}$ for $i = 1, 2, \ldots, k$. This, of course, gives $P(S) = \sum_{i=1}^{k} 1/k = 1$, and it is easy to see that the other axioms are satisfied as well. With such an assignment for the single-element events, the probability for any event $B \subset S$ becomes simply $P(B) = m/k$, where m is the number of elements belonging to B. This discussion is summarized in the following definition.

DEFINITION 1.9

An experiment with finite sample space $S = \{1, 2, \ldots, k\}$ is said to have *equally likely outcomes* if the single-element event probabilities are all equal. In this case the probability of occurrence for any event $B \subset S$ is simply $P(B) = m/k$, where m is the number of elements belonging to B.

Example 1.14

Assume that two fair dice are rolled one time. A simple sample space for the experiment of rolling the two dice is given by $S = T \times T$, where $T = \{1, 2, \ldots, 6\}$; the assumption that the two dice are fair then implies that the $6 \times 6 = 36$ two-tuples in S are equally likely to occur. This, in turn, says that each single-element event, like $\{(1, 1)\}$, has a probability of occurrence equal to $\frac{1}{36}$.

Many popular dice games hinge on the value for the sum of the two numbers rolled with a pair of fair dice; the different values for this sum are not equally likely to occur. Let B_j, where $j = 2, 3, \ldots, 12$, represent the event that the sum is j; thus B_2 is the event that the sum observed is 2, while B_7 is the event that the sum observed is 7. Table 1.1 lists the 2-tuples belonging to $B_2, B_3, \ldots, B_{12}$.

The resulting probabilities for $B_2, B_3, \ldots, B_{12}$ are given in Table 1.2.

Using the values in this table, it is easy to compute probabilities that the sum of the two numbers rolled satisfies certain requirements. For example, if we define A to be the event that the sum of the two numbers rolled is at least 8, we have

$$P(A) = P(B_8) + P(B_9) + \cdots + P(B_{12}) = \frac{5 + 4 + 3 + 2 + 1}{36} = \frac{5}{12}.$$

Table 1.1

Elements of B_j	$j =$
(1,1)	2
(1,2),(2,1)	3
(1,3),(2,2),(3,1)	4
(1,4),(2,3),(3,2),(4,1)	5
(1,5),(2,4),(3,3),(4,2),(5,1)	6
(1,6),(2,5),(3,4),(4,3),(5,2),(6,1)	7
(2,6),(3,5),(4,4),(5,3),(6,2)	8
(3,6),(4,5),(5,4),(6,2)	9
(4,6),(5,5),(6,4)	10
(5,6),(6,5)	11
(6,6)	12

Table 1.2

Event	B_2	B_3	B_4	B_5	B_6	B_7	B_8	B_9	B_{10}	B_{11}	B_{12}
Probability	$\frac{1}{36}$	$\frac{2}{36}$	$\frac{3}{36}$	$\frac{4}{36}$	$\frac{5}{36}$	$\frac{6}{36}$	$\frac{5}{36}$	$\frac{4}{36}$	$\frac{3}{36}$	$\frac{2}{36}$	$\frac{1}{36}$

If we let C be the event that the sum is an odd number, then

$$P(C) = P(B_3) + P(B_5) + \cdots + P(B_{11}) = \frac{2+4+6+4+2}{36} = \frac{1}{2}.$$

If you let D be the event that the sum of the two numbers observed is even, what is the value for $P(D)$?

Equally likely sample spaces can be useful in testing claims of clairvoyants; a simple case is discussed in the next example.

Example 1.15

Ms. $\mathcal{G}$ claims she can predict or control the face that will land uppermost when a fair coin is flipped. To test this claim, we will flip a fair coin three times; let $T = \{r, w\}$ represent the two outcomes, right (she correctly identified the face to occur) and wrong, and use $S = T \times T \times T$ as the sample space for the three flips. If she is merely guessing which face will occur on each flip, then the elements of S should be equally likely; since there are $2 \times 2 \times 2 = 8$ three-tuples in S, the event $A = \{(r,r,r)\}$, that she is right on all three flips has a probability $\frac{1}{8}$ of occurring. If we found A did in fact occur, we might tend to believe her claim. (In actual fact, we would probably want to see her performance on a larger number of flips before making any judgment.)

Example 1.16

Table 1.3

Day	Sunday	Monday	Tuesday	Wednesday	Thursday	Friday	Saturday
Count	687	685	685	687	684	688	684

What is the probability that the 13th day of a randomly chosen month falls on Friday? Since there are seven days in a week, an easy answer is $\frac{1}{7}$. A more precise (and surprising) answer is derived from reviewing the structure of our calendar system, which repeats every 400 years. Every fourth year is a leap year, unless it ends in "00"; in this case it is a leap year only if the first two digits of the year are also divisible by 4. Thus, 1700, 1800, and 1900 were not leap years, but 2000 will be a leap year. In one full period of our calendar there are 97 leap years, 4800 months, and a total of $97(366) + 303(365) = 146,097$ days or exactly 20,871 weeks. Each of these 4800 months has a 13th day; as pointed out in the solution to a problem posed in the *American Mathematical Monthly* (1933, vol. 40, p. 607), an enumeration of the days of the week on which this date falls for the 4800 months gives the results in Table 1.3.

If we select one of these 4800 months at random, the probability of finding the 13th day on a Friday is $688/4800 > \frac{1}{7}$; Friday is the most likely day of the week for the 13th to fall on. If the 13th day of a month falls on a Friday, then the first of that month falls on a Sunday; thus the most likely day for the first day of a randomly chosen month is Sunday.

Probability theory is concerned with modeling the effects of chance; behind-the-scenes manipulation may totally obliterate the effects of randomness on the outcome observed.

Example 1.17

A sly individual wants to take advantage of human gullibility. Specifically, she will convince one or more people that she is able to correctly predict the outcomes of professional football games and, having convinced them, then ask for payment for her predictions for subsequent games. To see how she might accomplish this, assume she selects 64 (which is 2^6) people at random and sends each of them a letter at the start of the season. The first week of the season features an important game with team $\mathcal{A}_1$ versus team $\mathcal{B}_1$; the letter states her prediction of which of the two teams will win. In half (32) of the letters she says the winner will be $\mathcal{A}_1$ and in the other 32 she says the winner will be $\mathcal{B}_1$.

After the first week is over, she has made the correct prediction of who would win the important game of that week, for 32 persons. She then sends each of these 32 people another letter predicting the outcome of the big game of week 2, which features $\mathcal{A}_2$ versus $\mathcal{B}_2$; again in half (now 16) of these she picks $\mathcal{A}_2$, and in the other half she picks $\mathcal{B}_2$. At the end of week 2 she has made two correct predictions in a row for each of 16 persons. By now the scheme should be fairly clear; each of these 16 will receive a letter from her about who will win the big game in week 3. Eight of these persons will be told the winner will be A_3 and the remaining eight will be told the winner will be B_3, and after the third week she has made three correct predictions for each of 8 persons. At this time she might try to solicit money from each of these people, to learn her prediction for week 4, based on her extraordinary performance. Without knowing anything about the game, she can deliver very good predictions (to some people) with this scheme.

The fairness of two dice does not extend to fair (equally likely) sums of the two numbers to occur. However, the following example indicates that fairness at one level may extend to another.

Example 1.18

Four people, $A, B, C,$ and D, enter a single-elimination tournament, involving a game that cannot end in a tie. Such a tournament requires a total of three games; game 1 pits A versus B, while game 2 features C versus D. The third and final game will involve the winners of games 1 and 2; the person winning this game is the tournament winner. If all four are of equal ability, it would seem logical that each of them has probability $\frac{1}{4}$ of winning the tournament. That this must be the case can be seen from the following type of analysis.

Since the tournament will consist of a total of three games, a reasonable sample space has 3-tuples as elements; the first entry in each 3-tuple gives the winner of game 1, while the second and third entries give the winners of the other two games. That is, let us use

$$S = \{(A, C, A), (A, D, A), (B, C, B), (B, D, B),$$
$$(A, C, C), (A, D, D), (B, C, C), (B, D, D)\}.$$

Here we make the assumption of equally likely outcomes, since the players are assumed to be of equal ability. Let D be the event that D wins the tournament; that is,

$$D = \{(A, D, D), (B, D, D)\},$$

so $P(D) = \frac{2}{8} = \frac{1}{4}$. It is easy to see that this is the probability of winning for each of the players.

There are many other assignments of probabilities to these single-element events that also give $\frac{1}{4}$ as the probability for each player winning the tournament; assuming the 3-tuples used in this sample space to be equally likely implies a probability of $\frac{1}{4}$ for each player winning the tournament, but not vice versa. You will be asked to explore this in Exercises 1.3.

For most popular games involving two players and where one player starts (as in tic-tac-toe, chess, or checkers), it is commonly the case that the person moving first has an advantage. This need not be true for all games, as illustrated in the following example.

Example 1.19

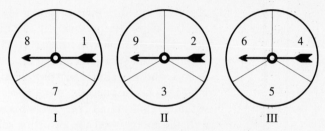

Figure 1.9 Three spinners for a game

Consider a game involving three spinners; the person to start selects one of the spinners, and then the opponent selects one of the two remaining. Each player spins her pointer, which eventually stops on a number; the winner is the one who spins the larger number. If the spinners used in Fig. 1.9 are employed, then the person selecting first can be forced to have a probability of winning less than $\frac{1}{2}$.

For any pair there are $3^2 = 9$ different pairs of numbers that may occur. These are all equally likely with equal-size sectors. Notice that if *I* and *II* are selected, then there are five such pairs for which the number from *II* exceeds the number from *I*; if *II* and *III* are selected, there are six pairs for which the *III* number exceeds the *II* number; and if *III* and *I* are selected, there are six pairs for which the *I* number exceeds the *III* number. Thus, if the person starting selects a spinner, the second person can always choose one for which the probability of producing the larger number is $\frac{5}{9}$ or $\frac{6}{9}$.

EXERCISES 1.3

1. A die is rolled one time, as in Example 1.13, except that now the die is weighted in such a way that faces 1, 2, and 3 have equal probabilities of occurrence, as do faces 4, 5, and 6, but the faces with the larger numbers of spots are each 4 times as likely to occur as are the faces with the smaller numbers of spots. Evaluate the probabilities of occurrence of the following:
   a. the single-element events   b. the sum equals an odd number

2. The sectors of the circle below span angles proportional to the numbers. What are the implied probabilities of the spinner stopping in the various sectors?

3. A die is rolled one time. Let *C* be the event that an odd number of spots occur, let *D* be the event that an even number of spots occur, and assume that $P(C) = 2P(D)$. Does this give enough information to completely specify the probability function?

4. $\mathcal{A}$ rolls a fair die; then $\mathcal{B}$ rolls a fair die. The person rolling the higher number wins (no one wins if the numbers are the same). What is the probability that $\mathcal{A}$ wins? that $\mathcal{B}$ wins? that there is no winner?

5. A hat contains six slips of paper, each bearing one of the integers 1, 2, 3, 4, 5, and 6. (Thus each slip has one number on it and each of these integers occurs once on some slip.) Two slips are selected at random, without replacement (i.e., two slips are selected simultaneously).
   a. If the integers on the two slips selected are added together, what are the possible values for the sum?
   b. Evaluate the probabilities of occurrence of these different sums.

6. There are 20,871 Fridays in one (400-year) period of our calendar. Enumeration shows that 643 of these fall on the 29th day of some month, while 629 fall on the 30th and 399 fall on the 31st. If we select a Friday at random, what are its most likely dates?

7. If Ms. $\mathcal{G}$, of Example 1.15, were presented with eight coin flips, and she really only guesses the outcome for each, what is the probability she would be right on all eight flips?

8. If a person wanted to establish a track record for predictions of major game outcomes by correctly predicting eight in a row for each of k people using the scam mentioned in Example 1.17, how many people must be contacted initially with a prediction for the first game?

9. For the single-elimination tournament, with four contestants, mentioned in Example 1.18, what is the probability that any one of the contestants does *not* win the tournament? What is the probability that any one of them loses the first game?

10. Five different-colored bowls containing the same dog food are laid out in a row. If a dog chooses one of these bowls at random from which to eat, what is the probability he chooses the blue bowl (exactly one of the bowls is blue)? If a second dog is used, what is the probability that he chooses the blue bowl? What is the probability that both dogs choose the blue bowl? What is the probability that both dogs choose the same bowl (regardless of color)?

11. For the single-elimination tournament discussed in Example 1.18, use the same sample space, with the elements in the order listed there. However, instead of assuming the single-element events are equally likely, assume these single-element probabilities to be $\frac{1}{20}$ times 1, 2, 3, 4, 4, 3, 2, 1, for the order given in the example. Evaluate the probability of each of the individuals winning the tournament; also evaluate the probability that each one wins her first game.

12. For the single-elimination tournament of Example 1.18, use the same sample space and give an assignment of probabilities to the single-element events that is not equally likely but which gives $\frac{1}{4}$ as the probability for each of them winning. With your assignment, what is the probability of each entrant winning his first game?

13. A round-robin tournament calls for each entrant to play every other entrant one time; the winner of the tournament, if there is one, is the one to win the most games. Assume $\mathcal{A}$, $\mathcal{B}$, and $\mathcal{C}$ enter such a tournament,

requiring then a total of three games to be played: A versus B, A versus C, and B versus C. Assuming all entrants are of equal ability, what is the probability each one of them will be the winner? What is the probability there will be no winner (i.e., they each win one game)?

14. A fair coin is flipped three times. With the sample space $S = \{0, 1, 2, 3\}$, where $x \in S$ indicates the number of heads that occur, is it reasonable to assume these elements are equally likely to occur? Can you suggest more reasonable values for the probabilities of the single-element events in S?

15. There are, of course, many collections of three spinners, each with three sectors including each of the integers 1 to 9 exactly once. For most of these, the person who first selects a spinner has the advantage. What is an obvious distribution of $1, 2, \ldots, 9$ into three partitions of size 3 that gives the starter a probability of 1 of getting the larger number?

16. *Continuation of Exercise 15* Complete enumeration shows that there are only five partitions of $1, 2, \ldots, 9$ into three parts that give the *second* chooser an advantage over the first. Can you find another one?

1.4 | Counting Techniques

In modeling an experiment with a finite sample space and equally likely outcomes, the probability of occurrence for any event B is the ratio of the number of elements in B to the number of elements in S. Thus, evaluation of probabilities for these cases requires the ability to count the numbers of elements in S and the number of elements in any $B \subset S$ of interest. We shall study some simple, widely applicable counting procedures in this section, and see their application in solving problems.

With finite sample spaces, every event is itself a finite set; one crude way to count the number of elements in such a set is simply to list all the elements and then count the number of items in the list. This procedure, however, could be very time-consuming because the word *finite* is not synonymous with *small*. Perhaps equally important, the construction of the list of elements that belong to a set is not necessarily error-free; it is very easy to forget items that should be on the list, and in complicated lists to include the same item more than once, leading to errors in the count.

There are a number of concepts and procedures that are useful in counting the numbers of elements belonging to a set. The simplest of these is referred to as the multiplication principle (see page 28).

This principle is directly employed in counting the number of elements in the Cartesian product of two sets; as was mentioned earlier, if A has m elements and B has n elements, then $A \times B$ has mn elements (2-tuples). Constructing the list of all 2-tuples in $A \times B$ requires two *operations*: selecting an element from A for each first position and selecting an element from B for

---

MULTIPLICATION PRINCIPLE

According to the multiplication principle, if a first operation can be performed in m ways and, for each of these, a second operation can be performed in n ways, then the two operations can be performed together in mn ways.

---

each second position. Since there are m elements in A and n elements in B, m and n give the number of ways to perform the first and second operation, respectively. Thus the full list of 2-tuples in $A \times B$ contains mn items, the total number of ways that the two operations can be performed.

Many persons find tree diagrams useful in enumerating all the possible combinations embodied in the multiplication principle. Suppose a personal computer vendor carries system units made by three different manufacturers ($\mathcal{D}$, $\mathcal{N}$, $\mathcal{C}$) and display monitors of four different types (**1**, **2**, **3**, **4**). If you are purchasing a personal computer (consisting of the system unit plus display), there are $3 \times 4 = 12$ different combinations from which you could choose. These are illustrated in the tree diagram presented in Fig. 1.10.

The basic idea of this type of diagram is simple: First list the different ways the first operation (choosing one of the system units) can be performed. Then, for each of these, list the possible choices of monitor display. The result at the bottom of the tree is the 12 pairs that are possible for your purchase.

The multiplication principle generalizes in the obvious way. If k operations are to be performed, and operation i can be performed in n_i ways regardless of how any of the other operations is performed, then the number of ways that all k operations can be performed is given by the product $\prod_{i=1}^{k} n_i = n_1 n_2 \cdots n_k$. Tree diagrams can be extended for any number k of levels, but as k increases they quickly involve an unwieldy number of branches.

The following example employs this multiplication principle to count the number of different automobile license plates that can be made, if certain conventions are followed.

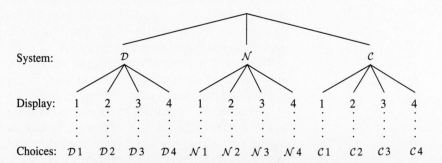

Figure 1.10   Tree diagram for computer choices

Example 1.20

Many states use automobile plates that contain six positions, the positions being filled with letters and digits according to some pattern. The multiplication principle can be employed to count the total number of plates possible, for any given pattern. Suppose a license plate contains three digits followed by three capital letters. Then if all possible combinations are employed, the total number of plates possible can be enumerated by counting the number of ways that $k = 6$ operations can be performed; the six operations consist of choosing the particular item to be placed into positions 1 through 6.

Since the first position must be a digit, the first operation can be performed in 10 ways (because the digit placed in the first position could be 0, 1, ... , or 9). Similarly, no matter what digit is placed into the first position, the second position can also be filled in 10 ways, as can the third. Whichever three-digit number occurs in the first three positions, the fourth position can be filled in 26 ways, using any of the 26 capital letters in the Roman alphabet, as can the fifth and sixth positions. Thus the chore of enumerating all the possible plates would result in a list having $10 \cdot 10 \cdot 10 \cdot 26 \cdot 26 \cdot 26 = 17,576,000$ members, which indeed would take a long time to complete, although it is a finite list.

In actual fact, not all possible combinations are employed; assume that the first letter on a plate can be neither I or O, since these might be mistaken for digits. The number of plates possible then would be $10 \cdot 10 \cdot 10 \cdot 24 \cdot 26 \cdot 26 = 16,224,000$, since now the fourth operation can be done in only 24 ways. Assume further that there are 196 remaining three-letter combinations that may be objectionable to one or more persons or groups. Since each three-letter combination would appear on $10^3 = 1000$ plates, not using any plates with these combinations would remove 196,000 plates from the list. With the first letter neither I nor O, and not allowing the use of the 196 three-letter combinations, the resulting number of plates possible is $16,224,000 - 196,000 = 16,028,000$.

Many counting problems hinge on being able to count the number of ways distinct (or different) objects can be placed in a row. One word frequently used in these applications is *permutation*.

DEFINITION 1.10

A *permutation* of n distinct objects is an arrangement of those objects in a definite order. Any change in ordering gives a different permutation.

Clearly, two objects, say a and b, can be permuted (or listed) in a total of two ways, ab and ba; ab and ba are each permutations of a and b. Three objects, say a, b, and c, give rise to six different permutations: *abc*, *acb*, *bac*, *bca*, *cab*, and *cba*. A permutation can be thought of as an n-tuple; two different 2-tuples can be constructed using two different objects, and six different 3-tuples can be constructed using three different objects.

How may n-tuples can be constructed from the first n integers, where each integer is used only once? The number of different n-tuples (or permutations) that can be constructed from these n objects can be easily counted by envisioning the n operations necessary to make up the full list of n-tuples. To begin making up this list, we select an object for placement in the first (leftmost) position; this first operation can be done in n ways. Now one of the remaining $n - 1$ objects must be selected for the second position, an operation that can be done in $n - 1$ ways. After the first two positions have been filled, we select one of the remaining $n - 2$ objects for the third position, which can be done in $n - 2$ ways. We proceed in this way through all of the positions. By the time we have reached the $(n - 1)$st position, $n - 2$ objects have been put into place and there are $n - (n - 2) = 2$ objects remaining, either of which may be selected for this position; this operation can be done in two ways. Finally, when the nth (right-most) position is reached, there is one object left, so the final operation can be performed in one way. Thus, the total number of ways that all n operations can be performed, which is the same as the total number of n-tuples or permutations possible, is $n(n - 1)(n - 2) \cdots 2 \cdot 1$, the product of the first n integers.

This product (or function) occurs frequently in counting problems and various other areas of mathematics; it is called *n-factorial* (or the *factorial of n*) and is denoted by $n!$. Thus, as noted earlier, there are $2! = 2$ different permutations of two objects and $3! = 3 \cdot 2 \cdot 1 = 6$ different permutations of three objects. Four different objects give rise to $4! = 4 \cdot 3 \cdot 2 \cdot 1 = 24$ different permutations, while five different objects give rise to $5! = 120$ different permutations. The value of $n!$ increases very rapidly as n increases.

Example 1.21

The same six school children line up daily for lunch each school day. The distinct number of ways they can arrange themselves in a line on any given day, which is the same as the number of 6-tuples possible, is $6! = 720$. If there are 180 school days in the school year, and all six of these students remain together throughout their school years, they could line up in a different order (permutation) every school day for four years without a repetition.

When considering questions of probability it is useful to be aware of various ways of restricting the ordering. Suppose these six children are represented by the letters a, b, c, d, e, and f, and assume that they line up totally at random for lunch each day; then our experiment consists of the order they assume in line for a given day. The sample space for this experiment is

$$S = \{(x_1, x_2, \ldots, x_6) : x_i = a, b, c, d, e, f, \ x_i \neq x_j, \ i = 1, 2, \ldots, 6\}.$$

The number of elements in S is $6! = 720$, and if the students truly line up at random on this day, each single-element event has probability $\frac{1}{720}$.

Let A be the event that b immediately follows a in line on this particular day. The lineup to occur might be (c, d, a, b, f, e), for example, or *any other* arrangement that lists a right before b. The probability of A occurring is given by first counting the number of elements in S that have a immediately in front of b in the 6-tuple and then dividing by 720. This type of counting problem can be easily solved by considering the number of possible arrangements of the symbols $\underline{ab}, c, d, e, f$; that is, we treat $\underline{ab}$ as a single symbol, together with the four other, single letters. Tying a and b together like this assures that the arrangements counted will have b immediately following a. This strategy gives a total of five symbols that can be arranged in any order. The total number of such arrangements is $5! = 120$, so $P(A) = \frac{120}{720} = \frac{1}{6}$.

For event B, in which b will immediately precede a in line, we now count the number of arrangements of $\underline{ba}, c, d, e, f$. Again the number of possible arrangements is $5! = 120$, and $P(B) = \frac{1}{6}$. Now let C be the event that a and b are together in line (in either order). Since the set of 6-tuples with a before b is disjoint from the set of 6-tuples with b before a, the number of 6-tuples belonging to C is $120 + 120 = 240$ (the sum of those with a preceding b plus those with b preceding a). Thus $P(C) = \frac{1}{3}$.

Example 1.22

A secretary uses a computer program to type the same form letter to six persons, each of which is personally addressed; let 1, 2, 3, 4, 5, and 6 represent the different addresses. The secretary then uses this program to print envelopes for the letters. The printer to which these commands are sent is in a different work area, and the output is then delivered back to the secretary by an office aide. Assume that the letters are brought back in their original order but, because the aide tripped and fell, the envelopes have been randomly permuted when delivered. The secretary simply places the first letter into the first envelope, second letter into second envelope, and so forth, and then mails them. What is the probability that the letter addressed to 3 is in the envelope addressed to 3?

The sample space for this experiment consists of the 6! different orderings of the envelopes, assumed to be equally likely; of these, 5! contain 3 in the third (correct) position. Thus the probability that the letter to 3 is in the correct envelope is the ratio $\frac{5!}{6!} = \frac{1}{6}$. This is also the probability that the letter to 1 or 4 or any single one of them is in the correct envelope.

Since $(n-1)!$ is the product of the first $n-1$ integers, $n(n-1)!$ is the product of the first n integers; that is,

$$n! = n(n-1)! . \tag{1}$$

If we set $n = 1$ in Eq. (1), we get $1! = 1 \cdot (1-1)! = 1 \cdot 0!$; since $1! = 1$, it follows that $0! = 1$, which may not seem natural. You may take this result as a convention if desired. These factorial numbers coincide with particular values of the *gamma function*, which also gives $0! = 1$. It is important to remember that $0! = 1$.

The word *permutations* is used in a second way in the following definition.

DEFINITION 1.11

Given n distinct objects, the *number of permutations of n things r at a time*, where $r \leq n$, is the number of r-tuples that can be made from the n objects. This number is denoted by $_nP_r$.

It is easy to evaluate $_nP_r$; using the first n integers, $1, 2, 3, \ldots, n$, the total number of r-tuples possible is

$$_nP_r = n(n-1)(n-2)\cdots(n-(r-1)),$$

since the left-most position can again be filled by any of the n objects, the second position can then be filled by any of the remaining $n-1$ objects, and so on. Any of the remaining $n-(r-1)$ objects can be selected for the final rth position (each integer occurs at most one time in any r-tuple). Since

$$n! = n(n-1)\cdots(n-r+1) \times (n-r)(n-r-1)\cdots 2 \cdot 1$$
$$= {}_nP_r \times (n-r)!$$

it follows that

$$_nP_r = \frac{n!}{(n-r)!}.$$

Example 1.23

Numbers of different arrangements of distinct letters in a row provides an illustration of this use of permutations. Consider the eight letters in the word *computer*. Call any arrangement of letters a "word," whether it actually spells a word in some language or not. There are $8! = 40{,}320$ different eight-letter words that could be made from the letters in *computer*, allowing no repetitions of letters. In the same manner, there are $_8P_5 = 8!/3! = 6720$ different five-letter words that can be constructed with these same letters, allowing no repetitions of letters.

It is also straightforward to count words taking certain types of restrictions into account. Since the word *computer* contains three vowels and five consonants, $5! = 120$ of the five-letter words contain no vowels, so the difference, $6720 - 120 = 6600$, is the number of five-letter words possible which contain one or more vowels.

Suppose five of these letters are selected (no repetitions) at random and then placed in a row. What is the probability that the word so constructed contains no vowels? that it contains at least one vowel? If the five letters are selected at random from the eight, then an equally likely sample space S has as elements the full list of 6720 5-tuples that can be constructed from the eight letters. Let A represent the event that the word constructed has no vowels. Then the number of elements in A is 120 and $P(A) = 120/6720 = \frac{1}{56} = .018$; $\overline{A}$ is the event that the word has one or more vowels, so $P(\overline{A}) = \frac{55}{56} = .982$.

If repetitions of letters are allowed, a total of $8^8 = 16{,}777{,}216$ different eight-letter words are possible using the letters in *computer*, since each of the different positions could be filled by any of the eight letters, and $8^5 = 32{,}768$

five-letter words can be constructed. Of these, $5^5 = 3125$ contain no vowels, and $32,768 - 3,125 = 29,643$ contain one or more vowels. If the five letters are selected at random from the eight, this time with repetition allowed, the natural sample space S with equally likely outcomes contains all 32,768 five-letter selections possible. Letting A represent the event that the word constructed contains no vowels, now $P(A) = 3125/32,768 = .095$. Note that this is larger than its value when no repetitions are allowed and that $P(\overline{A}) = 29,643/32,768 = .905$ is smaller by the same amount.

In counting permutations, order within the n-tuple is important since two different orderings of the same objects are counted as being different. It is also of interest to be able to evaluate the numbers of items in certain lists where order within the item does not count, that is, to count items as being different only if they do not contain exactly the same symbols, regardless of order. This distinction is the same as that made between sets and n-tuples; two sets are equal if they contain the same elements, regardless of order, whereas two n-tuples are different if they contain the same elements in a different order. The word *combinations* is used for unordered items, as introduced in the following definition.

DEFINITION 1.12

The *number of combinations of n things taken r at a time*, written $\binom{n}{r}$, is the number of distinct subsets, each of size r, that can be constructed from a set with n elements where $r \leq n$.

Example 1.24

The three symbols a, b, c generate $_3P_2 = 3!/1! = 6$ different permutations of size 2: ab, ac, ba, bc, cb, and ca. The *set* $\{a, b, c\}$, though, has only $\binom{3}{2} = 3$ subsets of size 2: $\{a, b\}$, $\{a, c\}$, and $\{b, c\}$. Each distinct subset of two objects gives rise to two different permutations, as is apparent from the previous lists.

The discussion in Example 1.24 gives the relationship between permutations and combinations for any sizes of n and r. Any subset of r objects generates $r!$ different permutations. Thus

$$\binom{n}{r} r! = {}_nP_r = \frac{n!}{(n-r)!},$$

which gives

$$\binom{n}{r} = \frac{n!}{r!(n-r)!}.$$

The quantity $\binom{n}{r}$ is called a *combinatorial coefficient*, or a *binomial coefficient*, since it also occurs in the statement of the binomial theorem.

Example 1.25

Many common card games use a standard 52-card deck, which contains 13 cards in each of four suits: hearts ($\heartsuit$), diamonds ($\diamondsuit$), spades ($\spadesuit$), and clubs ($\clubsuit$). The first two of these suit symbols are red (and the cards bearing them are called red cards), while the last two suit symbols are black. Each of these suits contains 13 denominations: ace, 2, 3, ..., 10, jack, queen, king, so the deck contains four cards of each of these denominations, one in each suit.

Most card games involve each player receiving a subset of these cards, without regard to the order in which the cards are received. Thus, counting the number of different possible selections (called *hands*) that could be dealt in five-card draw poker requires the use of combinations, not permutations; the number of such distinct hands (subsets of size 5) is $\binom{52}{5} = 52!/5!47! = 2,598,960$. If five cards are selected at random from such a deck, then each of these possible subsets of size 5 is equally likely to occur (and a sample space with equally likely outcomes would have these 2,598,960 subsets as elements).

Suppose five cards are selected at random from this deck of 52. Let R be the event that all the cards selected are red, B that all the cards selected are black, H that all the cards selected are hearts. (No two of these events are complementary.) The sample space S then has as elements the list of 2,598,960 subsets of size 5 that could be chosen. Of the 52 cards in the deck, 26 are red (hearts and diamonds) and the other 26 are black. There are $\binom{26}{5} = 65,780$ of these subsets that contain only red cards (and an equal number that contain only black cards); thus $P(R) = P(B) = 65,780/2,598,960 = .025$. The number of subsets of size 5 that contain only hearts is $\binom{13}{5} = 1287$, so $P(H) = 1287/2,598,960 = .000495$; it is quite rare to get all 5 cards from the same suit if they are selected at random from the 52.

Example 1.26

A Little League baseball team has 15 players; the number of nine-player lineups that the coach could use to start the game, disregarding the assignment of persons to positions, is $\binom{15}{9} = 5005$. If the coach selects the persons to start the game for his team at random, a sample space S with these 5005 subsets of size 9 as elements is appropriate.

Two of the players available are named Joe and Hugh, respectively. Let J be the event that Joe is one of the players selected to start a given game and let H be the event that both Joe and Hugh are selected to start the game. The probabilities of these two events, then, can be evaluated by enumerating the numbers of subsets of size 9 that include only Joe and those that include both Joe and Hugh. The number of subsets of size 9 that include Joe is $\binom{14}{8} = 3003$, since counting the subsets that include this one individual is equivalent to first listing his name in each subset, and then choosing any eight of the remaining players to complete the nine-member team. Thus $P(J) = 3003/5005 = .6$, if the selection is made at random. The number of subsets of size 9 that can be selected from the original 15 players and includes both Joe and Hugh is $\binom{13}{7} = 1716$, since the remaining seven members could be chosen from any of the 13 remaining persons available; thus $P(H) = 1716/5005 = .343$.

These combinatorial coefficients have a number of properties that are simple to derive. For any integer $n > 0$,

$$\binom{n}{0} = \binom{n}{n} = 1,$$

a fact that is easily seen by writing out the factorials involved, or by realizing that a set with n elements has one subset of size 0, $\emptyset$, and one subset of size n, the whole set. Similarly,

$$\binom{n}{1} = \binom{n}{n-1} = n,$$

since any set of size n has n different subsets of size 1, as well as n subsets of size $n - 1$; to get a subset of size $n - 1$ we must select all the elements except one. The one left behind could be any of the n elements in the set. This reasoning holds for any size subset; that is, in selecting a subset of size r, say, exactly $n - r$ elements must be left behind. There is a one-to-one correspondence between subsets of size r and subsets of size $n - r$. Thus it is necessary that

$$\binom{n}{r} = \binom{n}{n-r} = \frac{n!}{r!(n-r)!},$$

for any size r.

Granted the set $S = \{1, 2, \ldots, n\}$, we have seen that S has exactly $\binom{n}{r}$ subsets of size r. Clearly, every subset of size r either does, or does not, include the particular element 1. The number of subsets of size r that include 1 as an element is $\binom{n-1}{r-1}$; this is true since the remaining $r - 1$ elements in the subset can be any from the other $n - 1$ in the set. The number of subsets of size r, each of which do *not* include 1 as an element, is $\binom{n-1}{r}$, the number of ways that r elements could be selected from $\{2, 3, \ldots, n\}$. The sum of these two numbers must equal the total number of subsets of size r that can be selected from S; that is,

$$\binom{n-1}{r-1} + \binom{n-1}{r} = \binom{n}{r}. \tag{2}$$

This relation is also easily verified by writing out the two combinatorial coefficients in their factorial form, putting them over a common denominator, and canceling common factors.

Equation (2) is the basis for *Pascal's triangle*, a procedure that allows us to build up the values of combinatorial coefficients by addition rather than multiplication. Before the widespread availability of fast digital computers this was of more than passing interest. Since $\binom{n}{0} = \binom{n}{n} = 1$, for any n, and in particular, when $\binom{1}{0} = \binom{1}{1} = 1$, the triangle displayed in Fig. 1.11 is easy to construct. The first row gives the combinatorial coefficients for $n = 1$, the second gives the values for $n = 2$, the third for $n = 3$, and so forth, with the final row giving the values of $\binom{7}{r}$ for $r = 0, 1, \ldots, 7$. Each row begins and ends with 1, while the interior values are given by the sum of the two values bracketing it in the preceding row. You could use this procedure to

$$1 \quad 1$$
$$1 \quad 2 \quad 1$$
$$1 \quad 3 \quad 3 \quad 1$$
$$1 \quad 4 \quad 6 \quad 4 \quad 1$$
$$1 \quad 5 \quad 10 \quad 10 \quad 5 \quad 1$$
$$1 \quad 6 \quad 15 \quad 20 \quad 15 \quad 6 \quad 1$$
$$1 \quad 7 \quad 21 \quad 35 \quad 35 \quad 21 \quad 7 \quad 1$$

Figure 1.11 Pascal's triangle

evaluate $\binom{n}{r}$, no matter how large n and r might be, but the danger of errors in addition, which are propagated for all later rows, certainly increases with n.

As already noted (and as is evident from Fig. 1.11), $\binom{n}{r}$ and $\binom{n}{n-r}$ are equal, with the smallest values occurring at the extremes of $r = 0$ and $r = n$; the value for $\binom{n}{r}$ increases symmetrically as r and $n - r$ approach $n/2$. The two middle values (with $r = (n-1)/2$ and $(n+1)/2$) are the largest if n is odd, while the single middle value (with $r = n/2$) is the largest if n is even.

Equation (2) also provides an easy, inductive proof of the binomial theorem. If x and y are any two real numbers, we can easily see by removing the parentheses that

$$(x+y)^2 = y^2 + 2xy + x^2 = \binom{2}{0}x^0 y^{2-0} + \binom{2}{1}x^1 y^{2-1} + \binom{2}{2}x^2 y^{2-2}$$

$$= \sum_{i=0}^{2} \binom{2}{i}x^i y^{2-i}.$$

Assume this result is true for an arbitrary positive integer n, that

$$(x+y)^n = \sum_{i=0}^{n} \binom{n}{i}x^i y^{n-i};$$

then it must also be true for the next larger integer $n+1$, as can be seen as follows:

$$(x+y)^{n+1} = y(x+y)^n + x(x+y)^n$$

$$= \binom{n}{0}y^{n+1} + \binom{n}{1}xy^n + \binom{n}{2}x^2 y^{n-1} + \cdots + \binom{n}{n}x^n y$$

$$+ \binom{n}{0}xy^n + \binom{n}{1}x^2 y^{n-1} + \cdots + \binom{n}{n-1}x^n y + \binom{n}{n}x^{n+1}$$

$$= \quad y^{n+1} + \binom{n+1}{1}xy^n + \binom{n+1}{2}x^2 y^{n-1} + \cdots + \binom{n+1}{n}x^n y + x^{n+1}$$

$$= \sum_{i=0}^{n+1} \binom{n+1}{i}x^i y^{n+1-i}.$$

Here Pascal's triangle has been used to combine pairs of terms that have the same powers of x and y. Since the proposition is true for $n = 2$, it must also be true for $n = 3$, which in turn implies it is true for $n = 4$, and so on. This proves the *binomial theorem*, stated below.

> **THEOREM 1.6** If x and y are any two real numbers and n is any positive integer, then
>
> $$(x + y)^n = \sum_{i=0}^{n} \binom{n}{i} x^i y^{n-i}.$$

The binomial theorem plays a role in many different areas in mathematics. It is the basis for the name for the *binomial random variable*, as we will see in Chapter 3.

At the start of this section, it was stated that $S = \{1, 2, \ldots, k\}$ has 2^k subsets; the following example establishes this result in two ways, one of which employs the binomial theorem.

Example 1.27

Let $S = \{1, 2, \ldots, k\}$; the number of subsets of size r that S has is $\binom{k}{r}$, as we saw earlier. Thus, if $\binom{k}{r}$ is summed for all possible values of r, from 0 to k, we arrive at the total number of subsets of S:

$$\sum_{i=0}^{k} \binom{k}{i} = \sum_{i=0}^{k} \binom{k}{i} x^i y^{k-i}, \quad \text{where } x = y = 1,$$

$$= (x + y)^k = (1 + 1)^k = 2^k.$$

A second way to establish this same result is to actually count the number of items in the list of all subsets of S. In making up this list of all subsets, k operations must be performed; we must go to each individual item in turn and either include it in the roster for the subset or not. That is, each operation can be performed in two ways, so the total number of ways to do all k operations is the product $2 \cdot 2 \cdots 2 = 2^k$.

Example 1.28

As in Example 1.22, assume that six letters (addressed to 1, 2, 3, 4, 5, and 6) are randomly assigned to their six envelopes. Let A_i, where $i = 1, 2, \ldots, 5$, be the event that the letter to i is paired with the correct envelope. (Note that if five of the letters are correctly assigned, so is the sixth.) Then the union $B = A_1 \cup A_2 \cup \cdots \cup A_5$ is the event that at least one letter is assigned to the correct envelope.

As seen earlier, $P(A_i) = \frac{1}{6}$ for each i; the probability that both 1 and 2 are assigned correctly is $P(A_1 \cap A_2) = (6-2)!/6! = 4!/6!$ since there are 4! 6-tuples with both 1 and 2 in their correct positions. Clearly, this is also the value for $P(A_i \cap A_j)$, for all $i \neq j$. Similarly, $P(A_1 \cap A_2 \cap A_3) = (6-3)!/6!$ since there are 3! 6-tuples with each of 1, 2, 3 in the correct positions; this is the probability for the intersection of any three of $A_1, A_2, \ldots, A_5$. In short, the probability of the intersection of any k of these events is $(6-k)!/6!$ for $k = 2, 3, 4, 5$.

Recall that $P(B) = P(A_1 \cup A_2 \cup \cdots \cup A_5)$ is given by the sum of the probabilities of all the individual events, minus the sum of the probabilities of all pairs, plus the sum of the probabilities of all triples, and so on. There are $\binom{6}{k}$ different intersections of k of these events, so

$$P(B) = \binom{6}{1}\frac{(6-1)!}{6!} - \binom{6}{2}\frac{(6-2)!}{6!} + \binom{6}{3}\frac{(6-3)!}{6!} - \binom{6}{4}\frac{(6-4)!}{6!}$$
$$+ \binom{6}{5}\frac{(6-5)!}{6!}$$
$$= 1 - \frac{1}{2!} + \frac{1}{3!} - \frac{1}{4!} + \frac{1}{5!} = .63333.$$

The probability of this union is approximately equal to the alternating series

$$1 - e^{-1} = 1 - \left(1 - 1 + \frac{1}{2!} - \frac{1}{3!} + \cdots\right) = .63212;$$

thus, if we assign n letters at random to their envelopes, the probability that at least one letter will be assigned the correct envelope is essentially $1 - e^{-1}$, no matter how large n may be.

In discussing permutations, it was assumed that the items being permuted were distinguishable, or different; interchanging the positions of any two of them gave rise to a new permutation. How might we evaluate the number of different four-digit numbers that can be made from the digits in 1221? Or perhaps evaluate the number of different "words" that can be constructed from the letters in *Tennessee*? Both of these examples involve counting the numbers of permutations of things that are not all different. The two 1's (and the two 2's) in 1221 are not distinguishable from each other, nor are the e's, n's, or s's in *Tennessee*. With the correct point of view, counting arrangements of things that are not all different is quite straightforward.

To have a general notation, assume we have n objects in total, of k different kinds. Let n_1 be the number of objects of the first kind, n_2 the number of objects of the second kind, and so forth, with n_k the number of objects of the kth kind. Then it must be true that $n = \sum_{j=1}^{k} n_j$; that is, if we sum the numbers of repetitions for all the kinds of objects, we get the total number of objects available. For example, the word *Tennessee* contains nine letters ($n = 9$) of which $n_1 = 1$ is T, $n_2 = 4$ are e's, $n_3 = 2$ are n's, and $n_4 = 2$ are s's, so $k = 4$ and $n = n_1 + n_2 + n_3 + n_4 = 1 + 4 + 2 + 2 = 9$.

Any permutation of these nine letters can be identified (or defined) by simply recording the position number(s) occupied by the letters of each type. One of these possible permutations is *sene Tseen*, in which T occupies position 5, the e's occupy positions 2, 4, 7, and 8, the n's occupy positions 3 and 9, and the s's occupy positions 1 and 6. Different permutations can then be distinguished by changes in the position numbers for the types of objects. This gives a simple way of counting the total number of permutations possible.

Using this discussion, we can determine the number of permutations possible with our general notation by evaluating the number of ways we can choose positions for the first type of object, followed by the number of ways we can choose positions for the second type of object, and so on, to counting the number of ways we can choose positions for the kth (final) type of object. Then we evaluate the product of these numbers using the multiplication principle. Thus, the number of ways to choose positions for the first type of object is $\binom{n}{n_1}$. At this point there are then $n - n_1$ positions still available, any n_2 of which could be chosen for the second type of object; that is, the number of ways to choose positions for the second type of object is $\binom{n-n_1}{n_2}$. There are then $\binom{n-n_1-n_2}{n_3}$ ways to choose positions for the third type of object. We proceed in this way until, having chosen positions for the first $k - 1$ types of object, we are left with $n - n_1 - n_2 - \cdots - n_{k-1} = n_k$ positions for the n_k objects of type k, in other words things come out just even, as they must. The total number of permutations possible, then, is the product of these numbers, which, after canceling common factors, is

$$\binom{n}{n_1}\binom{n-n_1}{n_2}\cdots\binom{n-\sum_{i=1}^{k-1} n_i}{n_k} = \frac{n!}{n_1!n_2!\cdots n_k!}.$$

If all n objects were different, then $n_i = 1$ for each i and $k = n$, the number of permutations possible is $n!$. If there are indistinguishable objects in the group, this number is simply divided by the product of the factorials of the numbers of repetitions of the various types of objects, since any permutation among them would result in an arrangement indistinguishable from the original. The quantity

$$\binom{n}{n_1 \; n_2 \; \cdots \; n_k} = \frac{n!}{n_1!n_2!\cdots n_k!}$$

is called a *multinomial coefficient*; in some senses it is an extension of the binomial (or combinatorial) coefficient used earlier. It appears in the distribution for a multinomial random variable (or vector), as will be seen in Section 5.3 and in the multinomial theorem (see exercise 31 below.)

Example 1.29

The number 1221 has $n = 4$ digits and contains $k = 2$ types of objects: $n_1 =$ two 1's and $n_2 =$ two 2's. Thus the number of different (distinguishable) four-digit numbers that could be constructed from these four symbols is $\binom{4}{2 \; 2} = 4!/2!2! = 6$; these 6 four-digit numbers are 1122, 1212, 1221, 2211, 2121, and 2112.

Consider the letters in the word *Tennessee*; this gives $n = 9$ objects in total: $n_1 = 1$ is T, $n_2 = 4$ are e's, $n_3 = 2$ are n's, and $n_4 = 2$ are s's. Then the total number of permutations of these nine letters is $\binom{9}{1\ 4\ 2\ 2} = 9!/1!\,4!\,2!\,2! = 3780$, rather than $9! = 362,880$, which would be possible with nine different letters.

EXERCISES 1.4

1. A state issues automobile license plates that bear all possible combinations of three digits, followed by three lowercase letters. How many different plates are possible? If they use all combinations of three digits, followed by *either* three lowercase or three capital letters (for all positions), how many plates are possible? If they use all combinations of three digits, followed by three letters, where each letter could be either lowercase or capital, how many plates are possible? If they use plates with six symbols, each of which may be a digit or a letter (and each letter either lower- or uppercase), how many plates are possible?

2. Six letters (addressed to 1, 2, 3, 4, 5, and 6) are randomly allocated to their six envelopes; what is the probability that the letters to 1 and 2 get into the correct envelopes?

3. Consider again the six school children assumed to line up at random for lunch in Example 1.21.
   a. What is the probability that a and b are *not* beside each other in line?
   b. What is the probability that a precedes b in line (whether immediately or not)? (*Hint*: Consider a sample space of 2-tuples, where the first element gives a position number for a and the second gives the position for b.)

4. Use the sample space of 2-tuples mentioned in part b of Exercise 3 to evaluate the probability that a and b are side by side in line (in either order).

5. Assume that 14 automobiles, bearing the numbers 1 through 14, enter a race; the first three across the finish line will receive prizes (of three different values). How many different ways could the prizes for first, second, and third place be distributed among the entrants? How many of these include first prize going to automobile 13?

6. Many computer outlets will mix or match components for customers. One of these outlets allows a customer to choose an Intel 80386 or 80486 processor, to have 2, 4, 6, or 8 MB of memory, with a monochrome, VGA, or superVGA display; in addition, the customer may choose to have no hard disk drive, or a hard disk drive with 80, 160, or 320 MB of storage. How many different computers can be assembled by this outlet?

7. A pizza parlor offers its customers their choice(s) of six different toppings, each of which may or may not be placed on the basic pizza, which has a crust, tomato sauce, and mozzarella cheese. How many different pizzas are available?

8. The word *algorithm* contains nine different letters. With no repetitions of letters, how many different nine-letter words can be made with these letters (where a word is an arrangement of the letters, whether or not it is meaningful in some language)? How many six-letter words with no repetitions are possible ? How many of these contain one or more vowels? If one of these six-letter words is selected at random, what is the probability that it contains no vowels?

9. The owner of a small business employs 15 people, of whom five are male and 10 are female. Due to economic difficulties the owner must lay off five people. If the five to be laid off were selected totally at random, what is the probability they would all be female? all male? What is the probability that among those laid off is at least one male and at least one female?

10. *For poker players* Count the number of five-card hands that contain the following:
    a. a full house (three cards of one denomination and 2 of another)
    b. a flush (all five cards from one suit)
    c. a straight (five cards in sequence, starting with ace, 2, 3, ... , or 10, regardless of suit)
    d. three of a kind (three cards from one denomination, plus two more of different denominations)
    e. two pairs (two cards from one denomination, two from a second, plus one more card of a different denomination)
    f. a pair (two cards from one denomination, plus three more, each of different denominations)

11. Reconsider the Little League team with 15 members, two of whom are named Joe and Hugh (Example 1.26). To determine the starting lineup the coach puts 15 slips of paper into a box, numbered 1 through 15. Each team member draws one of these slips, and those selecting the numbers 1 through 9 will be on the starting lineup. What is the probability Joe is in the starting lineup? that both Joe and Hugh start the game?

12. Assume a small high school has nine boys out for basketball.
    a. Ignoring the assignment of individuals to positions, how many different five-man starting teams could be fielded by this school?
    b. What is the probability that any specific individual would be on the starting team, if the selection of members is made at random?
    c. Let the letters a, b, c, d, e, f, g, h, and i represent the nine players available. What is the probability the starting team includes a, c, e, and g if the selection of members is made at random?

13. A California municipal court district assembles a panel of 45 persons; they are to be available to serve as jurors for trials scheduled during a three-week period. When a trial is scheduled, an initial collection of 18 of these persons will be selected at random from the panel of 45 for questioning by the judge and the attorneys for the two sides of the case. If you are one of the 45 persons on the panel, what is the probability that you will be among the initial 18 selected? Suppose both you and your neighbor are on the panel; what is the probability that both of you will

be among the initial 18? that at least one of you two will be included in the initial selection?

14. Assume that a professional football team is allowed to carry 44 players on its traveling roster.

   a. How many different 11-man starting groups could be fielded by such a team? (Ignore position assignments.)

   b. Assume that the starting group will always include seven particular players, and that the remaining four are selected at random from 10 of the other travelers (called $a, b, c, d, e, f, g, h, i$, and j). What is the probability that h, i, and j will start a given game?

   c. Use the information in part b to determine the probability that a, b, c, and d will start the game.

15. A box contains four white pills, indistinguishable to the eye; two tablets are aspirin, the other two are cold medication. Doug will select one pill at random to ingest, and then Hugh will select one at random. Since there are two of each type of pill, it is possible that both will select an aspirin. Let D be the event that Doug selects an aspirin and let H be the event that Hugh selects an aspirin.

   a. Using $S = \{aa, ac, ca, cc\}$ as a sample space, where aa indicates that both select an aspirin, with equally likely outcomes, evaluate the probabilities of occurrence for D, H, and $H \cap D$.

   b. Now use $S = \{(x_1, x_2) : x_i = 1, 2, 3, 4, x_i \neq x_j\}$, where the aspirin tablets are numbered 1, and 2 and the cold medication tablets are numbered 3 and 4, and assume equally likely outcomes. Reevaluate the probabilities asked for in part a.

   Why are these different?

16. Extend Pascal's triangle, presented in the text, to include the values for $\binom{10}{r}$.

17. *Vandermonde convolution*   If m_1, m_2 are positive integers, show that

$$\sum_{j=0}^{n} \binom{m_1}{j} \binom{m_2}{n-j} = \binom{m_1 + m_2}{n},$$

where $n = \min(m_1, m_2)$, the smaller of m_1 and m_2. (*Hint:* Let $m_2 > m_1$. Each subset of size m_1 selected from $S = \{1, 2, \ldots, m_1, m_1 + 1, \ldots, m_1 + m_2\}$ must contain j elements from $\{1, 2, \ldots, m_1\}$ and $m_1 - j$ elements from $\{m_1, m_1 + 1, \ldots, m_1 + m_2\}$.)

18. Is it possible to define a set S that has 11 subsets in total?

19. Consider the ratio $\binom{n}{r} / \binom{n}{r+1}$; if the ratio is less than 1, then $\binom{n}{r} < \binom{n}{r+1}$. Compare the value for this ratio with 1 to find the largest possible value for $\binom{n}{r}$, where n is fixed.

20. Suppose $S = \{1, 2, \ldots, k\}$; how many subsets of S have 1 as an element?

21. There are n people in the same room, each of whom has a birthdate (month and day of month, not including year). Assume that each year has only 365 dates (ignore leap years) and that each day is equally likely to be the birthdate for any particular person. What is the smallest value

for n such that $P(A) > \frac{1}{2}$, where A is the event that two or more people in the room share the same birthdate?

22. Each of n women toss their hats into the air in a strong wind; the hats are subsequently returned at random to the women. What is the probability that none of them receives her own hat?

23. How many different four-digit numbers are there in our usual system of recording numbers? (Remember, the first digit cannot be 0.)

    a. Suppose one of these four-digit numbers is selected at random from the totality possible. Evaluate the probability that the number selected contains no 3's.

    b. Again suppose one of these four-digit numbers is selected at random, and evaluate the probability that the selected number contains exactly two 5's.

    c. Once again suppose one of these four-digit numbers is selected at random, and evaluate the probability that the selected number contains (exactly) one 8 and (exactly) one 9.

24. Evaluate the number of 10-letter words (meaningful or not) that can be made with the letters in *statistics*.

    a. Count the number of these 10-letter words that have all three s's together; thus also evaluate the probability that all three s's are side by side if these 10 letters are placed at random in a row.

    b. What is the probability that all three s's occur together *and* all three t's occur together, in the same word?

25. A display tub in a retail store contains 200 packages of light bulbs, each package containing two bulbs. Fifteen of the packages contain one or more defective bulbs, and the other 185 packages each contain two good bulbs. You select three of the packages at random from the tub to purchase.

    a. Describe a reasonable sample space of equally likely outcomes for this experiment, and evaluate the number of elements it has.

    b. Using the sample space in part a, evaluate the probability that all the bulbs you purchase are good.

    c. Again using the sample space from part a, evaluate the probability that you select three bad packages. Is this event the complement of the event in part b?

26. Suppose you have a deck of 20 cards, all of which have identical backs; 10 of the fronts are green, while the remaining 10 are red. A "clairvoyant" claims he can correctly identify the color of at least 18 of these cards without seeing their faces; that is, with all 20 cards lying face down in front of him, he will split them into two groups of 10, claiming that all those in one group are red and all those in the second group are green. What is the probability he will correctly identify at least 18 if he has no special talent and is merely guessing the color for each card?

27. Frequently advertisers of modern products mention how well a product does in practice; generally these statements describe only part of the situation. Suppose a major racing car event has 30 cars entered. Of these

30 cars, assume that 20 use motor oil A, while six use B, and four use C.

a. Keeping track only of the motor oil used by the cars, how many different orderings could be observed as the cars pass the finish line of the race?

b. Assume that all the cars and all the motor oils are equally good, so that all the orders listed in part a are equally likely. What is the probability that the winning car uses oil A?

c. Using the assumptions of part b, find the probability that both of the first two finishers use oil A.

d. What is the probability that all of the first three finishers use oil C?

28. *Only for the foolhardy*   Consider the letters in the word *proper*, and count the total number of words that can be constructed using only these letters, each at most once. (Sum the number of words possible with one letter, two letters, and so on up to six letters.)

29. Given a sports league with nine teams as members, how many different pairs of teams are possible — that is, what is the number of games necessary for each team to play every other one time? If $\mathcal{A}$ is one of these teams, how many games does it play in? If each team is to play every other team five times in the season, how many games are required?

30. We have seen two different ways to show that a set with k elements has 2^k subsets, one of which involves the binomial expansion of $(1+1)^k = 2^k$; examine the expansion of $(1-1)^k = 0$ to show that the sum of every other term in

$$\sum_{i=0}^{k} \binom{k}{i},$$

starting with either the $i = 0$ or $i = 1$ term, gives 2^{k-1}.

31. The multinomial coefficients occur in the expansion of the sum of three or more terms raised to the nth power. That is, if n is a positive integer, and x, y, and z are any real numbers, then

$$(x+y+z)^n = \sum_i \sum_j \sum_k \binom{n}{i\,j\,k} x^i y^j z^k.$$

where the sum is over all 3-tuples (i, j, k) for which $i = 0, 1, 2, \ldots, n$, $j = 0, 1, 2, \ldots, n$, $k = 0, 1, 2, \ldots, n$, and $i + j + k = n$. Show this result to be true.

32. Suppose there are 100 registered voters in a small town, 60 of whom favor a bond issue and the remaining 40 do not. The town treasurer selects 10 of these registered voters at random and asks each one whether or not he or she supports the bond issue.

a. How many different samples of 10 voters could be selected?

b. What is the probability that six of those polled are in support of the bond issue?

c. What is the probability that a majority (six or more) of those polled are in support of the bond issue?

1.5 | Conditional Probability

Conditional probability provides a powerful tool that is useful in many circumstances. It reflects the power of conditional reasoning, employed quite naturally by most people, although most might not bother to categorize it as such.

Suppose you are a contestant on a television game show and are allowed to pick one of three envelopes (one red, one green, and one blue). You will be paid the number of dollars written on the piece of paper contained in the envelope you choose. The papers in two of the envelopes have "100" written on them, and the third has "1,000,000" written on it. If you choose one of the envelopes at random, the probability you will receive the $1,000,000 payment is $\frac{1}{3}$.

If the master of ceremonies selects the red envelope first and shows you that it contains a paper with "100" written on it before you make your choice, the probability that you will win the $1,000,000 payment will then be $\frac{1}{2}$, not $\frac{1}{3}$ as before, since you will not then select the red envelope. This latter value is different, of course, precisely because the circumstances have changed: You are now selecting one of two, not one of three, envelopes to determine your prize; this is an example of a *conditional probability* computation, the result of already knowing that the red envelope does not contain the $1,000,000 paper.

Let us introduce the notation employed for conditional probability in the context of this example. Let A represent the event that you will win the $1,000,000 payment; with no further information beyond the fact that one of the three envelopes contains the 1,000,000 number, you know that $P(A) = \frac{1}{3}$ as already noted. Let B represent the event that the red envelope is shown to contain the number 100. We shall denote the probability that you will win the big prize, given that B occurred, by $P(A \mid B)$, which is read "the conditional probability of A occurring given that B has occurred" or the probability of A given B. In this case, $P(A \mid B) = \frac{1}{2}$, which is greater than $P(A)$.

In this example it is easy to see that we have two different values for the probability of A occurring, one conditional and one not. Now let us discuss the general definition for the conditional probability of an event A occurring given that B is known to have occurred. Suppose we have an experiment with sample space S and that A and B where $(A \subset S, B \subset S)$ are two events that may occur. If B has already been observed, then the particular element of S that was the actual outcome must have been an element of B. Therefore conditional reasoning tells us that the only possible elements to have occurred are those that belong to B. (In a sense B becomes the sample space.) If A is to occur given B has occurred, it follows that the final outcome observed would have to be an element of $A \cap B$, since *both* A and B occurred. Thus a rational measure of this conditional probability of A occurring is provided by the ratio $P(A \cap B)/P(B)$, the proportion of the time that both occur relative to the proportion of the time that B occurs. This is in fact the definition of the conditional probability of A, given B has occurred, as given in the following definition.

DEFINITION 1.13

In an experiment with sample space S, let B be any event such that $P(B) > 0$. Then the *conditional probability of A occurring, given that B has occurred*, is

$$P(A \mid B) = \frac{P(A \cap B)}{P(B)}, \qquad (1)$$

for any $A \subset S$.

Note in this definition that the probability of the conditioning event B must be positive. This is necessary, of course, since division by 0 is undefined. For finite sample spaces S, events with positive probability coincide with events that might in fact be observed. (There would be no use in conditioning on any other type of event.) One of the exercises at the end of the section asks you to verify that this definition for a conditional probability, denoted $P(A \mid B)$, satisfies the Kolmogorov axioms.

Granted that conditional probabilities satisfy the axioms, it follows then that all their consequences hold true; in particular, given B has occurred, the conditional probability that A did not occur is 1 minus the probability it did occur: $P(\overline{A} \mid B) = 1 - P(A \mid B)$. As we will see, the conditional probability that A occurs, given that B has occurred, may be larger than, equal to, or smaller than $P(A)$, the unconditional probability of A occurring.

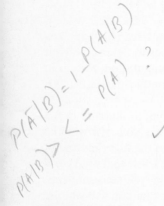

Example 1.30

Conditional reasoning is useful in solving many problems. Suppose a person approaches the locked door to his home in the dark; any one of the four keys on the key ring in his pocket may be the correct one to unlock the door. If he selects one of them at random to try in the door, the probability that it will be the correct one is $\frac{1}{4}$. Assuming he will not try the same key twice, the conditional probability that the second key he tries is the correct one, given the first was not, should be given by $\frac{1}{3}$.

Let us formally derive this value using Definition 1.13. Assume the keys are numbered 1 through 4 and that key number 1 is the correct key for this door. Since the owner will not try the same key twice, a reasonable sample space S contains the $4! = 24$ different orderings in which he might attempt to insert the keys into the door lock. If he is really in the dark and cannot identify the correct key, this sample space has equally likely outcomes. Let B be the event that he does *not* select the correct key on the first try, and let A be the event that he *does* select the correct key on the second try. Then B contains all the 4-tuples in S that do not contain 1 (the correct key) in the first position; there are $3! = 6$ permutations with 1 in the first position, leaving 18 permutations that do not have 1 in the first position. Thus $P(B) =$

$\frac{18}{24} = \frac{3}{4}$. To compute $P(A\,|\,B)$ we need to find $P(A \cap B)$. $A \cap B$ contains all permutations that have 1 in position 2 and do not have 1 in position 1; this is equivalent to those permutations with 1 in position 2 (since 1 cannot be used twice). Thus $A \cap B$ contains six elements of S, and $P(A \cap B) = \frac{6}{24} = \frac{1}{4}$. Thus $P(A\,|\,B) = P(A \cap B)/P(B) = \frac{1}{3}$ as expected.

If we know $P(A \cap B)$ and $P(B)$, then their ratio gives $P(A\,|\,B)$, the conditional probability for A occurring. A very common and powerful use of conditional probability is given by rewriting Eq. (1) as

$$P(A \cap B) = P(B)P(A\,|\,B).$$

Since

$$P(B\,|\,A) = \frac{P(B \cap A)}{P(A)}$$

and $B \cap A = A \cap B$, the probability of the intersection of A and B can also be written $P(A \cap B) = P(A)P(B\,|\,A)$.

Example 1.31

Sampling without replacement can be modeled quite easily with conditional probability. Suppose the milk case in your supermarket has 40 cartons of the milk you buy, and 10 of them are spoiled (which is not evident from the outside of the cartons). You purchase two cartons. What is the probability that you don't get any spoiled cartons? Let A be the event that the first carton you select is not spoiled, and let B be the event that the second carton is not spoiled. Assuming your selections are made at random, $P(A) = \frac{30}{40}$. Given A (the first carton you selected was not spoiled) has occurred, there are then 39 cartons of milk remaining in the display, of which 10 are still spoiled, so $P(B\,|\,A) = \frac{29}{39}$. The intersection, $A \cap B$, is the event that you get no spoiled cartons; thus $P(A \cap B) = P(A)P(B\,|\,A) = \left(\frac{30}{40}\right)\left(\frac{29}{39}\right) = \frac{87}{156} = .558$. If you select a third carton, the probability that none of the three is spoiled is, in the same way, $\left(\frac{30}{40}\right)\left(\frac{29}{39}\right)\left(\frac{28}{38}\right) = 2436/5928 = .411$, the product of one unconditional probability with two conditional probabilities.

This experiment illustrates a fact that is true in general. It is also counterintuitive for many persons, at least when first encountered. Let B be defined as before, that the second carton you select is not spoiled. We have seen that the probability the first one selected is unspoiled is $P(A) = \frac{3}{4}$. What is the value for $P(B)$? Recall that we can write $B = (B \cap A) \cup (B \cap \overline{A})$, since B can happen either with or without A. Also, since $A \cap \overline{A} = \emptyset$, these two intersections making up B are mutually exclusive, and $P(B) = P(B \cap A) + P(B \cap \overline{A})$. Since $P(B \cap A) = P(A)P(B\,|\,A) = \left(\frac{30}{40}\right)\left(\frac{29}{39}\right)$ and $P(B \cap \overline{A}) = P(\overline{A}) \times P(B\,|\,\overline{A}) = \left(\frac{10}{40}\right)\left(\frac{30}{39}\right)$, then $P(B) = \left(\frac{30}{40}\right)\left(\frac{29}{39}\right) + \left(\frac{10}{40}\right)\left(\frac{30}{39}\right) = \frac{30}{40} = P(A)$. Thus the unconditional probability that the second carton selected is unspoiled is the same as the probability that the first is unspoiled.

Notice that the value for $P(B)$ is an average of the two conditional probabilities $P(B\,|\,A)$ and $P(B\,|\,\overline{A})$; the value the average takes is the same as the value of the unconditional probability that the first carton is not spoiled. If we think of probabilities as being relative frequencies, this relationship says that if this experiment were repeated a large number of times, the proportion of time that the second carton selected is unspoiled (unconditionally) is the same as the proportion of time that the first is unspoiled. In the same way, the unconditional probability that the third carton purchased is unspoiled (if a third one were taken) is also $\frac{30}{40}$, as is the probability for the fourth, fifth, and so on, up to the 40th. Equivalently, if you think of the 40! different orderings in which these cartons could be removed from the case, the number of 40-tuples in which the first carton is unspoiled is equal to the number of 40-tuples with the second one unspoiled.

The representation of the probability of an intersection as a product is easily extended to any number of events. If A, B, and C are any three events, we have

$$P(A \mid B \cap C) = \frac{P(A \cap B \cap C)}{P(B \cap C)}.$$

It follows by multiplication that

$$P(A \cap B \cap C) = P(B \cap C)P(A \mid B \cap C);$$

since $P(B \cap C)$ can be written $P(C)P(B \mid C)$, then

$$P(A \cap B \cap C) = P(C)P(B \mid C)P(A \mid B \cap C).$$

The probability of the intersection of k events can always be written as the product of k terms: the unconditional probability of any one of them, times the conditional probability of a second of them given the first, and so forth, with the final term being the conditional probability of the last of them given *all* the preceding events have occurred. That is, if A_1, A_2, ..., A_k are k events, then

$$P(A_1 \cap A_2 \cap \cdots \cap A_k) = P(A_1)P(A_2 \mid A_1) \cdots P(A_k \mid A_1 \cap A_2 \cap \cdots \cap A_{k-1}).$$

A sequence of events $A_1, A_2, \ldots, A_k, \ldots$ is said to partition the sample space S if the following definition is satisfied.

DEFINITION 1.14

Let $A_1, A_2, \ldots$ be a sequence of events, defined for an experiment with sample space S. These events are called a *partition of S* if

1. $A_i \cap A_j = \emptyset$ for all $i \neq j$.
2. $A_1 \cup A_2 \cup \cdots \cup A_k \cup \cdots = S$.

In short, a group of events partition S if they cut it into nonoverlapping "pieces" and together account for the whole sample space S (not leaving out any portion of S); from the third Kolmogorov axiom then it follows

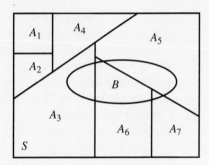

Figure 1.12   Partition of S

that $\sum_{i=1}^{\infty} P(A_i) = 1$. If S has k elements, then the k single-element events themselves provide a partition of S. Figure 1.12 pictures a sample space S and a group of events that partition S into seven pieces; note they also partition event B.

If $A_1, A_2, \ldots,$ partition the sample space S, they also partition *any* event B: $B = (A_1 \cap B) \cup (A_2 \cap B) \cup \cdots,$ and these intersections are mutually exclusive. Since a probability function is additive for unions of mutually exclusive events, it follows that $P(B) = \sum_{i=1}^{\infty} P(A_i \cap B)$; recall that $P(A_i \cap B) = P(B \mid A_i)P(A_i)$ so we have $P(B) = \sum_{i=1}^{\infty} P(B \mid A_i)P(A_i)$. Formally stated as follows, this is called the *theorem of total probability*.

THEOREM 1.7   If $A_1, A_2, \ldots$ is a partition of S, and B is any event, then $P(B) = \sum_{i=1}^{\infty} P(B \mid A_i)P(A_i)$.

The unconditional probability $P(B)$ is a weighted average of the conditional probabilities $P(B \mid A_i)$, with weights $P(A_i)$. This result was used earlier and is illustrated once again in the following example.

Example 1.32

Suppose a calculator manufacturer buys her main processor chip from three different suppliers; call these suppliers s_1, s_2, and s_3. From past experience, 1% of the circuits supplied by s_1, 4% of the circuits supplied by s_2, and 2% of the circuits supplied by s_3 have been defective. Granted that this manufacturer buys 30% of her circuits from s_1, 10% from s_2, and the rest from s_3, we can use the theorem of total probability to compute the probability that one of these circuits is found to be defective when checked before being assembled into a calculator. Let B be the event that the circuit is found to be defective, and let A_i be the event that the circuit was supplied by s_i, for $i = 1, 2, 3$; it is clear that A_1, A_2, A_3 partition S (since the circuit must have been produced by one of these suppliers) and that they partition B as well. From the information given, then, $P(A_1) = .3$, $P(A_2) = .1$, and $P(A_3) = .6$, while $P(B \mid A_1) = .01$, $P(B \mid A_2) = .04$, and $P(B \mid A_3) = .02$. Thus

$$P(B) = (.01)(.3) + (.04)(.1) + (.02)(.6) = .019,$$

from the theorem of total probability.

One of the early results in probability theory, published more than 200 years ago, is named after the Reverend T. Bayes. This theorem expresses the way one must modify initial values of probabilities (called their *prior values*) after having observed an event (or observed data) to yield the subsequently implied values for the same probabilities (called their *posterior values*). It provides the basis for Bayesian approaches to statistical inference.

From the theorem of total probability, and granted that $A_1, A_2, \ldots$ form a partition of S, we have $P(B) = \sum_{i=1}^{\infty} P(B \mid A_i) P(A_i)$. It follows that

$$P(A_i \mid B) = \frac{P(A_i \cap B)}{P(B)}$$

$$= \frac{P(A_i) P(B \mid A_i)}{\sum_{i=1}^{\infty} P(B \mid A_i) P(A_i)}$$

for any event A_i in the partition $A_1, A_2, \ldots$. This result is known as *Bayes' theorem*. Notice it expresses the relationship between conditional probabilities, where the roles of the events have been reversed, showing how $P(B \mid A_1)$, say, is related to $P(A_1 \mid B)$. The following two examples illustrate the use of this theorem.

Example 1.33

Two friends carefully examine a coin; A_1 says the coin is fair, that the probabilities of getting a head or a tail are equal. A_2 says the coin is biased, that the probability of getting a head is 3 times as likely as getting a tail. Let us agree there is a 50% chance that either of these opinions is correct, and treat these opinions as being events, so $P(A_1) = P(A_2) = \frac{1}{2}$; $P(A_1)$ and $P(A_2)$ are called the prior probabilities of these two events (or opinions). If B is the event that a head occurs on one flip of the coin, we have $P(B|A_1) = \frac{1}{2}$, and $P(B \mid A_2) = \frac{3}{4}$.

The coin is flipped one time and a head occurs, thus B occurs. We can then use Bayes' theorem to reevaluate the probabilities of A_1 and A_2 in the light of B occurring. We find that $P(B) = \left(\frac{1}{2}\right)\left(\frac{1}{2}\right) + \left(\frac{1}{2}\right)\left(\frac{3}{4}\right) = \frac{5}{8}$ and $P(A_1 \mid B) = P(A_1)P(B \mid A_1)/P(B) = \left(\frac{1}{2}\right)\left(\frac{1}{2}\right)\left(\frac{8}{5}\right) = \frac{2}{5}$, while $P(A_2 \mid B) = \frac{3}{5}$. This evidence (one head in one flip) says we should now give A_1 only a 40% chance of being right, with A_2 having a 60% chance.

Bayes' theorem also allows us to evaluate the probabilities of A_1 and A_2 given a tail $(\overline{B})$ occurred on the coin flip. As may not be immediately intuitive, this updating of the probabilities of A_1 and A_2 is not symmetric; you will find that $P(A_1 \mid \overline{B}) = \frac{2}{3}$. The occurrence of a tail on flipping the coin has a greater impact on the posterior probability of A_1 than does the occurrence of a head on the posterior probability of A_2.

Example 1.34

Assume that the probability is .95 that the jury selected to try a criminal case will arrive at the correct verdict whether innocent or guilty. Further, suppose that the local police force is quite diligent in performing its function, that 99% of the people brought to trial are in fact guilty. Given that a jury finds

Table 1.4

Results for 10,000 Cases

		Defendant Is		
		Guilty	*Innocent*	*Total*
Jury	*Guilty*	9,405	5	9,410
decides	*Innocent*	495	95	590
	Total	9,900	100	10,000

a defendant innocent, what is the probability that she is in fact innocent? To use the notation of Bayes' theorem, let A_1 be the event that the defendant is guilty, and let $A_2 = \overline{A}_1$ be the event that she is innocent. These two events then partition the sample space, the full collection of possibilities. Further, let B be the event that the defendant is found innocent (and then $\overline{B}$ is the event she is found guilty). We want to evaluate $P(A_2 \mid B)$, the probability the defendant is innocent, given that the jury finds her innocent.

The information we are given is $P(B \mid A_2) = P(\overline{B} \mid A_1) = .95$, $P(A_1) = .99$, which then also gives $P(\overline{B} \mid A_2) = P(B \mid A_1) = .05$ and $P(A_2) = 1 - P(A_1) = .01$. Thus

$$P(A_2 \mid B) = \frac{P(A_2)P(B \mid A_2)}{P(A_1)P(B \mid A_1) + P(A_2)P(B \mid A_2)}$$

$$= \frac{(.01)(.95)}{(.99)(.05) + (.01)(.95)} = .161.$$

It may strike you that this value is too low in light of the information given; it is easy to compare this probability with the wrong quantity. Before this person went to trial the probability she was innocent was assumed to be .01; after the innocent verdict was given, this value has increased to .161. Table 1.4 presents the results of 10,000 cases, in the proportions used for this example. The column totals of 9900 and 100 represent the fact that 99% of the people brought to trial are guilty; each of these groups is then split into those found guilty and innocent by the jury (the 95%–5% split). The second row shows that a total of 590 persons are found innocent, of whom 95 ($95/590 = .161$) are in fact innocent while the rest are actually guilty.

Conditional probability can at times involve subtle points that can be easily overlooked. The following example discusses a problem, presented in a nationwide Sunday newpaper magazine supplement, that apparently generated a great deal of interest and caused much confusion as well. This same example had been the subject of a fairly lengthy series of letters published in *The American Statistician* some 17 years earlier.

Example 1.35

An occasional columnist for a national Sunday newspaper magazine supplement presents questions and answers in her column. Recently she presented a probability problem involving a television game show, similar in content to the discussion at the start of this section. The problem was posed as follows.

You are a contestant on a television show and will be allowed to select one of three closed doors; you will keep what you find behind the door you choose. The master of ceremonies tells you that an expensive new automobile is behind one of the doors, while each of the other two open to a goat. You select your door (without opening it), and the master of ceremonies then opens one of the two remaining doors and shows you the goat behind his door. You are then allowed to keep the door you initially selected or shift to the other door that has not been opened. The question posed by the columnist was "Should you keep the door you initially selected or switch to the other unopened door?" She also stated that the contestant should definitely make the switch of doors.

Several weeks later the columnist reported that this question had brought forth a great deal of mail, all of it essentially saying there would be no point in switching doors since the (conditional) probability of getting the automobile is $\frac{1}{2}$ with either door, after the master of ceremonies opens a door. The columnist then went on to claim that the person should switch, and that by so doing the contestant's chance of winning the automobile increases from $\frac{1}{3}$ to $\frac{2}{3}$. She based her reasoning on the following: Suppose the doors are numbered 1, 2, 3, and you choose door 1. The true situation then must be one of the three lines in the following table:

Door 1	Door 2	Door 3
Automobile	Goat	Goat
Goat	Automobile	Goat
Goat	Goat	Automobile

Notice that only one of these three original possibilities includes the contestant winning the automobile by staying with door 1, whereas two out of these three have the automobile behind either door 2 or door 3. Once the master of ceremonies has opened a door (always one with a goat), then two times out of three the contestant would benefit by changing to the other door. The automobile is equally likely to be found behind any door; the subsequent fact that you will be shown a goat behind one of the doors you did *not* choose then makes it more likely that the automobile is behind the remaining door you did not choose.

This reasoning might be more easily understood if looked at from the perspective of the producer of the show. If this choice is offered 300 times (remember that the master of ceremonies always opens a door to a goat) and the contestant always stays with his or her initial choice, the automobile would be won $\frac{1}{3}$ of the time (so 100 automobiles are given away). With this strategy it follows that no automobile is given away the remaining 200 times. If the contestant were to switch doors every time, these two numbers would be reversed.

For any event A with $0 < P(A) < 1$, the theorem of total probability assures us that

$$P(B) = P(B \mid A)P(A) + P(B \mid \overline{A})P(\overline{A}),$$

where B is any event; since the probabilities of A and $\overline{A}$ are positive and sum to 1, $P(B)$ is an average of the two conditional probabilities $P(B \mid A)$ and $P(B \mid \overline{A})$ and thus must lie between their two values. Conditional probabilities also satisfy the Kolmogorov axioms, so all of their consequences follow; in particular, this averaging effect is also true for conditional probabilities, *so long as the conditioning is on the same event.* If we manipulate conditional probabilities, where the conditioning events are not the same, some subtle and possibly nonintuitive results can occur. This phenomenon is illustrated in the following example, which discusses a set of observed data reported in a sociology journal.

Example 1.36

In 1981, M. Radelet reported the following observed data in the *American Sociological Review* (vol. 46, pp. 918–927). These data record the results of 326 murder trials held in 20 Florida counties in the years 1976 and 1977, in each of which the defendant was convicted; his article was concerned with the imposition of the death penalty in murder trials for white defendants versus black defendants. The data reported are given in Table 1.5.

Let us use these observed data as the basis for the probability model of the penalty imposed on a convicted murderer. To illustrate some conditional probability computations, define the events

D_W : The defendant is white. $\qquad D_B$: The defendant is black.
V_W : The victim is white. $\qquad V_B$: The victim is black.
$\quad\ Y$: Death penalty is imposed. $\qquad N$: Death penalty is not imposed.

Note that three pairs of these events are complementary. With the data given in Table 1.5 specifying the model for relative frequencies (probabilities), then, we have

$$P(Y) = \tfrac{36}{326}, \qquad P(N) = \tfrac{290}{326}, \qquad P(D_W) = \tfrac{160}{326}, \qquad P(V_B) = \tfrac{112}{326},$$

and so on. Conditional probabilities can now be evaluated directly from partial tables of these data; for example, by adding the rows for the two types of victims, we arrive at Table 1.6.

We can then see that

$$P(Y \mid D_W) = \tfrac{19}{160} > P(Y \mid D_B) = \tfrac{17}{166};$$

Table 1.5
Imposition of Death Penalty

Race of		Death Penalty		
Defendant	*Victim*	*Yes*	*No*	*Total*
White	White	19	132	151
	Black	0	9	9
Black	White	11	52	63
	Black	6	97	103
	Totals	36	290	326

Table 1.6
Imposition of Death Penalty

| | **Death Penalty** | | |
Defendant	*Yes*	*No*	*Total*
White	19	141	160
Black	17	149	166
Totals	36	290	326

the conditional probability that a white defendant will receive the death penalty is larger than the conditional probability that a black defendant will. If we look back to the original data in Table 1.5 we notice that

$$P(Y \mid D_W \cap V_W) = \tfrac{19}{151} < \tfrac{11}{63} = P(Y \mid D_B \cap V_W),$$
$$P(Y \mid D_W \cap V_B) = \tfrac{0}{9} < \tfrac{6}{103} = P(Y \mid D_B \cap V_B).$$

The conditional probability that a white defendant will receive the death penalty is smaller than the probability that a black defendant will, granted the victim is white; this is also true if the victim is black, which may not seem possible since $P(Y \mid D_W) > P(Y \mid D_B)$. What causes this apparent reversal?

The value for $P(Y \mid D_W)$ is an average of $P(Y \mid D_W \cap V_W)$ and $P(Y \mid D_W \cap V_B)$; the value for $P(Y \mid D_B)$ is an average of $P(Y \mid D_B \cap V_W)$ and $P(Y \mid D_B \cap V_B)$. Since these two averages are taken with respect to two different pairs of probabilities, reversals in value like this may occur. This is an example of a phenomenon that has come to be known as *Simpson's paradox*, named after the author of an article written in 1951, although it had apparently been described by earlier writers.

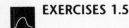 **EXERCISES 1.5**

1. A regular 52-card deck is shuffled well; the top two cards are then turned over, one at a time. Evaluate the probability that
   a. The top card is red.
   b. The second card is red, given that the first was red.
   c. Both of the first two cards are black.
   d. The first two cards are of different colors.
2. A fair coin is flipped three times. What is the probability that the second flip gives a head, given that a head occurred on the first flip?
3. A fair coin is flipped three times. What is the probability that the second flip gives a head, given that (exactly) one head occurred on the three flips?
4. A fair coin is flipped three times. What is the probability that the second flip gives a head, given that at least one head occurred on the three flips?
5. Given an experiment with sample space S, show that

$$P(A \mid B) = \frac{P(A \cap B)}{P(B)}$$

satisfies the Kolmogorov axioms for a probability function.

6. Show that the probability of the union of two events can be written
$$P(A \cup B) = P(A) + P(B)P(\overline{A} \mid B).$$

7. An urn contains one red ball and one blue ball. A ball is selected at random from the urn. If the ball is red, it is simply put back into the urn; if the ball selected is blue, then it is put back, along with another blue ball (so the urn then contains three balls). Then another ball is drawn at random from the urn. What is the probability that the second ball drawn is red? that it is blue?

8. A display case contains 100 individually packaged 60-watt light bulbs, two of which are defective and will not operate. You observe two people in front of you remove two bulbs each; you then remove one yourself to purchase.
   a. What is the probability that all of the bulbs available are good when you make your selection?
   b. What is the probability that the bulb you select is defective?

9. A small chest has three drawers, labeled A, B, and C, each of which contains a single coin. The coin in drawer A is fair: One face is called heads, the other tails, and each is equally likely to occur if the coin is flipped. The coin in drawer B has heads on both faces, and the coin in drawer C has tails on both faces. One drawer is selected at random; the coin within is selected and flipped one time. What is the probability that a head is observed?

10. Consider again the case of two friends A_1, and A_2 who disagree on the probability of a head occurring (event B) when a coin is flipped. Assume that $P(A_1) = P(A_2) = \frac{1}{2}$ (i.e., they are equally likely to be correct) and that the conditional probabilities of observing B are $P(B \mid A_1) = p_1$ and $P(B \mid A_2) = p_2$, respectively. Show that $P(A_2 \mid B) = P(A_1 \mid \overline{B})$ (i.e., that the posterior probability of A_2 given a head is the same as the posterior probability of A_1 given a tail) requires $p_1 + p_2 = 1$ (i.e., that their initial opinions of the probability of a head occurring are symmetric about $\frac{1}{2}$).

11. Return to the television show, with the doors, goats, and automobile (Example 1.35). Suppose again that you choose door 1, but this time you are also told that if the master of ceremonies has a choice of which door to open, he will flip a coin in your presence to decide which of the two doors he will open. (He will not flip a coin if he has no choice.)
    a. Should you switch if you see him flip a coin?
    b. Should you switch if he opens door 2 without flipping a coin?

12. Assume that among married couples with exactly two children there are equal numbers with (boy, boy), (girl, girl), (girl, boy) (boy, girl), where the order in the 2-tuple indicates the order of birth of the children. One of these families is selected at random.
    a. What is the probability that the selected family has two girls, given that it has at least one girl?
    b. Suppose you are currently married and have one child (a girl) and are expecting your second child; clearly, after the birth of your second child you will have at least one girl in your family of two children. Is the answer to part a the probability that your second child will be a girl?

13. In a given state, the conditional probability that a voter selected at random is registered as a Republican, given he is male, is smaller than the probability he is registered as a Democrat. The same situation holds for females; that is, the conditional probability that a voter is registered Republican is smaller than the probability she is registered as a Democrat. Does it follow that the (unconditional) probability that a voter selected at random is registered as a Republican is smaller than the probability of being registered as a Democrat?

14. Sixteen teams (including your favorite) enter a single-elimination tournament; thus a total of four rounds will be required to determine the winner. Assume that the probability of your favorite team winning its first game is .9, while the probabilities of its winning succeeding games, given it gets to play them, are .8, .7, and .6, respectively.

    a. What is the probability your favorite team wins the tournament?

    b. What is the probability your favorite team is eliminated in the first round? the second round?

15. Fifteen balls are in an urn, of which 10 are white. You select four balls at random from the urn, without replacement. Evaluate the probability that

    a. All four balls are white.   b. None of the balls is white.

    c. Exactly one of the balls selected is white.

16. An urn contains 15 balls, of which five are red and the rest are white. One ball is selected at random from this urn; if the ball selected is red, it is returned to the urn, together with an additional 10 red balls. If the ball selected is white, it is not returned to the urn, nor are any additional balls added. A second ball is then drawn from the urn. What is the probability that this second ball is red?

17. Suppose the two-stage experiment described in Exercise 16 was in fact performed, and that the second ball selected from the urn was found to be white. What is the conditional probability that the first ball selected was also white?

18. Show the following:

    a. $A \subset B$ implies $P(A \mid B) = P(A)/P(B)$.

    b. $P(\overline{A} \mid B) = 1 - P(A \mid B)$.

    c. if $P(A) = P(B) = 2/3$, then $P(A \mid B) \geq 1/2$.

    d. $P(A \mid B) > P(A)$ implies $P(B \mid A) > P(B)$.

19. Assume that medical science has a test for cancer which is 90% accurate for both those who do and those who do not have cancer. Also assume that 5% of the population tested actually has cancer. If an individual takes this cancer test and the result is positive (i.e., the test says the person has cancer) what is the probability the person really does have cancer?

20. Assume that the probability is .95 that the jury selected to try a criminal case will arrive at the correct verdict of innocent or guilty. Further, suppose that the local police force is quite diligent in performing its function, that 99% of the people brought to trial are in fact guilty. Given that a jury finds a defendant guilty, what is the probability that he is in fact guilty?

21. Suppose a calculator manufacturer buys the same integrated circuit from three different suppliers, s_1, s_2, and s_3. In the past, 1% of the circuits supplied by s_1, 4% of the circuits supplied by s_2, and 2% of the circuits supplied by s_3 have been defective. Assume that this manufacturer buys 30% of her circuits from s_1, 10% from s_2, and the rest from s_3. Now suppose that all of these circuits are tossed into the same bin, without regard for the name of the supplier. If one of these circuits is selected at random from the bin and after testing is found defective, what is the probability that it came from s_2?

22. All students are required to take a specific course in government and history, sometime before graduation. Typically this course has a mixture of students from both the lower division (freshmen and sophomores) and the upper division (juniors and seniors). One term this class had 260 students, each of whom completed the course and received a final grade. A description of the final grades given is presented in the following table.

		Final Grade		
Sex	*Division*	*B+ or better*	*B or less*	*Total*
Male	Lower	42	57	99
	Upper	29	16	45
Female	Lower	47	14	61
	Upper	33	22	55
	Totals	151	109	260

a. If one person were selected at random from this class, what is the probability that this person received a grade of $B+$ or better?

b. What is the conditional probability that the person in part a received a grade of $B+$ or better, given the person is male? female?

c. Do the entries in this table exhibit Simpson's paradox (see Example 1.36)? (Compare the probabilities of getting a grade of $B+$ or better, given sex, in both divisions.)

1.6 | Independent Events

Granted an experiment with events $A, B \subset S$, one of the following must be true:

$$P(A) > P(A\,|\,B), \qquad P(A) < P(A\,|\,B), \qquad \text{or } P(A) = P(A\,|\,B).$$

The last case is quite special, since it says the conditional probability of observing A, given that B has occurred, is the same as the unconditional probability of A occurring. Think of what this says in probabilistic terms. If we have no information about whether B has occurred, our measure of the chance of A occurring is simply $P(A)$, its unconditional probability. On the other hand, if we are given that B has occurred, our measure of the chance of A occurring is $P(A\,|\,B)$, its conditional probability of occurrence. If these two measures are equal, $P(A) = P(A\,|\,B)$, then the occurrence of B has provided us with no information about whether A has also occurred (because the measure of the chance that A has occurred is the same in both cases).

Recall that $P(A\,|\,B) = P(A \cap B)/P(B)$. Thus $P(A\,|\,B)$ will equal $P(A)$ if and only if the probability of the intersection of A and B is given by the product of the individual probabilities, that is, $P(A \cap B) = P(A)P(B)$. If two events satisfy the latter relationship, they are called independent events, as given in the following definition.

DEFINITION 1.15 ———————————————————

If $A \subset S$ and $B \subset S$ are any two events with nonzero probabilities, A and B are called *independent* if and only if

$$P(A \cap B) = P(A)P(B).$$

Showing that two events are independent simply means computing the probabilities involved $P(A)$, $P(B)$, and $P(A \cap B)$ and verifying that Definition 1.15 is satisfied.

Example 1.37

If a fair coin is flipped twice, it seems intuitive that the outcome of the first flip should have no effect or influence on the outcome of the second flip; the results of the two flips should be independent. This is easily verified. Since the coin is fair, our sample space is

$$S = \{(h,h), (h,t), (t,h), (t,t)\};$$

the four single-element events are equally likely and have probability $\frac{1}{4}$. Now let $A = \{(h,h), (h,t)\}$, the event that we get a head on the first flip, and let $B = \{(h,h), (t,h)\}$, the event that we get a head on the second flip. Thus $A \cap B = \{(h,h)\}$, and we find

$$P(A \cap B) = \tfrac{1}{4} = \left(\tfrac{1}{2}\right)\left(\tfrac{1}{2}\right) = P(A)P(B);$$

these two events are independent and our intuition is borne out.

On the other hand, if we again let A be the event that a head occurs on the first flip and let $C = \{(h,t), (t,h)\}$, the event that exactly one head occurs, it is not clear whether A and C are independent. To see whether they are requires also evaluating $P(C)$ and $P(A \cap C)$ and then checking to see if the definition is satisfied. It is easy to see that $P(C) = \frac{2}{4}$ and $P(A \cap C) = \frac{1}{4} = P(A)P(C)$, and these two events are also independent.

Let D be the event that two heads occur, and let E be the event that two tails occur. (A is still the event that a head occurs on the first flip.) Intuitively, it should seem that A and D are *not* independent, since we cannot get two heads unless we got a head on the first flip (as well as the second). The event D can't occur unless A occurs, so D is a subset of A; it is easy to see that $P(A \cap D) \neq P(A)P(D)$, so they indeed do not satisfy the definition of independence. In like manner, it doesn't seem as if A and E should be independent because, if we know A occurred, then we also know E did *not* occur. It is easy to verify that A and E do not satisfy the definition of independent events.

If A and B are independent, then $P(B \mid A) = P(A)P(B)/P(A) = P(B)$; that is, the conditional probability of B occurring given that A has occurred is equal to the unconditional probability of B occurring. As you are asked to show in Exercises 1.6, the independence of A and B implies that $(A, \overline{B})$, $(\overline{A}, B)$, and $(\overline{A}, \overline{B})$ are also independent pairs of events.

In common English usage, the word *independent* and the phrase *mutually exclusive* have some meanings that border on being synonymous. In probability theory, though, there is no overlap in meaning at all, since the truth of one of these properties denies the possibility of the other. Suppose A and B are two events, such that $P(A) > 0$ and $P(B) > 0$; if A and B are independent, then $P(A \cap B) = P(A)P(B) > 0$. This implies that $A \cap B \neq \emptyset$ since, if the intersection were empty, its probability would have to be 0. Thus A and B cannot be mutually exclusive. On the other hand, if A and B are mutually exclusive, their intersection is empty and $P(A \cap B) = 0 \neq P(A)P(B)$, so they cannot be independent. If A and B are mutually exclusive, they cannot happen together, so the occurrence of one of them gives us information about the other, specifically, that this second event certainly did *not* occur. If A and B are independent, then it must be true that both of them could occur at the same time. It is wise to realize the sharp distinction between these two properties.

The extension of this concept of independence to three or more events may not be immediately obvious. Recall that A and B are independent if the occurrence of B gives no information about the occurrence of A, and vice versa, that is, $P(A \mid B) = P(A)$ and $P(B \mid A) = P(B)$. If we are to call three events A, B, and C independent, we would expect any pair of them to be independent. In addition, we should insist that the joint occurrence of two of them, say $B \cap C$ occurs, also gives no information about whether A has occurred; that is, we should also insist that $P(A \mid B \cap C) = P(A)$, which in turn implies that we should have $P(A \cap (B \cap C)) = P(A)P(B \cap C) = P(A)P(B)P(C)$. Thus three events A, B, and C are independent only if

$$P(A \cap B) = P(A)P(B),$$
$$P(A \cap C) = P(A)P(C),$$
$$P(B \cap C) = P(B)P(C),$$
$$P(A \cap B \cap C) = P(A)P(B)P(C).$$

The first three of these equations might seem to imply the truth of the last (and vice versa), but that is not true, as is illustrated in the following example.

Example 1.38

Assume that the probability of rain being reported in a small town on April 15, in any given year, is $\frac{1}{2}$; also assume that the occurrences of rain on this date are independent from one year to another. Let A be the event that rain is reported this year on this date, let B be the event that rain is reported on this date next year, and let C be the event that the same thing (either it rains or it doesn't) happens on April 15 on *both* years. The sample space for the experiment of observing whether rain occurs on this date for these two years, then, can be taken as $S = T \times T$, where $T = \{r, n\}$, with r representing the occurrence of rain and n representing the occurrence of no rain; the position in the 2-tuple indicates the year.

The event $A \cap B = (r, r)$ is assigned probability $P(A)P(B) = \frac{1}{4}$ since we are assuming these events are independent. Similarly, the event $A \cap \overline{B} = (r, n)$ is assigned probability $P(A)P(\overline{B}) = \frac{1}{4}$ because of the assumption that occurrences of rain on this date are independent from one year to the next. In the same way, the remaining 2-tuples are also assigned probabilities of $\frac{1}{4}$. (Thus this is an equally likely sample space.) Then we have $A = \{(r, n), (r, r)\}$, $B = \{(r, r), (n, r)\}$, and $C = \{(r, r), (n, n)\}$, which give $P(A) = \frac{2}{4} = \frac{1}{2} = P(B) = P(C)$. It also follows that

$$P(A \cap B) = \tfrac{1}{4} = P(A)P(B),$$
$$P(A \cap C) = \tfrac{1}{4} = P(A)P(C),$$
$$P(B \cap C) = \tfrac{1}{4} = P(B)P(C),$$

so each pair of events is independent. These are called *pairwise independent events*. However, $P(A \cap B \cap C) = \frac{1}{4} \neq P(A)P(B)P(C)$, so A, B, and C are *not* independent events. In fact, if we know that A and B occurred, we also know that C occurred, $P(C|A \cap B) = 1$, which is the reason the three events are not independent.

If a group of k events, say $A_1, A_2, \ldots, A_k$, are independent, then for all pairs of events with $i \neq j$,

$$P(A_i \cap A_j) = P(A_i)P(A_j),$$

for all triples of events with $i \neq j \neq l$

$$P(A_i \cap A_j \cap A_l) = P(A_i)P(A_j)P(A_l),$$

and so on, up to

$$P(A_1 \cap A_2 \cap \cdots \cap A_k) = P(A_1)P(A_2) \cdots P(A_k).$$

Thus a total of $\binom{k}{2} + \binom{k}{3} + \cdots + \binom{k}{k} = 2^k - k - 1$ equations are true, granted that the k events are independent. Note that independence of subsets of groups of independent events is inherited: If k events $A_1, A_2, \ldots, A_k$ are independent, then the events in any subset of them are also independent. However, if $A_1, A_2, \ldots, A_k$ are independent, and A_{k+1} is independent of *each* of them, this does not imply that $A_1, A_2, \ldots, A_k, A_{k+1}$ must be independent. (In Example 1.38, with $k = 3$ events A, B, and C, we saw that A and B were independent, and that C was independent of each of them, but that the three events were not independent.)

Independence is frequently assumed in building a probability model for an experiment; this assumption then allows easy assignment of probabilities to the intersections of these events, by simply multiplying together the individual probabilities involved.

Example 1.39

Assume the probability that any U.S. citizen, at age 20, will live to age 65 (or beyond) is .72. If we assume that three 20-year-old friends will or will not each reach age 65 independently, we can then easily evaluate the probabilities of several events. Let L_1 be the event that the first of the three lives to age 65 (or beyond) and L_2 and L_3 be the events that persons 2 and 3 will as well. Thus $P(L_1) = P(L_2) = P(L_3) = .72$. The independence assumption gives the probability that all three of them live to age 65 (or beyond) as $P(L_1 \cap L_2 \cap L_3) = (.72)^3 = .3732$. The probability that none of them survives to this age is $P(\overline{L}_1 \cap \overline{L}_2 \cap \overline{L}_3) = (1 - .72)^3 = .0220$, and the probability that at least one of them lives to this age is $1 - .0220 = .9780$. That exactly one of them lives to this age is an event that can happen three ways: Only the first, only the second, or only the third survives. With our assumptions, each of these has the same probability, namely $(.72)(.28)(.28) = .056448$. Thus the probability that exactly one of them lives to this age is $3(.72)(.28)^2 = .169344$; it is also easy to evaluate the probability that exactly two of them live to this age.

A wide variety of experiments can be broken down into simpler parts, called *trials*. A trial can thus be thought of as a very simple experiment. The two preceding examples have this property. In Example 1.38 we modeled the occurrence or nonoccurrence of rain in two successive years; the experiment included the possibilities that could occur in the two years, each of which we can think of as being a trial. Example 1.39 discussed the experiment consisting of modeling whether each of three 20-year-old friends would survive to age 65; the separate lifetimes of the three friends are each trials. Other examples include rolling five fair dice (each die being a trial), purchasing four new tires for your car (each tire being a trial), selecting a sample of potential voters (each voter being a trial), and observing the performance achieved by an individual taking four courses in a given academic semester (each course being a trial). If the results of an experiment are determined by the outcomes of n simpler trials, where n is some positive integer, we can generally model the experiment more easily by recognizing this structure.

If the outcome for an experiment is in fact determined by the results of some number n of trials, it then is natural to use a sample space S that is the Cartesian product of n sets; these n sets are the sample spaces for the n individual trials. As a simple example, suppose a student is faced with a surprise exam in probability class one day. This exam contains $n = 3$ questions; the first two are true-false selections, and the last is multiple choice with four options presented. Granted that one and only one response is correct for each question and that we are interested in modeling the student's performance on the exam, a good sample space for the experiment (performance on the exam) would be given by $S = T \times T \times T$, where $T = \{r, w\}$ (r means the answer was right and w that it was wrong). This S is the Cartesian product of $n = 3$ sets, representing the outcomes for the respective questions, and the elements of S are 3-tuples.

In many experiments that consist of n trials, it is frequently reasonable to assume that the outcomes of the trials are independent, in other words, that what happens on any single trial has no effect on the outcome of any other. More exactly, this assumption says that if A_i is an event whose occurrence depends only on the outcome of trial i, for $i = 1, 2, \ldots, k \leq n$, then the events in any subset of these A_i's are independent. This independence of trials then translates into a simple rule for assigning probabilities to the single-element events (which are all n-tuples): The probability assigned is the product of n terms, representing the probabilities of observing the appropriate result on the corresponding trial.

To illustrate this relationship, return to the case of the student faced with a surprise exam with three questions. Let us also assume that she is totally unprepared and will guess at the correct response for each question. If this were true, the outcomes of the three trials (answers guessed for the three questions) should be independent; whether she happens to correctly guess the answer to any given question has no effect on whether she also correctly guesses the answer for any other. The probability of guessing the right answer (outcome r) for either of questions 1 or 2 is $\frac{1}{2}$ (since there are two options available), while the probability of guessing the right answer to question 3 is $\frac{1}{4}$ (since it presents four choices). Then the probability assigned every 3-tuple in S is a product of three terms, each of which is determined by the result on the corresponding question. Thus the single-element events $\{(r, r, r)\}$ and $\{(r, w, r)\}$ are each assigned a probability of $\left(\frac{1}{2}\right)\left(\frac{1}{2}\right)\left(\frac{1}{4}\right) = \frac{1}{16}$, while the single-element event $\{(r, w, w)\}$ is assigned a probability of $\left(\frac{1}{2}\right)\left(\frac{1}{2}\right)\left(\frac{3}{4}\right)$; it is easy to see the probability assignments for the five remaining single-element events using this rule. The probability she would guess the correct answers to at least two questions is the sum of the probabilities of the four single-element events, each of which contains at least two r's; this value is $\frac{3}{16}$.

Example 1.40

Transistor radios are produced on an assembly line, and each radio either is defective in some manner or is not defective. Assume the probability is .05 that any radio is defective (and thus .95 it is nondefective). Further assume that the defective radios occur independently. This means, for example, that any specific radio being defective has no effect on whether any subsequent or preceding radio is also defective. Events whose occurrences hinge only on whether specific radios are defective are all independent.

Let us consider the experiment that consists of examining some number, say 4, of these radios for defects. This experiment thus consists of four independent trials. We shall use $S = T \times T \times T \times T$ as our sample space, where $T = \{d, n\}$, with d representing a defective radio and n a nondefective one. From the information given, the appropriate probability measure for each individual trial, with sample space T, is $P(\{d\}) = .05$, $P(\{n\}) = .95$. Since the four trials are independent, the probability measure for S is defined by the probabilities of occurrence for its single-element events. For example,

$$P(\{(n,n,n,n)\}) = (.95)(.95)(.95)(.95) = (.95)^4,$$
$$P(\{(d,d,d,d)\}) = (.05)(.05)(.05)(.05) = (.05)^4,$$
$$P(\{(d,n,d,n)\}) = (.05)(.95)(.05)(.95) = (.05)^2(.95)^2,$$
$$P(\{(d,d,d,n)\}) = (.05)(.05)(.05)(.95) = (.05)^3(.95).$$

Notice that the probability for each single-element event is the product of four terms; each term is either .05 or .95, depending on whether the corresponding radio is defective or nondefective. The probability of occurrence for any other event (any union of single-element events) is given by the sum of the single-element event probabilities for those elements belonging to it. Thus, if A is the event that exactly one of the four radios is defective, then $A = \{(d,n,n,n),(n,d,n,n),(n,n,d,n),(n,n,n,d)\}$ is the union of four single-element events; *each* of these single-element events has a probability of $(.05)(.95)^3$, giving $P(A) = 4(.05)(.95)^3 = .171$.

The formal definition of an experiment consisting of n independent trials is given below.

DEFINITION 1.16 ——————————————————————

An experiment is said to consist of n *independent trials* if and only if

1. The sample space S can be written as a Cartesian product of n sets: $S = T_1 \times T_2 \times \cdots \times T_n$.
2. For every $(x_1, x_2, \ldots, x_n) \in S$,

$$P(\{(x_1, x_2, \ldots, x_n)\}) = P_1(\{x_1\})P_2(\{x_2\}) \cdots P_n(\{x_n\}),$$

where $P_i(\{x_i\})$ is the probability of $x_i \in T_i$ occurring on trial i.

In many cases the sample spaces for the individual trials $T_1, T_2, \ldots, T_n$ are identical (meaning each trial has the same set of possible outcomes) and the probability measure employed is the same for each trial, as in Example 1.40; this need not be true in general.

Example 1.41

Assume that Rachel is taking a psychology class; on Monday of one week the class is given a short surprise quiz consisting of four true-false questions and three multiple-choice questions. One response is correct for each question. Suppose Rachel has not had time to keep up her studies and is totally unprepared for all the questions; consequently she will guess in choosing her answer for each question. In modeling how well she might do on this quiz, we shall assume this is an experiment with seven independent trials; that is, whether she happens to guess the correct answer to question 3, say, should have no effect on whether she will also guess the correct answer to any other question. Thus, we shall use $T = \{c, i\}$, where c represents a correct answer and i represents an incorrect answer, for *each* of the questions (trials); the sample space for the experiment (her performance on the exam) then is $S = T \times T \times \cdots \times T$, a collection of 7-tuples.

Assume that questions 1 through 4 are true-false and that questions 5 through 7 are multiple-choice, with three choices for each, exactly one of which is the correct response. Since Rachel is guessing, she has probability $\frac{1}{2}$ of answering each true-false question correctly and probability $\frac{1}{3}$ of answering each multiple-choice question correctly. Thus the probability measures for the individual trials are specified by

$$P_i(\{c\}) = \tfrac{1}{2}, \quad \text{for } i = 1, 2, 3, 4,$$
$$= \tfrac{1}{3}, \quad \text{for } i = 5, 6, 7.$$

The probabilities for $\{(x_1, x_2, \ldots, x_7)\} \subset S$ then are products of the values for the probabilities of the appropriate entries occurring in the 7-tuples. Thus, for example,

$$P(\{(c, c, c, c, c, c, c)\}) = \left(\tfrac{1}{2}\right)^4 \left(\tfrac{1}{3}\right)^3 = \tfrac{1}{432} = .00231$$
$$P(\{(c, c, i, i, c, i, i)\}) = \left(\tfrac{1}{2}\right)^2 \left(\tfrac{1}{2}\right)^2 \left(\tfrac{1}{3}\right) \left(\tfrac{2}{3}\right)^2 = \tfrac{4}{432} = .00926$$
$$P(\{(i, i, i, i, i, i, i)\}) = \left(\tfrac{1}{2}\right)^4 \left(\tfrac{2}{3}\right)^3 = \tfrac{8}{432} = .0185.$$

Rachel must answer at least six questions correctly to pass the exam. Let A be the event that she passes the exam; this event then contains all 7-tuples having at least 6 c's. There are eight 7-tuples belonging to A, seven with one i (which can occur in any of the seven positions) and one 7-tuple containing all c's (no i's). $P(A)$ is the sum of the probabilities of these eight single-element events. Thus

$$P(A) = 4\left(\tfrac{1}{2}\right)^4 \left(\tfrac{1}{3}\right)^3 + 3\left(\tfrac{1}{2}\right)^4 \left(\tfrac{1}{3}\right)^2 \left(\tfrac{2}{3}\right) + \left(\tfrac{1}{2}\right)^4 \left(\tfrac{1}{3}\right)^3 = \tfrac{11}{432} = .0255.$$

A person who had time to prepare should expect to have a higher probability than this of passing the exam.

The fact that an experiment consists of n trials does not imply that the trials must be independent; simple random sampling without replacement provides an example of dependent trials, as discussed in the following example.

Example 1.42

Twenty people attend a party at which two prizes are to be given "by lot" (at random). Each attendee is given a slip of paper bearing a number from 1 through 20 (and each number occurs exactly one time). The two numbers that determine the prize winners then are drawn at random, without replacement, from a hat containing slips of paper, each with a number from 1 to 20. The first number drawn is equally likely to be any of the 20, while the second number drawn is equally likely to be any of the remaining 19. There are $\binom{20}{2} = 190$ different pairs of numbers that could be drawn (where order of selection does not count), and $_{20}P_2 = 380$ ordered pairs that might be observed, all of which are equally likely to occur. We shall use this latter collection of ordered pairs for our sample space S.

This is also an example of an experiment with two trials, corresponding to the two numbers drawn. The Cartesian product $S = T \times T$, where $T = \{1, 2, \ldots, 20\}$, could be used as the sample space, with the understanding that any 2-tuple that has the same number in the two positions will receive probability 0 (and that the remaining 2-tuples are equally probable). If we let A be the event that the first number drawn is 1, and B the event that the second number drawn is 1, we have $P(A) = P(B) = \frac{19}{380} = \frac{1}{20}$ since there are equal numbers of 2-tuples in S with the number 1 in the first or second position. Of course, $P(B \mid A) = 0$ since the number 1 couldn't be drawn twice if the selection is made without replacement; events A and B are not independent in this case. Equivalently, these two *trials* are not independent since the range of possibilities for the second draw depends on the first number selected.

Doing the drawing without replacement assures that the same person cannot get both prizes. However, if the selection of winning numbers were done with replacement, then the same number could be drawn twice. Defining A and B as before, we again have $P(A) = P(B) = \frac{1}{20}$, but now $P(B \mid A) = \frac{1}{20} = P(B)$, so these two events are independent; the two trials are independent, because the outcome of the first draw has no effect on the second draw.

Building a probability model for an experiment with dependent trials demands more information than is required if the trials are independent. If the trials are independent, the probability of observing any possible outcome on the fifth trial, say, is the same, no matter what occurs on the preceding or succeeding trials. If the trials are not independent, then the model must take the dependencies into account. The following example discusses a simple situation of modeling dependent trials.

Example 1.43

The abilities of major league baseball players are summarized or described in many ways; one of these is the batting average, which expresses the ratio of the number of hits made divided by the number of times at bat for all the games in the current year. Suppose a particular player, in a given game, will be at bat three times; that is, he will have three opportunities to score a hit. Also suppose his batting average coming into the game is .300. His three times at bat in this game can be modeled as an experiment with three trials. The simplest possible model, in many ways, would be the one that assumes these trials are independent. (That is, whether he gets a hit his second time at bat, say, does not depend on how he did on his first chance or on other aspects of the game.) However, this assumption may not be realistic since

Table 1.7

Single-Element Event Probabilities

Event	Probability
$\{(h,h,h)\}$	$(.3)(.4)(.4) = .048$
$\{(h,h,n)\}$	$(.3)(.4)(.6) = .072$
$\{(h,n,h)\}$	$(.3)(.6)(.2) = .036$
$\{(h,n,n)\}$	$(.3)(.6)(.8) = .144$
$\{(n,h,h)\}$	$(.7)(.2)(.4) = .056$
$\{(n,h,n)\}$	$(.7)(.2)(.6) = .084$
$\{(n,n,h)\}$	$(.7)(.8)(.2) = .112$
$\{(n,n,n)\}$	$(.7)(.8)(.8) = .448$

it ignores the psychology of the game and various dynamics that may be in place.

To illustrate one way to model his three times at bat (trials) that does *not* make the independence assumption, let us arbitrarily assume that the probability of his scoring a hit his first time at bat is .3, his current batting average. Rather than saying the probabilities of his subsequent hits are also all given independently by .3, let us assume his probability of getting a hit is .4 *if he made a hit on his previous opportunity* and is .2 *if he did not*. We shall use the sample space $S = T \times T \times T$, where $T = \{h, n\}$ with h representing a hit and n representing no hit. These assumptions again allow easy specification of the probabilities of occurrence for the single-element events, with the slight complication that the probability for the second or later component depends on the preceding entry. Table 1.7 lists the eight single-element events for this experiment and their probabilities of occurrence, granted these assumptions.

The probabilities of these single-element events sum to 1, as they must. We can then evaluate the probabilities of occurrence of any events of interest by summing the appropriate single-element event probabilities. For example, if A is the event that the player gets (exactly) one hit, then

$$P(A) = .144 + .084 + .112 = .340,$$

while if B is the event that he gets (exactly) two hits, we have

$$P(B) = .072 + .036 + .056 = .164.$$

You will be asked in Exercises 1.6 to reconsider this probability model using an independent trials structure.

EXERCISES 1.6

1. A fair coin is flipped three times; let A be the event that a head occurs on the first flip and B the event that (exactly) one head occurs. Are A and B independent?

2. A coin is flipped three times; let A be the event that exactly one head

occurs and *B* the event that exactly two tails occur. Without recourse to computations, can you see that these two events are not independent?

3. A professional baseball player is at bat three times in a given game. Assume that the probability he gets a hit each time at bat is .3 and that his three times at bat are independent trials. Find the following probabilities:
   a. He gets three hits in this game.   b. He gets no hits in this game.
   c. He gets exactly two hits in this game.

4. Three friends, Joe, Hugh and Rae, will individually and independently attempt to solve the same problem. The probabilities that each of them is successful are .9, .8, and .7 respectively.
   a. What is the probability the problem will be solved by at least one of them?
   b. What is the probability that the problem is solved by Rae alone?
   c. Granted the problem has been solved, what is the probability that it was solved only by Joe?

5. Assume that events *A* and *B* are independent. Show that each of the pairs $(A,\overline{B})$, $(\overline{A},B)$, and $(\overline{A},\overline{B})$ is also independent.

6. You place an order, by mail, for three items; the mail-order firm has a new clerk handling your order. This person has probability .2 of making a mistake (independently) on each item placed into the box to be sent to you. What is the probability that the items you receive are the items you ordered? (Assume that a mistake on any item results in sending something you did not order.)

7. Insurance companies employ mortality tables to set the rates they will charge to insure the lives of people. These tables are based on actual experience (historical records of births and deaths) and typically give the proportions of people born in a given year who survive for an additional year, at various ages. You could, for example, consult such a table and find that a person at age 20 has probability .7 (roughly) of still being alive at age 65. Suppose that two friends are currently 20 years old and that their remaining life spans are independent (e.g., they will not both die during the same epidemic or die in the same car crash). Also assume that .7 accurately reflects the probability that each of them will survive to age 65 (or beyond).
   a. What is the probability that *both* of them will survive to age 65?
   b. What is the probability that exactly one of them will survive to age 65?

8. A local police department allows home burglar alarms to be wired into their headquarters; if the alarm is sounded, the department agrees to send a police officer out to investigate. Assume that the probability of a false alarm (i.e., the burglar alarm incorrectly sounds a warning) is .1 for each alarm the police receive and that these signals are independent. If this police department receives four burglar signals in one 24-hour period, what is the probability that they are all real (i.e., that none are false alarms)? What is the probability that exactly one of the four is a false alarm?

9. A total of 40 automobiles crossed a bridge, in the north to south direction,

during a 30-minute period. For each automobile crossing the bridge, the sex of the driver (male M or female $\overline{M}$), the origin of the automobile (domestic D or foreign $\overline{D}$), and an indicator of the automobile's color (red R or not-red $\overline{R}$) were noted. The counts observed are given in the following table:

| Sex | Origin | Color of Car | | |
		Red	Not Red	Total
Male	Domestic	5	7	12
	Foreign	2	6	8
Female	Domestic	7	1	8
	Foreign	6	6	12
	Totals	20	20	40

Use these counts as the probability model for selecting one automobile at random crossing this bridge from north to south.

  a. Define the events $M, D,$ and R as suggested by the previous discussion, and evaluate their probabilities.

  b. Show that no two of these events are independent but the probability of the intersection of all three events is given by the product of the individual probabilities.

10. Three teams, call them A, B, and C, enter a round-robin tournament in which each team plays every other team one time; no ties are allowed in this game. Assume that $P(A \text{ beats } B) = .4$, $P(B \text{ beats } C) = .5$, and $P(C \text{ beats } A) = .6$ and that the outcomes of all the games are independent. Evaluate the probabilities of the following:

  a. A wins the tournament   b. B wins the tournament.

  c. No one wins the tournament.

11. In one morning an Avon saleswoman calls at 16 homes. Assume the probability that she makes a sale at any single home is .1, and that these events are independent. What is the probability that she makes at least one sale on a given morning?

12. A surprise examination contains 10 true-false questions. If a student is guessing in answering each question and will pass by correctly answering seven or more questions, what is the probability that the student will pass the exam?

13. An eight-question exam is multiple-choice, with four choices per question, of which exactly one is correct. If passing requires six or more questions to be answered correctly, evaluate the probability that a person who is guessing will pass.

14. A surprise examination contains eight questions, each multiple-choice with four possible answers presented, of which exactly one is correct. A student must answer six of the questions correctly to pass. If Pat knows the answers to questions 1, 3, 5, and 6 but will have to guess at the answers to the other four, what is the probability she will pass the exam?

15. For the baseball situation of Example 1.43, assume his three times at bat are independent trials and that the probability of getting a hit is .3 for each trial. Evaluate the probabilities of his getting

a. no hits   b. one hit   c. two hits   d. three hits

in this game.

16. Recall the baseball player with three times at bat discussed in Example 1.43. With the model discussed there, evaluate the probability that he makes a hit his first time at bat. Do the same for his second and third times at bat.

17. A "natural" way to model the performance of a baseball player is to *always* use his current batting average as the probability of getting a hit. Consider again a player with three times at bat in a certain game, where he has had 60 previous times at bat in this season, of which 18 resulted in his scoring a hit. (Thus his batting average is $\frac{18}{60} = .3$ coming into the game, and the probability of his getting a hit the second time at bat is $\frac{19}{61}$ or $\frac{18}{61}$, depending on whether he got a hit on his first attempt.) Evaluate the probabilities of his getting

a. no hits   b. one hit   c. two hits   d. three hits

in this game.

1.7 | Summary

A set is any collection of objects, typically specified by the roster method (a listing of all elements) or the rule method (in which the elements must satisfy a given rule). $a \in A$ signifies that a is an element of A while $a \notin A$ signifies a is not an element of A. The empty set, $\emptyset$, contains no elements. A is a subset of B (written $A \subset B$) if and only if $x \in A$ implies that $x \in B$; the empty set is a subset of A (written $\emptyset \subset A$) for any set A. A and B are equal (written $A = B$) if and only if $A \subset B$ and $B \subset A$.

The union of A and B (written $A \cup B$) contains all the elements that belong to either of A or B. The intersection of A and B (written $A \cap B$) contains only those elements that belong to both A and B. Two distributive laws link these operations:

$$A \cup (B \cap C) = (A \cup B) \cap (A \cup C),$$

$$A \cap (B \cup C) = (A \cap B) \cup (A \cap C).$$

The complement of a set A is denoted $\overline{A}$ and contains all the elements in the sample space S that are *not* in A. DeMorgan's laws state that $\overline{A \cup B} \equiv \overline{A} \cap \overline{B}$ and $\overline{A \cap B} \equiv \overline{A} \cup \overline{B}$. An n-tuple is an ordered list of n elements, written $(x_1, x_2, \ldots, x_n)$. The Cartesian product of n sets $A_1, A_2, \ldots, A_n$ is denoted $A_1 \times A_2 \times \cdots \times A_n$; it contains as elements all possible n-tuples $(x_1, x_2, \ldots, x_n)$ where $x_i \in A_i$, for $i = 1, 2, \ldots, n$.

The probabilistic model for an experiment consists of the set S of possible outcomes, the collection of events $A \subset S$ and a rule that specifies $P(A)$, the probability of observing A for every event A. Two events A and B are mutually exclusive if $A \cap B = \emptyset$. To satisfy the properties of relative frequencies the probability function must satisfy the Kolmogorov axioms:

KOLMOGOROV AXIOMS

1. For any event $A \subset S$, $P(A) \geq 0$.
2. $P(S) = 1$.
3. If $A_1, A_2, \ldots$ are mutually exclusive events, then

$$P(A_1 \cup A_2 \cup \cdots) = P(A_1) + P(A_2) + \cdots .$$

Some consequences that follow from these axioms:

CONSEQUENCES

1. $P(\emptyset) = 0$.
2. $P(\overline{A}) = 1 - P(A)$.
3. $P(A \cup B) = P(A) + P(B) - P(A \cap B)$.
4. If $A \subset B$, then $P(A) \leq P(B)$.

A finite sample space has k different possible outcomes and there are then k different single-element events, with probabilities $p_1, p_2, \ldots, p_k$, where $\sum_{i=1}^{k} p_i = 1$. The probability of any event A is given by $P(A) = \sum_{i \in A} p_i$. If these single-element events are equally likely, then each must have a value $\frac{1}{k}$, and the probability of any event A ($A \subset S$) is simply $n(A)/k$, where $n(A)$ counts the number of elements belonging to A.

Counting procedures are useful in enumerating the numbers of elements belonging to certain sets. The multiplication principle states that the number of ways two operations can be jointly performed is mn, granted that the first can be done in m and the second in n ways. A permutation of n distinct objects is an arrangement of them in a row, in a definite order (the number of n-tuples that can be constructed); the total number of such permutations is $n! = n(n-1)\cdots 1$, where $0! = 1$. The number of permutations of n things r at a time (the number of r-tuples possible) is $_nP_r = n!/(n-r)!$. The number of combinations of n things taken r at a time (the number of distinct subsets of size r) is $\binom{n}{r} = n!/r!(n-r)!$. Three useful facts about combinatorial co-efficients are that $\binom{n}{0} = 1$, $\binom{n}{n-1} = n$, and $\binom{n}{r} = \binom{n}{n-r}$. Pascal's triangle is built from the relation

$$\binom{n-1}{r-1} + \binom{n-1}{r} = \binom{n}{r}.$$

BINOMIAL THEOREM

If x and y are any two real numbers and n is any positive integer, then

$$(x+y)^n = \sum_{i=0}^{n} \binom{n}{i} x^i y^{n-i}.$$

Given n objects, of which n_j are of type j, $\sum_{j=1}^{k} n_j = n$, the number of permutations possible is

$$\binom{n}{n_1 \, n_2 \cdots n_k} = \frac{n!}{n_1! n_2! \cdots n_k!};$$

this number is also called a multinomial coefficient.

The conditional probability of A occurring, given that B has occurred, is the ratio $P(A\,|\,B) = P(A \cap B)/P(B)$, from which it follows that the probability of an intersection of events is given by $P(A \cap B) = P(B)P(A\,|\,B) = P(A)P(B\,|\,A)$.

PARTITION

The events $A_1, A_2, \ldots, A_k, \ldots$ are called a partition of S if

1. $A_i \cap A_j = \emptyset$ for all $i \neq j$.
2. $A_1 \cup A_2 \cup \cdots \cup A_k \cup \cdots = S$.

THEOREM OF TOTAL PROBABILITY

If $A_1, A_2, \ldots$ is a partition of S, and $B \subset S$ is any event, then $A_1, A_2, \ldots$ also partition B, and

$$P(B) = P(B\,|\,A_1)P(A_1) + P(B\,|\,A_2)P(A_2) + \cdots$$

BAYES' THEOREM

If $A_1, A_2, \ldots$ is a partition of S, and $B \subset S$ is any event, then

$$P(A_i\,|\,B) = \frac{P(A_i \cap B)}{P(B)}$$

$$= \frac{P(A_i)P(B\,|\,A_i)}{P(A_1)P(B\,|\,A_1) + P(A_2)P(B\,|\,A_2) + \cdots}$$

A and B are independent events if and only if $P(A \cap B) = P(A)P(B)$, which in turn implies that the occurrence of one of them gives no information about whether the other has also occurred. $A_1, A_2, \ldots, A_k$ are independent if and only if the probability of the intersection of any number of them is equal to the product of the corresponding probabilities.

INDEPENDENT TRIALS

An experiment is said to consist of n independent trials if and only if

1. $S = T_1 \times T_2 \times \cdots \times T_n$.
2. For every $(x_1, x_2, \ldots, x_n) \in S$,

$$P(\{(x_1, x_2, \ldots, x_n)\}) = P_1(\{x_1\})P_2(\{x_2\}) \cdots P_n(\{x_n\}),$$

where $P_i(\{x_i\})$ is the probability of $x_i \in T_i$ occurring on trial i.

 **EXERCISES 1.7**

1. Let $A = \{x : -1 < x < 1\} = B$ and find the region $A \times B$.
2. Three dogs are each presented (individually and independently) with the same dog food, in four different colored bowls (red, blue, green, and yellow). Each dog picks one of the four bowls (colors) to eat from first. Define a reasonable sample space for describing the possible selections by the dogs. Let A be the event that they all choose the same color, and express A as a subset of S.
3. Define $A = \{(x, y) : x + y = 4\}$, $B = \{(x, y) : -x + 2y = 0\}$, and find $A \cap B$.
4. An individual purchases three lottery tickets, each of which may or may not win some prize. The probability at least one ticket wins a prize is .3, the probability (exactly) one ticket wins a prize is .26, and the probability at most two tickets wins some prize is .99. What are the four single-element probabilities, using $S = \{0, 1, 2, 3\}$?
5. Suppose four people, $\mathcal{A}$, $\mathcal{B}$, $\mathcal{C}$, and $\mathcal{D}$, enter a single-elimination tournament involving a game that cannot end in a tie. Such a tournament requires a total of three games; game 1 pits $\mathcal{A}$ versus $\mathcal{B}$, while game 2 features $\mathcal{C}$ versus $\mathcal{D}$. The third and final game involves the winners of games 1 and 2; the person winning this game is the tournament winner. Since the tournament consists of a total of three games, use the sample space

$$S = \{(\mathcal{A}, \mathcal{C}, \mathcal{A}), (\mathcal{A}, \mathcal{D}, \mathcal{A}), (\mathcal{B}, \mathcal{C}, \mathcal{B}), (\mathcal{B}, \mathcal{D}, \mathcal{B}),$$
$$(\mathcal{A}, \mathcal{C}, \mathcal{C}), (\mathcal{A}, \mathcal{D}, \mathcal{D}), (\mathcal{B}, \mathcal{C}, \mathcal{C}), (\mathcal{B}, \mathcal{D}, \mathcal{D})\},$$

where the entry indicates the winner of the corresponding game. Assume the single-element probabilities are in the proportions 3:1:3:1:3:1:3:1 for the order listed in S. What are the single-element probabilities? Evaluate the probabilities of each individual being the tournament winner.
6. A fair die is rolled three times; the sum of the three numbers to occur then lies between 3 and 18. Work out the probability that each integer between 3 and 18 is the value for the sum.

7. With n persons in a room, let A_n be the event that two or more of them share the same birth week; assume there are 52 weeks in the year, each of which is equally likely to occur as the birth week for each person. What is the smallest value for n such that $P(A_n) > \frac{1}{2}$?

8. How many 11-letter "words" can be made with the letters in *probability*? How many of these begin with p and end with y?

9. Granted a tire dealer has in stock 15 of the tires you want to buy, of which three have a defect in the sidewall, what is the probability that you get at least one tire with a defective sidewall when you buy four of these tires? That you get three tires with defective sidewalls?

10. An island telephone exchange uses five-digit telephone numbers, each of which must start with 45. How many different telephone numbers are possible?

11. Show that if $P(A \mid B) > P(A \mid \overline{B})$, then $P(A \cap B) > P(A)P(B)$.

12. Five fair dice are rolled one time. Evaluate the probability that five different faces occur.

13. A small chest contains two drawers; drawer 1 contains three silver and two gold coins, while drawer 2 contains two silver and five gold coins. Two coins are selected at random, without replacement, from drawer 1 and placed in drawer 2. Two coins are then drawn from drawer 2 without replacement. If the last two coins are both silver, what is the probability that both coins transferred from drawer 1 were silver?

14. Six fair dice are rolled one time. Evaluate the probability that each of the six faces occurs.

15. Two people are hired by the same corporation at the same time. The probability that at least one of them is still working for the corporation one year later is .7, while the probability that exactly one of them is still working for the corporation one year later is .4. Evaluate the probability that
    a. neither of them   b. both of them
    is still working for the corporation one year later.

16. Each lottery ticket purchased has probability $\frac{1}{9}$ of winning a prize. What is the probability of winning (at least) one prize if you purchase five such tickets? if you purchase 9 tickets?

17. The two owners of a small shop do not agree on the probability that a person entering their business will make some purchase. The first owner says this probability is .4, while the second says it is .25. Let A_1 and A_2 represent these two opinions, and assume they are equally likely to be correct, so that $P(A_1) = P(A_2) = \frac{1}{2}$. If two customers come in and neither makes a purchase (independently of each other), what are the resulting posterior probabilities of A_1 and A_2 being correct?

18. Let A_1, A_2 be a partition with prior probabilities $P(A_1)$, $P(A_2)$, respectively, and let B be an event with conditional probabilities $P(B \mid A_1)$, $P(B \mid A_2)$. Show that the posterior probability of A_1 (i.e., $P(A_1 \mid B)$) is larger than $P(A_1)$ if and only if $P(B \mid A_1) > P(B)$.

CHAPTER 2

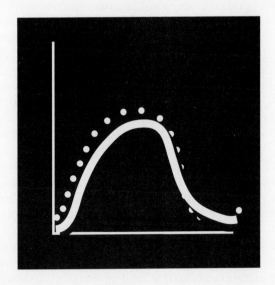

RANDOM VARIABLES

One of the most frequent uses of probability is modeling the behavior of "random variables," a colorful name that is meant to be descriptive. The word *variable* means a numeric quantity; the word *random* is used to indicate that chance or randomness plays a role in determining the value of the variable. Thus the term *random variable* describes a numeric variable whose value is not perfectly predictable; its value could be thought of as being subject to, or determined by, chance to some degree.

Suppose you are to take a standardized test of some sort, for example, the Scholastic Aptitude Test. Let X be the score you will receive when your test is graded. Before the fact, even if you have taken this exam before, the value of X is not known; it is an example of a random variable. We can, however, develop procedures for modeling X, based on the reasoning we have employed in modeling the outcomes of experiments: Before you take this exam, we can conceive of the list of *possible* scores you might receive and thus the range of possible values that X might equal. After you have taken the exam, your grade is determined to be x, which is a member of the list of possible scores assumed. We call x the *observed value of the random variable X*.

Additional examples of random variables include the number of patients N to be helped by a specific medical treatment, the number of sales D of a given item to be made by a department store in the coming month, the number of traffic fatalities T to occur in a given area during a holiday weekend, the time of birth Y of the first baby delivered in a given year at a hospital, the number of miles M a car will travel on a full tank of gasoline under certain conditions. We shall use capital letters as names for random variables. The observed value (the actual value observed or realized for the random variable) could be any in a range of possibilities and is generally denoted by the lowercase of the name of the random variable.

This chapter introduces the idea of a random variable and explores procedures that have proven useful in modeling random variables and their behavior. In Sections 2.1 and 2.2 we shall see that there are two distinct types of random variables, which are very frequently employed in practical problems, called discrete and continuous random variables. Of the cases mentioned in the previous paragraph, note that $N, D,$ and T are counting variables, so the possible observed values are whole numbers, counts to be observed; these are examples of *discrete* random variables. In contrast, the random variables Y (a time) and M (a distance) are not restricted to the counting numbers (integers) since in theory, times and distances can equal any positive real number, not just the separated, distinct integers. Thus the collections of possible observed values should contain all points in an interval. Y and M are examples of *continuous* random variables. Slightly different approaches are required in modeling discrete versus continuous random variables.

In Section 2.3 we shall discuss the mean, variance, and standard deviation of a random variable, the most commonly employed summary measures. Section 2.4 discusses the use of quantiles (or percentiles) as summary measures for continuous random variables.

2.1 | Discrete Random Variables

A random variable X is a numeric quantity whose actual observed value x may be any value in set of real numbers. Consider Example 1.14, where the language of random variables was not employed. Our experiment consists of one roll of a pair of fair dice; the sample space for the experiment then is $S = T \times T$, where $T = \{1, 2, \ldots, 6\}$ represents the number of spots uppermost on a single die. Define X to be the sum of the two numbers to occur on this roll of the dice; thus the observed value of X will be the sum of the two numbers in the 2-tuple that is observed. X is an example of a random variable. Table 2.1 lists the possible 2-tuples that could be observed, together with the value

Table 2.1

Possible Observed Values for X, the Sum of Two Fair Dice

Outcome	Observed Value x
$(1,1)$	2
$(1,2),(2,1)$	3
$(1,3),(2,2),(3,1)$	4
$(1,4),(2,3),(3,2),(4,1)$	5
$(1,5),(2,4),(3,3),(4,2),(5,1)$	6
$(1,6),(2,5),(3,4),(4,3),(5,2),(6,1)$	7
$(2,6),(3,5),(4,4),(5,3),(6,2)$	8
$(3,6),(4,5),(5,4),(6,2)$	9
$(4,6),(5,5),(6,4)$	10
$(5,6),(6,5)$	11
$(6,6)$	12

for x, their sum. Formally, this random variable X has mapped the elements of S (which are 2-tuples) onto the integers 2, 3, ... , 12, the possible values for the sum of the two numbers rolled.

The set of possible sums of the two numbers observed when the dice are rolled is called the *range*. For this random variable X the range is $R_X = \{2, 3, \ldots, 12\}$. In addition to mapping the 2-tuples in S onto R_X, we can also transfer the probability measure from S to R_X, the range of X. Granted the two dice are fair, the 36 2-tuples in S are equally likely to occur, so they each have probability $\frac{1}{36}$. Since we will observe $X = 4$, for example, if any of the three 2-tuples $(1,3)$, $(2,2)$, or $(3,1)$ occurs, then $\frac{3}{36}$ (the sum of the probabilities for these single-element events) is the probability of observing the event $X = 4$. In fact, we can consider each $x \in R_X$ and evaluate the probability of finding $X = x$; this generates the probability function $P(X = x)$ shown in Table 2.2.

From our example we can summarize the characteristics of the probability function for a random variable X. The set of possibilities for the observed value of the random variable is called its range; it is labeled R_X, where the subscript identifies the random variable and is a set of real numbers. The generic symbol for an element of the range is a lowercase letter. Thus if a random variable X has range R_X, the elements of the set R_X are denoted by x; the value observed for X is x, one of the elements of the range for the random variable X. In modeling a random variable X, its range R_X plays the role that the sample space S has in modeling the outcome of an experiment. Events involving X are subsets of R_X, sets of real numbers.

Table 2.2

Probability Function for X, the Sum of Two Fair Dice

x	2	3	4	5	6	7	8	9	10	11	12
$P(X = x)$	$\frac{1}{36}$	$\frac{2}{36}$	$\frac{3}{36}$	$\frac{4}{36}$	$\frac{5}{36}$	$\frac{6}{36}$	$\frac{5}{36}$	$\frac{4}{36}$	$\frac{3}{36}$	$\frac{2}{36}$	$\frac{1}{36}$

For example, if we roll a pair of fair dice and let X be their sum, it could be that the total observed is less than or equal to 4. This statement defines an event (which may or may not be observed) expressed as a subset of R_X; this event is $\{2,3,4\}$ and its probability of occurrence is

$$P(X \leq 4) = \sum_{x=2}^{4} P(X = x) = \frac{6}{36} = \frac{1}{6}.$$

The event that the sum is an odd number (i.e., the value for X is one of the integers 3, 5, 7, 9, or 11) has probability

$$P(X \text{ is odd}) = \sum_{j=1}^{5} P(X = 2j + 1) = \frac{2 + 4 + 6 + 4 + 2}{36} = \frac{1}{2}.$$

Of course, $P(X \text{ is even}) = 1 - \frac{1}{2} = \frac{1}{2}$, as well.

The range of a random variable will always be a set of real numbers, its possible observed values. For our purposes, sets of real numbers are of two different types, which we shall call discrete and continuous. If the set of numbers consists only of separated, distinct points on the real line, it is discrete. Examples of this type of set include $R_X = \{2, 3, \ldots, 12\}$ (the range for the sum of two dice), the set of even numbers $\{2, 4, 6, \ldots\}$, the set of all integers $\{\ldots, -1, 0, 1, \ldots\}$. Any set that can be mapped onto a subset of the set of integers is discrete (and may have an infinite number of elements). A random variable Y whose range R_Y is discrete is called a discrete random variable. We shall discuss and illustrate some discrete random variables in this section.

A set that contains all the points in an interval on the real line, or a union of intervals, will be called continuous; these sets occur as ranges of continuous random variables. Examples include $\{x : 0 \leq x \leq 12\}$, $\{y : y > 0\}$, the set of all real numbers, and so on. A random variable X whose range R_X is continuous is called a continuous random variable; this type of random variable will be introduced in Section 2.2.

We have already discussed one discrete case at some length (the sum of two numbers rolled with a fair pair of dice). We derived $P(X = x)$, for $x \in R_X$, from the probability function used for the experiment. This function provides the *probability law* for X, the probability model for its behavior. An alternative notation for $P(X = x)$ as defined next, is frequently used.

DEFINITION 2.1

If X is a discrete random variable with range R_X, the *probability function* for X is $p_X(x) = P(X = x)$, which gives the probability of occurrence for each $x \in R_X$.

The following three examples discuss a few discrete random variables and their probability functions.

Example 2.1

Table 2.3

Probability Function for Y

y	$\frac{3}{2}$	2	$\frac{5}{2}$	3	$\frac{7}{2}$	4	$\frac{9}{2}$
$p_Y(y)$	$\frac{1}{10}$	$\frac{1}{10}$	$\frac{1}{5}$	$\frac{1}{5}$	$\frac{1}{5}$	$\frac{1}{10}$	$\frac{1}{10}$

Suppose a hat contains five slips of paper, each bearing one number; each of the integers 1 through 5 are used one time. Two of these slips of paper are selected at random (without replacement). If Y is the average of the numbers drawn (i.e., the sum divided by 2), what is the probability function for Y? The range for Y is specified by the total collection of possibilities for this average; at one extreme the two numbers drawn could be 1 and 2 (giving $y = \frac{3}{2}$), and at the other extreme the two numbers drawn could be 4 and 5 (giving $y = \frac{9}{2}$). The range for Y is $R_Y = \{\frac{3}{2}, 2, \frac{5}{2}, 3, \frac{7}{2}, 4, \frac{9}{2}\}$. Since there are five slips of paper available, there are $\binom{5}{2} = 10$ different pairs of slips that could be selected; each of these pairs then has probability $\frac{1}{10}$ of occurring, granted the selection is made at random. A single one of these pairs gives $y = \frac{3}{2}, 2, 4, \frac{9}{2}$ (pairs (1, 2), (1, 3), (3, 5), and (4, 5), respectively, while each of two pairs results in $y = \frac{5}{2}, 3, \frac{7}{2}$ (pairs (1, 4), (2, 3); (1, 5), (2, 4); and (2, 5), (3, 4), respectively). Thus the probability function for Y is as given in Table 2.3. The probability for any event involving the observed value for Y (subset of R_Y) then is given by summing the values of the probability function over the elements belonging to the event. For example, $P(Y \leq 3) = .6, P(2 \leq Y \leq 4) = .8, P(Y = 2, 3,$ or $4) = .4$.

Example 2.2

A small firm has three positions available, with equivalent requirements and responsibilities; eight persons, of whom five are female, apply for these positions. If the three persons to get the jobs are chosen at random from the eight, there are $\binom{8}{3} = 56$ different collections of three available; of these, $\binom{3}{3}\binom{5}{0} = 1$ collection contains only males, while $\binom{3}{2}\binom{5}{1} = 15$ collections contain two males, $\binom{3}{1}\binom{5}{2} = 30$ contain one male, and $\binom{3}{0}\binom{5}{3} = 10$ contain no males. Thus, if the selection is done at random and M is the number of males who get jobs, the probability function for M is

$$p_M(m) = \begin{cases} \binom{3}{m}\binom{5}{3-m} \Big/ \binom{8}{3}, & \text{for } m = 0, 1, 2, 3, \\ 0, & \text{otherwise.} \end{cases}$$

From this we have $P(M \geq 2) = \frac{16}{56}$; the probability is less than half of finding two or more males among those hired (if the assignment is made at random).

Let F be the number of females out of the eight applicants to receive jobs; then certainly $M + F = 3$, or $F = 3 - M$; that is, the value for F must be 3 less the number of males to receive jobs. Then $p_F(f) = P(F = f) = P(3 - M = f) = P(M = 3 - f) = p_M(3 - f)$. Thus the probability function for F has the same values as $p_M(m)$, except they occur at the translated values $m = 3 - f$. This probability function is

$$p_F(f) = \begin{cases} \dbinom{3}{3-f}\dbinom{5}{f} \bigg/ \dbinom{8}{3}, & \text{for } f = 0, 1, 2, 3, \\ 0, & \text{otherwise.} \end{cases}$$

The random variable F is a linear function of M; knowing the probability function for one of these two also immediately gives the probability function for the other.

Example 2.3

Many states run gambling games to raise revenue; in California the game lotto is popular. To play this game you select six integers (without replacement) from $\{1, 2, \ldots, 51\}$; the state then randomly selects six winning numbers from this same set. The amount you win (if anything) is determined by X, the number of values in your selected set that coincide with the state's set. The range for X is $R_X = \{0, 1, 2, \ldots, 6\}$; since the state selects its numbers at random, each of the $\binom{51}{6} = 18{,}009{,}460$ subsets of size 6 is equally likely to occur. Exactly x of your numbers will be matched in $\binom{6}{x}\binom{45}{6-x}$ of these selections, where $x \in R_X$. Thus the probability function for X is

$$p_X(x) = \begin{cases} \dbinom{6}{x}\dbinom{45}{6-x} \bigg/ \dbinom{51}{6}, & \text{for } x \in R_X, \\ 0, & \text{otherwise.} \end{cases}$$

You will win some amount of money if (and only if) $X \geq 3$ (and the bigger the value of X, the more you win). Thus the probability you win something, if you buy one ticket, is $P(X \geq 3) = \sum_{x=3}^{6} p_X(x) = 0.0166$; the probability you win nothing (and lose the \$1 paid) is $P(X < 3) = P(X \leq 2) = .9834$.

If we sum the value of the probability function for a random variable X over all the values in its (discrete) range R_X, the total must be 1 to satisfy the probability axioms. As done in the preceding examples, we shall take the probability function for a discrete random variable to equal 0 for $x \notin R_X$, giving a function defined over the whole real line; frequently it is convenient to explicitly define $p_X(x)$ only for $x \in R_X$, leaving the implied $p_X(x) = 0$ for $x \notin R_X$ not mentioned.

Any real-valued function, $p_X(x)$, of a real variable x can serve as a legitimate probability function for a discrete random variable if the following two requirements are met:

**REQUIREMENTS FOR THE PROBABILITY FUNCTION FOR
A DISCRETE RANDOM VARIABLE X**

1. $p_X(x) \geq 0$ for all real values of x.

2. $\sum_{x \in R_X} p_X(x) = 1$ for discrete R_X.

Many applications of probability and random variables are based on the structure of independent trials, discussed in Chapter 1. The following two examples are illustrations of this.

Example 2.4

Suppose you daily walk the same route, which takes you through three intersections with stoplights. Assume that the probability the light is red for your direction (so you must wait for it to turn green) is .4, independently for each intersection (each day). Let X be the number of red lights you encounter on a given day on this route. Immediately the range for X must be $R_X = \{0,1,2,3\}$, the only possibilities for the number of red lights you will encounter. The natural sample space to represent what may occur on a given day is the Cartesian product $S = T \times T \times T$, where $T = \{r,g\}$, where r and g represent the state of a given light when you arrive at the intersection. Then S has $2^3 = 8$ elements; the probability of any single-element event is the product of three terms, each of which is either .4 or .6, the assumed values for the probability of encountering a red or green light, respectively. The numbers of these 3-tuples that contain 0, 1, 2, and 3 rs are $\binom{3}{0}, \binom{3}{1}, \binom{3}{2}$, and $\binom{3}{3}$, respectively. Thus the probability function for X is

$$p_X(x) = \binom{3}{x}(.4)^x(.6)^{3-x}, \qquad \text{for } x = 0,1,2,3.$$

Then, the probability you will have to stop for at least one light is $P(X \geq 1) = \sum_{x=1}^{3} p_X(x) = .936$, while the probability you will have to stop for no more than one is $P(X \leq 1) = (.4)^3 + 3(.4)^2(.6) = .648$. This is an example of a binomial random variable, to be discussed in Chapter 3.

Recall that a discrete set consists of isolated points on the real line and is not necessarily finite; if X is discrete with infinite range R_X, then $\sum_{x \in R_X} p_X(x)$ is an infinite series that must converge to 1. The following example discusses a discrete random variable N whose range R_N is not finite.

Example 2.5

A fair coin is flipped until the first head occurs; N is the number of flips required. The results of the flips are assumed to be independent, so this is an example of an experiment with independent trials. The range for N is $R_N = \{1,2,3,\ldots\}$, the set of positive integers. $N = 1$ only if the first flip gives a head; thus $p_N(1) = \frac{1}{2}$, since the probability of getting a head is $\frac{1}{2}$ in flipping a fair coin. $N = 2$ if and only if the first flip gives a tail (so we have to flip again) and the second flip then gives a head; the probability of this occurring is $\frac{1}{2} \cdot \frac{1}{2} = \frac{1}{4}$, so $p_N(2) = \frac{1}{4}$. Continuing with this reasoning, $N = n \in R_N$ (i.e., the first head occurs on flip number n) if and only if we get $n - 1$ tails in a row followed by a head on flip number n; since the probability of this occurring with a fair coin is $\frac{1}{2}^{n-1}\left(\frac{1}{2}\right) = \frac{1}{2}^n$, it follows that the probability function for N is

$$p_N(n) = \begin{cases} 1/2^n, & \text{for } n = 1,2,3,\ldots, \\ 0, & \text{otherwise.} \end{cases}$$

For this to be a legitimate probability function it is necessary that

$$\sum_{n \in R_N} p_N(n) = \sum_{n=1}^{\infty} p_N(n) = 1.$$

The values for $p_N(n)$ form a geometric progression (see Appendix A), so

$$\sum_{n \in R_N} p_N(n) = \sum_{n=1}^{\infty} \frac{1}{2^n}$$

$$= \frac{1}{2} \sum_{n=0}^{\infty} \frac{1}{2^n} = \frac{1}{2} \left(\frac{1}{1 - \frac{1}{2}} \right) = 1,$$

and this function does satisfy the two rules for a probability function mentioned earlier.

The probability that N will equal an even number then is

$$P(N \text{ is even }) = \frac{1}{2^2} + \frac{1}{2^4} + \frac{1}{2^6} + \cdots$$

$$= \frac{1}{4} \left(1 + \frac{1}{4} + \frac{1}{4^2} + \cdots \right) = \frac{1}{3}.$$

Since the number of flips required to get the first head must be an odd number ($O = \{1, 3, 5, \dots\}$) or an even number ($E = \{2, 4, 6, \dots\}$), these two sets of outcomes are mutually exclusive and partition R_N; it follows then that $P(O) = 1 - P(E) = \frac{2}{3}$. It may seem curious that the probability of N equaling an odd number is twice as large as the probability of it equaling an even number. The fact that $P(\{1\}) = \frac{1}{2}$ with this assignment of probabilities immediately implies that $P(O) > \frac{1}{2}$.

The McLaurin series expansion of e^a gives

$$e^a = 1 + a + \frac{a^2}{2!} + \frac{a^3}{3!} + \cdots , \qquad \text{for any real } a. \tag{1}$$

Multiplying Eq. (1) by e^{-a} gives

$$1 = e^{-a} \left(1 + a + \frac{a^2}{2!} + \frac{a^3}{3!} + \cdots \right)$$

$$= e^{-a} \sum_{v=0}^{\infty} \frac{a^v}{v!}. \tag{2}$$

That is,

$$p_V(v) = \frac{a^v}{v!} e^{-a} \tag{3}$$

sums to 1 (the infinite series converges to 1), and all terms are nonnegative so long as $a > 0$, so Eq. (3) gives a legitimate probability function. (As will be seen in Chapter 3, $p_V(v)$ is an example of the probability function of a

Poisson random variable; the constant $a > 0$ is called a parameter.) A typical usage of this probability function is illustrated in the following example.

Example 2.6

Let V be the number of fatal automobile accidents, on a particular stretch of highway, over a one-month period. Then the set R_V of possible observed values would be discrete and consist of the counting numbers. So $R_V = \{0, 1, 2, \ldots\}$ where the integer $v \in R_V$ represents the number of fatal accidents to be observed in the given month. Quite arbitrarily at this point, let

$$p_V(v) = \frac{2^v}{v!} e^{-2}, \qquad \text{for } v = 0, 1, 2, \ldots . \tag{4}$$

Equation (4) is an example of the probability function given by Eq. (3), with $a = 2$.

Thus, this model gives

$$P(V \le 4) = e^{-2} \sum_{v=0}^{4} \frac{2^v}{v!}$$
$$= .9473.$$

The probability there will be at least six fatal accidents is

$$P(V \ge 6) = 1 - P(V \le 5)$$
$$= 1 - e^{-2} \sum_{v=0}^{5} \frac{2^v}{v!}$$
$$= 1 - .9834$$
$$= .0166.$$

Table 2.4 gives the values for the probabilities of 0 to 9 accidents occurring, rounded to four decimal places, with probability function (4); the probabilities of occurrence of the other $v \in R_V$, to four decimal places, are all 0. Since the sum of the values in this table is .9999, the probability of 10 or more accidents occurring is .0001 with this probability model.

Table 2.4

$v =$	0	1	2	3	4	5	6	7	8	9
$P(\{v\}) =$	.1353	.2707	.2707	.1804	.0902	.0361	.0120	.0034	.0009	.0002

Some discrete random variables occur frequently across many different applications and are given specific names. (The binomial, geometric, and Poisson variables already mentioned are among the most frequently used and will be discussed in detail in Chapter 3.)

One of the simplest standard discrete distributions is the discrete uniform probability law with positive integer parameter n. The word *parameter* is used to refer to any numeric constant occurring in a probability law; varying the parameter value(s) gives a new probability law of the same type. The name of the probability law is also frequently used in referring to the random variable itself; thus we shall refer to a discrete uniform random variable with integer parameter n, meaning the random variable that has the specified probability law and parameter value(s). The symbol used for the random variable bears no special significance; thus X or Y or, indeed, U could be used as the symbol for a discrete uniform random variable with parameter $n = 10$, say. The definition of the discrete uniform probability law with integer parameter n follows.

DEFINITION 2.2

A random variable X has the *discrete uniform probability law with integer parameter n* if

1. The range for X is $R_X = \{1, 2, \ldots, n\}$, where n is any positive integer.
2. The probability function for X is constant for $x \in R_X$; thus

$$p_X(x) = \frac{1}{n}, \qquad \text{for } x \in R_X.$$

This range R_X for a discrete uniform random variable contains n consecutive integers; thus the constant value $1/n$ for the probability function is necessary so that the sum of $p_X(x)$ over R_X gives 1. This particular discrete probability function is useful in modeling many types of fair gambling equipment; the following example discusses two such cases.

Example 2.7

We have used the example of a fair die several times; if X is the number of spots uppermost when a fair die is rolled, then X is discrete uniform with parameter $n = 6$; its probability function is

$$p_X(x) = \tfrac{1}{6}, \qquad \text{for } x = 1, 2, \ldots, 6.$$

The game of roulette involves spinning a horizontal wheel, with slots on its circumference; a small ball is rolled from the center of the wheel (or in an exterior ring) and eventually lands in one of the slots on the circumference. In European casinos, roulette wheels are subdivided into 37 sectors, with 37 slots on the circumference (one slot for each sector). These sectors are numbered 0 through 36 (0 is green in color, while half the remaining numbers are red and the other half black). If Y denotes the number on the slot into which the ball falls, for one play, then the range for Y is $R_Y = \{0, 1, 2, \ldots, 36\}$. Since this range starts at 0 rather than 1, Y is not a candidate for a uniform random variable as defined in Definition 2.2. However, $W = Y + 1$, which ranges through the integers from 1 to 37, does match this uniform range. The notation $W = Y + 1$ simply means we get the observed value for W by adding 1 to the observed value for Y; we call Y a *shifted uniform random variable*. Equivalently, we can write $Y = W - 1$, illustrating how the observed value for Y can be gotten from an observed value for W; this relation is called the *inverse transformation*. If the roulette wheel is fair (i.e., all spots occur with equal probability), W is then discrete uniform with parameter $n = 37$; its probability function is

$$p_W(w) = \tfrac{1}{37}, \qquad \text{for } w = 1, 2, \ldots, 37.$$

The probability function for Y, with this fairness assumption, is

$$p_Y(y) = \tfrac{1}{37}, \qquad \text{for } y = 0, 1, 2, \ldots, 36, \tag{5}$$

the same as the probability function for W except all the observed values are shifted to the left one unit (since $Y = W - 1$). The probability that Y equals an odd number is then given by summing this probability function, Eq. (5), over the odd numbers $(1, 3, \ldots, 35)$ in R_Y, yielding $18/37$ as the value for the probability. The probability that Y equals an even number (including 0 as an even number) is $19/37$.

In Chapter 3 we shall discuss several more standard discrete probability laws and the circumstances under which they occur.

EXERCISES 2.1

1. A hat contains 12 slips of paper, numbered consecutively from 1 to 12. One slip is removed at random, and Y is the number it bears. What is the probability function for Y?

2. Assume that one slip of paper is removed from the hat as in Exercise 1, and again Y is the number that it bears. Now suppose we add 5 to this number; that is, we define $X = Y + 5$. What is the probability function for X?

3. Assume that a hat contains 12 slips of paper, each of which bears the number 7. One slip is drawn from this hat at random; let X be the number on the slip drawn. What is the probability function for X?

4. Three men and two women apply for the same job; the interviewer will rank these five applicants and then choose the one ranked best. Assume that all possible rankings are equally likely.
   a. How many different arrangements can be made with three symbols of one type and two of another?
   b. For each of the possible rankings, we can let W denote the highest rank attained by a woman. For example, if the outcome is (m, w, m, m, w) (with the highest-ranked at the left), then $w = 2$. What is the range for w? Evaluate the probability function for w.
   c. What is the probability that a woman gets the job?

5. Two fair dice are rolled one time, one of which is red and the other is green. Let W be the *difference* (red number less green number) of the two numbers that occur.
   a. Find the probability function for W.
   b. Find the probability function for $X = |W|$, the magnitude of W.

6. A pair of fair dice is rolled one time. Let X be the larger of the two numbers to occur. Find the probability function for X. (If the two numbers are the same, the larger number is equal to their common value.)

7. Two fair dice are rolled one time. Let Y be the smaller of the two numbers to occur. Evaluate the probability function for Y. (If the two numbers are the same, the smaller number is equal to their common value.)

8. A bookstore has four different textbooks available on the same subject, priced \$48, \$52, \$44, and \$61, respectively. Let these four numbers represent R_X, the range of a discrete random variable X; also assume that the probability function for X is constant for all $x \in R_X$.
   a. What is the probability function for X?
   b. Describe an experiment for which this probability law occurs.

9. Each of two persons independently enter the bookstore mentioned in Exercise 8, and each chooses, at random, one of the four different textbooks available on the given subject. Let Y be the sum of the prices paid by these two persons for the books chosen by them. Evaluate R_Y and $p_Y(y)$.

10. An automated process is used to produce light bulbs; each bulb produced has probability .1 of being defective (not working), and defective bulbs occur independently. You buy four of these bulbs; let Y be the number of bulbs you buy that are defective. What is the range for Y and the probability function for Y? (*Hint:* Recall the earlier discussion of independent trials.)

11. A large supermarket display contains 100 individually wrapped light bulbs, 10 of which are defective. You select four of these bulbs at random (without replacement); let Y be the number of defective bulbs you select. What is the range for Y and the probability function for Y?

12. Each of three friends flips a single coin one time; assume that all three coins are fair, and let X be the number of heads to occur. What is the probability function for X?

13. Let X be the sum of the two numbers to occur on one roll of a pair of fair dice; show that the probability that the observed value for X is some multiple of k equals $1/k$, for $k = 2, 3, 4,$ and 6.

14. A fair die is rolled until the first 1 occurs; let N be the number of rolls required, and find $p_N(n)$, the probability function for N. Evaluate $P(N \leq 3)$ and $P(N > 1)$.

15. A fair die is rolled four times. Let X be the number of 1's to occur. Find the probability function for X, and evaluate $P(X \leq 2)$.

16. A random variable Y has probability function

$$p_Y(y) = \frac{r}{y!}e^{-1}, \qquad \text{for } y = 0, 1, 2, \ldots.$$

   a. What must be the value for r?
   b. Evaluate $P(1 \leq Y \leq 5)$ and $P(Y > 4)$.

17. A discrete random variable Z has the range $R_Z = \{1, 2, 3, 4\}$. Granted $p_Z(z) = z p_Z(1)$ for $z = 2, 3, 4$, find $p_Z(z)$ and evaluate $P(Z \geq 3)$.

2.2 | Cumulative Distribution Functions and Probability Density Functions

If X is a discrete random variable with range R_X, its probability function $p_X(x)$ defines the amount of probability at $x \in R_X$, where $\sum_{x \in R_X} p_X(x) = 1$; R_X and $p_X(x)$ define the *probability distribution* for X. An alternative way of defining the probability distribution for a random variable X is given by its cumulative distribution function, abbreviated *cdf*, which is defined below.

DEFINITION 2.3 —————————————————————

Let X be a random variable and let t be any real number; the *cumulative distribution function (cdf)* for X is $F_X(t)$, which gives the probability that the observed value for X will be less than or equal to t, for all real t:

$$F_X(t) = P(X \le t) \qquad \text{for } -\infty < t < \infty.$$

The *cdf* simply describes the accumulation of probability at any real number t.

When the probability function for X is given, it is easy to find its *cdf*; suppose a discrete random variable X has the probability function given in Table 2.5.

We could figuratively think of this X as having deposited three "chunks" of probability, of sizes .1, .6, and .3 at the integers 2, 5, and 6, respectively. To get $F_X(t)$ we can think of an accumulator starting at the left end of the real number line (at $t = -\infty$) and moving right. The accumulator $F_X(t)$ starts at 0 and keeps track of the accumulated probability at each value t; thus in this case $F_X(t) = 0$ until we reach $t = 2$, where $F_X(2) = .1$. The accumulated total stays constant at .1 until we reach $t = 5$, at which point $F_X(5) = .1 + .6 = .7$. Again the accumulation stays constant at .7 until we reach $t = 6$, with $F_X(6) = .1 + .6 + .3 = 1$, its maximum value. Thus the *cdf* for X is

$$F_X(t) = \begin{cases} 0, & \text{for } t < 2, \\ .1, & \text{for } 2 \le t < 5, \\ .7, & \text{for } 5 \le t < 6, \\ 1, & \text{for } t \ge 6. \end{cases}$$

Both $p_X(x)$ and $F_X(t)$ are graphed in Fig. 2.1.

Table 2.5

Probability Function for X

x	2	5	6
$P(X = x)$	.1	.6	.3

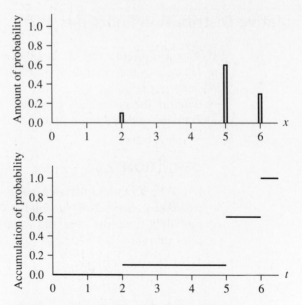

Figure 2.1 Two descriptions of the probability model for X given in Table 2.5

The *cdf* for any random variable X is defined over the entire real line; since $F_X(t) = P(X \leq t)$ is a probability, it is clear that $0 \leq F_X(t) \leq 1$; that is, the *cdf* must lie between 0 and 1 for all t. As illustrated in the preceding discussion, if X is a discrete random variable with probability function $p_X(x)$, then the probability that $X \leq t$ is given by summing the probability function over the values $x \in R_X$ (if there are any) which are no larger than t. If we use the symbolic notation "$x \leq t$" to indicate the more complete statement "$x \in R_X$ and $x \leq t$," the *cdf* for a discrete X can be written

$$F_X(t) = \sum_{x \leq t} p_X(x),$$

for all real t. In particular, if all of the elements of R_X exceed t (so that t is smaller than all $x \in R_X$), then this summation range is empty and $F_X(t) = 0$; if t is larger than all the elements in R_X, then the value for $F_X(t)$ is equal to the total sum of the probability function values, and $F_X(t) = 1$. This reasoning is illustrated again in the following example.

Example 2.8

Suppose a hat contains 10 slips of paper, of which one bears the number 1, two bear the number 2, three bear the number 3, and the remaining four each bear the number 4. If we select one of these slips of paper at random and let Y be the number it bears, the range for Y is $R_Y = \{1, 2, 3, 4\}$ and the probability function for Y is $p_Y(y) = y/10$, for $y \in R_Y$. Figure 2.2 pictures this probability function, as well as the *cdf* for Y.

This *cdf* is again gotten by simply describing the running accumulation of

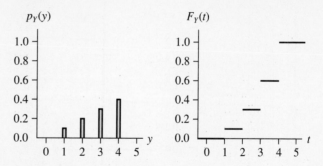

Figure 2.2   Probability function (*left*) and *cdf* (*right*) for Y in Example 2.8

probability for $-\infty < t < \infty$, which gives

$$F_Y(t) = \begin{cases} 0, & \text{for } t < 1, \\ .1, & \text{for } 1 \le t < 2, \\ .3, & \text{for } 2 \le t < 3, \\ .6, & \text{for } 3 \le t < 4, \\ 1, & \text{for } 4 \le t. \end{cases}$$

As is true for any discrete random variable, $F_Y(t)$ jumps in value for each $y \in R_Y$ (i.e., the accumulation instantaneously increases) and the size of the jump at $y \in R_Y$ is given by $p_Y(y)$, the amount of probability located at the point.

Every discrete random variable places positive mass at the elements of its range; this behavior, in turn, causes the *cdf* for a discrete random variable to be a step function, a discontinuous function which jumps in value at each element of the range of the random variable. This fact makes it easy to find the probability function for a discrete random variable, if its *cdf* has been specified, as illustrated in the following example.

Example 2.9

A college student living off-campus drives her car to school each day, passing through three intersections with stoplights on the way. As she approaches each of these intersections, the stoplight is either red or green. Let X be the number of these stoplights that are red on a given trip, and assume we are given that the *cdf* for X is

$$F_X(t) = \begin{cases} 0, & \text{for } t < 0, \\ \frac{27}{64}, & \text{for } 0 \le t < 1, \\ \frac{54}{64}, & \text{for } 1 \le t < 2, \\ \frac{63}{64}, & \text{for } 2 \le t < 3, \\ 1, & \text{for } t \ge 3. \end{cases}$$

This *cdf* is pictured on the left in Fig. 2.3; note that the jumps in $F_X(t)$ are at the integers 0, 1, 2, 3, which then must be the elements of R_X, the values with positive probability, so $R_X = \{0, 1, 2, 3\}$.

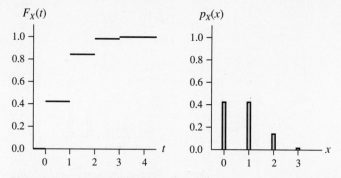

Figure 2.3   Cumulative distribution function (*left*) and probability function (*right*)

The size of the jump at any particular real number gives the amount of mass (probability) at that same value, since the jump indicates a discrete change in the accumulation of probability there. Thus the probability function for X is

$$p_X(x) = \begin{cases} \frac{27}{64} - 0 = \frac{27}{64}, & \text{for } x = 0, \\ \frac{54}{64} - \frac{27}{64} = \frac{27}{64}, & \text{for } x = 1, \\ \frac{63}{64} - \frac{54}{64} = \frac{9}{64}, & \text{for } x = 2, \\ 1 - \frac{63}{64} = \frac{1}{64}, & \text{for } x = 3, \\ 0, & \text{for any other } x, \end{cases}$$

as is also pictured in the figure. Since $F_X(t)$ jumps at only four points, there are only four elements in the range of X, as already indicated; the probability function must equal zero at all other values; otherwise there would have been further jumps in $F_X(t)$.

Cumulative distribution functions for discrete random variables are always discontinuous step functions; the jumps in value are caused by the discrete change in accumulation of probability each time an observed value of the random variable is encountered. Knowledge of the probability function $p_X(x)$ allows us to find the *cdf* $F_X(t)$ and vice versa.

If a is any real number, then the events $\{X \leq a\}$ and $\{X > a\}$ are complementary; thus $P(X \leq a) + P(X > a) = 1$, giving $P(X > a) = 1 - P(X \leq a) = 1 - F_X(a)$. Subtracting the value of the *cdf* (evaluated at any t) from 1 gives the probability that the random variable exceeds t.

Given two numbers c and d such that $c < d$, let A be the event $\{X \leq c\}$ while B is the event $\{X \leq d\}$; if A occurs (so the observed X is no larger than c), then certainly this observed X is also no larger than d (since $d > c$) and event B also occurs. Thus $A \subset B$ and $A \cup B \equiv B$. Recall also that $B = A \cup (\overline{A} \cap B)$ for this situation and that A and $\overline{A} \cap B$ are mutually exclusive. We then have $P(\overline{A} \cap B) = P(B) - P(A)$; that is,

$$P(c < X \leq d) = F_X(d) - F_X(c),$$

as pictured in Fig. 2.4. This result is illustrated in the following example.

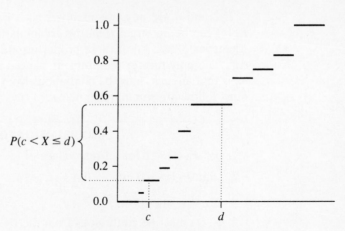

Figure 2.4 $P(c < X \le d) = F_X(d) - F_X(c)$

Example 2.10

You buy three scratch-off lottery tickets, each of which either does or does not win a prize. Let Y be the number of these tickets that win a prize; the range for Y then is $R_Y = \{0, 1, 2, 3\}$. Assume the *cdf* for Y is

$$F_Y(t) = \begin{cases} 0, & \text{for } t < 0, \\ .7023, & \text{for } 0 \le t < 1, \\ .9657, & \text{for } 1 \le t < 2, \\ .9986, & \text{for } 2 \le t < 3, \\ 1, & \text{for } t \ge 3. \end{cases}$$

What is the probability you will win more than one prize with these three tickets; that is, what is the value for $P(Y > 1)$? The two events $\{Y > 1\}$ and $\{Y \le 1\}$ are complementary, so $P(Y > 1) = 1 - P(Y \le 1) = 1 - F_Y(1) = .0343$. What is the value for $P(.5 < Y \le 2.5)$, that Y equals some value in $(.5, 2.5]$? Let $(a, b]$ represent the interval $a < t \le b$, which includes the right-hand endpoint (indicated by using the bracket) but does not include the left-hand endpoint (indicated by using an open parenthesis). We have $P(Y \le 2.5) = F_Y(2.5) = .9986$ and $P(Y \le .5) = F_Y(.5) = .7023$ from above. The difference in accumulation of probability $F_Y(2.5) - F_Y(.5)$ then must be the amount of probability contained in the interval $(.5, 2.5]$ (including 2.5 but excluding .5); thus $P(.5 < Y \le 2.5) = .9986 - .7023 = .2963$.

You should verify this by finding the probability function $p_Y(y)$; the interval $(.5, 2.5]$ contains two points (1 and 2) with positive probability. The sum $p_Y(1) + p_Y(2)$ thus also gives $P(.5 < Y \le 2.5)$.

As has been mentioned, it is necessary that $0 \le F_X(t) \le 1$ for all real values of t, since the *cdf* evaluated at any t gives a probability. Indeed, since the probability function sums to 1, the accumulation at $-\infty$ must be 0 and the accumulation at ∞ must be 1; that is,

$$\lim_{t \to -\infty} F_X(t) = 0, \qquad \lim_{t \to \infty} F_X(t) = 1.$$

If c and d are any two numbers such that $c < d$, then the accumulation $F_X(c)$ can be no larger than the accumulation $F_X(d)$; that is, $F_X(c) \leq F_X(d)$ whenever $c < d$. All of these properties are necessary for a function to be a *cdf*. The only further requirement, caused by our definition $F_X(t) = P(X \leq t)$, is that any *cdf* must be continuous from the right. The rules that any *cdf* must follow are summarized below.

REQUIREMENTS FOR $F_X(t)$

1. $0 \leq F_X(t) \leq 1$ for all real values of t.

2. $\lim\limits_{t \to -\infty} F_X(t) = 0$ and $\lim\limits_{t \to \infty} F_X(t) = 1$.

3. If $c < d$, then $F_X(c) \leq F_X(d)$.

4. $F_X(t)$ must be continuous from the right.

All of the *cdf*'s discussed to this point are in fact step functions: functions of t that remain constant, then jump to a new value at a particular point — because $p_X(t) > 0$ at that point — and remain constant, then jump and remain constant at another value, and so forth. In fact, this type of behavior for $F_X(t)$ shows that X must be a discrete random variable.

There are many continuous functions of a real variable t that satisfy the requirements for a *cdf*; for example,

$$F_X(t) = \begin{cases} 0, & \text{for } t \leq 0, \\ t, & \text{for } 0 < t < 1, \\ 1, & \text{for } t \geq 1, \end{cases} \tag{1}$$

is a continuous function that fills these requirements for a *cdf*. Since $F_X(0)$ (the accumulation is 0 at $t = 0$) and $F_X(1)$ (the accumulation is 1 at $t = 1$), all of the observed values for this X must lie between 0 and 1. If the probability distribution for X is given by this *cdf*, then

$$P\left(X \leq \tfrac{1}{3}\right) = F_X\left(\tfrac{1}{3}\right) = \tfrac{1}{3},$$
$$P\left(\tfrac{1}{4} < X \leq \tfrac{3}{4}\right) = F_X\left(\tfrac{3}{4}\right) - F_X\left(\tfrac{1}{4}\right) = \tfrac{3}{4} - \tfrac{1}{4} = \tfrac{1}{2},$$
$$P\left(X > \tfrac{2}{3}\right) = 1 - F_X\left(\tfrac{2}{3}\right) = 1 - \tfrac{2}{3} = \tfrac{1}{3}.$$

Any random variable whose *cdf* is a continuous function of t is called a continuous random variable. We shall define this concept more formally later.

If a random variable is continuous, then its *cdf* does not jump in value at any point, so its probability function is everywhere 0. This means that the probability for any individual point must be 0 with continuous random variables. However, for continuous random variables a probability of 0 does not imply that the given value cannot be an observed value. The continuous $F_X(t)$ given by Eq. (1) varies from 0 to 1 for $0 \leq t \leq 1$; thus some value in this interval must be the observed value for X (the range for X is $\{0 < x < 1\}$), but all these points have probability 0. The notion of continuity forces this conclusion; the individual points in (0,1) are too dense for them to have positive probability if the total of all the probability values is to be fixed at 1.

Since $F_X(t) = 0$ for $t \leq 0$, no negative value could be observed for this X, nor could any value greater than 1; since $F_X(t) = 1$ for $t \geq 1$, the accumulation of probability does not change. If $F_X(t)$ is continuous, those points in intervals over which $F_X(t)$ is increasing are the possible observed values for X, even though the probabilities for all these individual points must be 0.

The probability distribution for a discrete random variable can be defined by either its *cdf* or its probability function. For a continuous X the analog of the probability function is given by the derivative of the *cdf*:

$$f_X(t) = \frac{d}{dt} F_X(t),$$

the slope of the *cdf*, the rate at which the probability is accumulating. (For all the continuous *cdf*s used in this text, this derivative will exist at all but a finite number of points; since probabilities of individual points must be 0 in the continuous case, this poses no problem in modeling the behavior of continuous cases.) This derivative $f_X(t)$ is called the probability density function *(pdf)* for X. For the example discussed in the previous paragraph, the *pdf* for X is

$$f_X(t) = \begin{cases} \dfrac{d}{dt} F_X(t) = 1, & \text{for } 0 < t < 1 \\[2mm] \dfrac{d}{dt} F_X(t) = 0, & \text{for } t < 0 \text{ or } t > 1 \end{cases}.$$

The derivative does not exist at $t = 0$ or $t = 1$; we shall simply define $f_X(t) = 0$ for these two points.

This derivative $f_X(t)$ is 0 over all intervals where $F_X(t)$ is constant (slope 0) and is positive over intervals where $F_X(t)$ is increasing; at points for which the derivative does not exist, we define $f_X(t)$ to be any convenient value, typically 0. Then the range for a continuous X is $R_X = \{x : f_X(x) > 0\}$, the collection of values at which $F_X(t)$ is increasing; note that this is the same definition for the range R_X as for the discrete case, if $f_X(x)$ is replaced by the probability function $p_X(x)$. This discussion for the continuous case is summarized in the following definition.

DEFINITION 2.4

X is called a *continuous random variable* if its *cdf* is a continuous function of the argument t. The *probability density function (pdf)* for X is

$$f_X(t) = \frac{d}{dt} F_X(t);$$

the range for X is $R_X = \{x : f_X(x) > 0\}$ or, equivalently, the values for which the *cdf* is increasing.

Since integration is the opposite operation of differentiation, we also have $F_X(t) = \int_{-\infty}^{t} f_X(x)\, dx$; the accumulation of probability at t is given by the *area* under $f_X(x)$ to the left of t. Indeed, recalling that $P(c < X \le d) = F_X(d) - F_X(c) = \int_{-\infty}^{d} f_X(x)\,dx - \int_{-\infty}^{c} f_X(x)\,dx$ gives $P(c < X \le d) = \int_{c}^{d} f_X(x)\,dx$, the probability that the observed value for X falls within a finite length interval is simply given by the definite integral of $f_X(x)$ over the interval of interest (the area under $f_X(x)$). The following examples model the distributions for some continuous random variables.

Example 2.11

Assume you routinely take a public bus scheduled to leave your stop at 8:05 AM. Also assume that you always arrive at the bus stop between 8:00 AM and 8:05 AM, and let X represent your arrival time at this stop, on a given day, in minutes after 8:00 AM. Since you always arrive between 8:00 and 8:05, the range for X is $R_X = \{x : 0 < x < 5\}$ and the *pdf* for X is positive only for $0 < x < 5$. The particular shape of $f_X(x)$ over this interval controls the assignment of probabilities for X lying in subintervals of $(0,5)$. If every 1-minute interval between 8:00 AM and 8:05 AM is equally likely to contain your arrival time, we would take $f_X(x)$ to be constant for $0 < x < 5$ and use

$$f_X(x) = \begin{cases} \frac{1}{5}, & \text{for } 0 < x < 5, \\ 0, & \text{otherwise,} \end{cases}$$

as the *pdf* for X. This *pdf* is pictured in Fig. 2.5.

The height of the *pdf* must be $\frac{1}{5}$ to make the area under $f_X(x)$, over $(0, 5)$, equal to 1; the *pdf* is constant to reflect the assumption that every 1-minute subinterval is to have the same probability. This figure also indicates the area over the interval $(1, 3)$, which gives $P(1 \le X \le 3)$ for these assumptions. The probability of your arriving after 8:02 is, with this model, $P(X > 2) = \int_{2}^{5} \frac{1}{5}\,dx = .6$, and the probability of your being there by 8:04 is $P(X \le 4) = \int_{0}^{4} \frac{1}{5}\,dx = .8$. The *cdf* corresponding to this *pdf* is

$$F_X(t) = \int_{-\infty}^{t} f_X(x)\,dx = \begin{cases} \int_{-\infty}^{t} 0\,dx = 0, & \text{for } t \le 0, \\[2mm] 0 + \int_{0}^{t} \frac{1}{5}\,dx = \frac{t}{5}, & \text{for } 0 < t < 5, \\[2mm] 0 + 1 + \int_{1}^{t} 0\,dx = 1, & \text{for } t \ge 5. \end{cases}$$

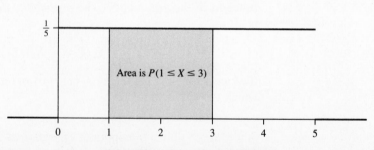

Figure 2.5 A constant *pdf* for X

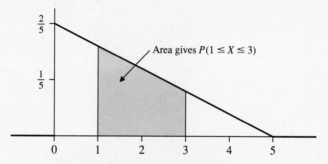

Figure 2.6 A triangular *pdf* for X

If you, like most people, want to make sure you don't miss the bus, it might be more realistic to assume a *pdf* $f_X(x)$ for your arrival time that gives a bigger probability of finding $0 < X < 1$ than of finding $4 < X < 5$. There are many shapes for $f_X(x)$ that accomplish this, one of the simplest being

$$f_X(x) = \begin{cases} \frac{2}{25}(5 - x), & \text{for } 0 < x < 5, \\ 0, & \text{otherwise,} \end{cases}$$

pictured in Fig. 2.6.

With this *pdf* we have

$$P(0 \leq X \leq 1) = \int_0^1 \tfrac{2}{25}(5 - x)\, dx = \tfrac{9}{25},$$

while

$$P(4 \leq X \leq 5) = \int_4^5 \tfrac{2}{25}(5 - x)\, dx = \tfrac{1}{25},$$

and

$$P(1 \leq X \leq 3) = \int_1^3 \tfrac{2}{25}(5 - x)\, dx = \tfrac{12}{25},$$

as pictured in Fig. 2.6. The *cdf* for X is

$$F_X(t) = \begin{cases} 0, & \text{for } t \leq 0, \\ 1 - \dfrac{(5 - t)^2}{25}, & \text{for } 0 < t < 5, \\ 1, & \text{for } t \geq 5, \end{cases}$$

as you can verify. These two *pdf*s provide two different probability models for X and correspond to two different assumptions about its behavior; which of the two may be better depends on the assumptions you wish to make.

Example 2.12

Let T represent the time a mechanic requires to diagnose and repair a certain automobile problem; take the *cdf* for T to be

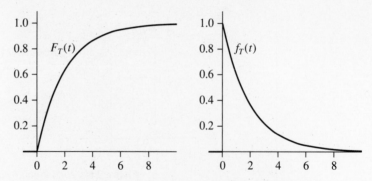

Figure 2.7 *cdf* and *pdf* for Example 2.12

$$F_T(t) = \begin{cases} 0, & \text{for } t \le 0, \\ 1 - e^{-t/2}, & \text{for } t > 0, \end{cases}$$

where T is measured in hours; this *cdf* is pictured in Fig. 2.7. Then the probability he will take no more than 2 hours is $P(T \le 2) = F_T(2) = 1 - e^{-1} = .6321$, while the probability he will take more than 1 hour is $P(T > 1) = 1 - F_T(1) = e^{-1/2} = .6065$. The *pdf* for T is given by the derivative of $F_T(t)$:

$$f_T(t) = \begin{cases} \frac{1}{2}e^{-t/2}, & \text{for } t > 0, \\ 0, & \text{for } t \le 0, \end{cases}$$

as pictured on the right in the same figure.

The probability that less than 1 hour is required to diagnose and repair the problem is

$$\int_0^1 \tfrac{1}{2}e^{-x/2}dx = -e^{-x/2}\Big|_{x=0}^{x=1} = 1 - e^{-.5} = .3935,$$

while the probability that it takes between 2 and 3 hours is

$$\int_2^3 \tfrac{1}{2}e^{-x/2}dx = -e^{-x/2}\Big|_{x=2}^{x=3} = e^{-1} - e^{-1.5} = .1447.$$

This distribution is an example of an exponential probability law, as we will see later.

To satisfy the probability axioms, any function $f_X(x)$ of a real variable x can be used as the *pdf* for a continuous random variable X as long as the following requirements are met.

REQUIREMENTS FOR THE *pdf* FOR A CONTINUOUS X

1. $f_X(x) \ge 0$ for all real x.

2. $\displaystyle\int_{x \in R_X} f_X(x) = 1.$

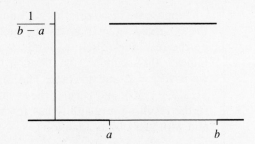

Figure 2.8 Continuous uniform *pdf* with parameters a, b

Notice the essential equivalence of these requirements for a *pdf* and the requirements for the probability function for a discrete X.

Just as with discrete random variables, there are certain continuous probability laws that occur frequently in applications; the continuous analog of the discrete uniform discussed in Section 2.1 is the continuous uniform probability law with parameters a, b where the interval (a, b) gives the range for X. Recall that the discrete uniform probability function is a constant for values in the range; the continuous uniform probability law shares this property in that the *pdf* is constant for values in the range. The continuous uniform probability law is defined below.

DEFINITION 2.5

A random variable X has the *continuous uniform probability law with parameters a, b* (where $a < b$) if
1. The range for X is $R_X = \{a < x < b\}$.
2. The *pdf* for X is constant for $x \in R_X$; thus the *pdf* is

$$f_X(x) = \begin{cases} \dfrac{1}{b-a}, & \text{for } a < x < b, \\ 0, & \text{otherwise.} \end{cases}$$

For the continuous uniform probability law the parameters a and b $(a < b)$ can be any two real numbers and are not restricted to being integers as in the discrete uniform case. This *pdf* is pictured in Fig. 2.8; because of its shape it is also sometimes called a *rectangular distribution*.

Example 2.13

If X has the continuous uniform distribution over the interval from 0 to 10, say, then its *pdf* is

$$f_X(x) = \tfrac{1}{10}, \qquad \text{for } 0 < x < 10.$$

Since the probability an interval, say $(3, 4)$, will contain the observed value for X is given by the area under the *pdf* over the interval, we have $P(3 < X < 4) = \tfrac{1}{10}$. In fact, *any* subinterval of the range of X, which has length 1, will have this same probability; the probability an interval contains the observed value is proportional to the length of the interval, with this probability law.

Suppose we *discretize* this continuous random variable by always rounding the observed value x up to the next larger integer, and suppose we let Y be the resulting (discrete) random variable. Formally, we write $Y = \lceil X$ to indicate this rounding up. Thus, if $0 < x \leq 1$, we have $y = 1$; if $1 < x \leq 2$, we have $y = 2$; and so forth. The range for Y then is the discrete set $R_Y = \{1, 2, 3, \dots, 10\}$, and each of these values has probability $\frac{1}{10}$ of occurring; that is, the resulting probability law for Y is discrete uniform with parameter 10 and with probability function

$$p_Y(y) = \int_{y-1}^{y} f_X(x)\, dx = \tfrac{1}{10}, \qquad \text{for } y = 1, 2, 3, \dots, 10.$$

The following example derives the *cdf* for the continuous uniform probability law with parameters a, b.

Example 2.14

Let X be a continuous uniform random variable with parameters a, b; then its *pdf* is

$$f_X(x) = \begin{cases} \dfrac{1}{b-a}, & \text{for } a < x < b, \\ 0, & \text{otherwise,} \end{cases}$$

and is thus a constant for all $x \in R_X = \{a < x < b\}$. The *cdf* for X gives the accumulation of probability on the real line up to and including the real number t. Using the discussion above, we have

$$F_X(t) = \int_{-\infty}^{t} f_X(x)\, dx = \begin{cases} \displaystyle\int_{-\infty}^{t} 0\, dx = 0, & \text{if } t \leq a, \\[2ex] \displaystyle\int_{-\infty}^{a} 0\, dx + \int_{a}^{t} \frac{1}{b-a}\, dx = \frac{t-a}{b-a}, & \text{if } a < t \leq b, \\[2ex] \displaystyle\int_{-\infty}^{a} 0\, dx + \int_{a}^{b} \frac{1}{b-a}\, dx + \int_{b}^{t} 0\, dx = 1, & \text{if } b < t. \end{cases}$$

This uniform *pdf* is pictured on the left, with the *cdf* pictured on the right, of Fig. 2.9. Note that this *cdf* is a continuous function of t, as is true for all

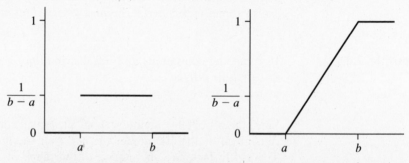

Figure 2.9 Uniform *pdf* (*left*) and *cdf* (*right*), with parameters a, b

the continuous random variables we shall employ, and has no jumps in value as seen in discrete *cdf*s.

For the continuous uniform probability law the *pdf* is a constant for $x \in R_X$; the accumulation of probability then must change linearly with t for $a < t < b$. This fact leads to the *cdf*, $F_X(t)$, being a *linear* function for $t \in R_X$. If X is uniform with $a = 0$ and $b = 1$, as discussed earlier, the *cdf* for X is

$$F_X(t) = \begin{cases} 0, & \text{for } t \leq 0, \\ t, & \text{for } 0 < t < 1, \\ 1, & \text{for } 1 \leq t, \end{cases}$$

The *cdf* for a random variable X fully describes the probability law for X, whether X is discrete or continuous. It proves a powerful tool for certain types of applications, as we shall see. The *cdf* is gotten from the probability function for a discrete X by summing values of its probability function $p_X(x)$; the *cdf* for a continuous random variable X is given by the integral of the *pdf*.

---

cdf FOR DISCRETE X

$$F_X(t) = \sum_{x \leq t} p_X(x).$$

cdf FOR CONTINUOUS X

$$F_X(t) = \int_{-\infty}^{t} f_X(x)\, dx.$$

Given the *cdf* for a random variable X, the probability law for X is discrete if $F_X(t)$ is a step function; the probability law for X is continuous if $F_X(t)$ is a continuous function of t. The probability function for X, or the *pdf* for X, can be derived from the *cdf*; for the discrete case the probability function is specified by the sizes of the jumps (and their locations) and in the continuous case the *pdf* is specified by the probability function (or slope) of the *cdf*.

PROBABILITY FUNCTION FOR DISCRETE X

$p_X(x)$ is the size of the jump in $F_X(t)$ at $t = x$.

pdf FOR CONTINUOUS X

$$f_X(t) = \frac{d}{dt} F_X(t).$$

The *cdf* for a random variable (either discrete or continuous) gives $P(X \leq t)$, the probability content of the interval $(-\infty, t]$. Simple differences of values of $F_X(t)$ give the probability content of finite-length intervals, including the right endpoint but not the left. Any interval either does or does not include its left- or right-hand endpoints, giving four possibilities: $(c,d), [c,d), (c,d]$, or $[c,d]$. For the case of the closed interval $[c,d]$, clearly

$$P(c \leq X \leq d) = P(X = c) + P(c < X \leq d)$$
$$= P(X = c) + F_X(d) - F_X(c).$$

It is equally easy then to see the results for the other possible cases:

$$P(c < X < d) = F_X(d) - F_X(c) - P(X = d)$$
$$P(c \leq X < d) = F_X(d) - F_X(c) - P(X = d) + P(X = c).$$

The difference $F_X(d) - F_X(c)$ always gives the probability on the half-open interval $(c,d]$. This distinction of including or not including endpoints has no effect if X is a continuous random variable, since all individual points then have probability 0; the probability content is the same, whether or not endpoints of the interval are included. That is, if X is any continuous random variable, then

$$P(c < X < d) = P(c \leq X \leq d)$$
$$= P(c \leq X < d)$$
$$= P(c < X \leq d)$$
$$= F_X(d) - F_X(c),$$

for any real numbers $c < d$. This, of course, is not true for all $c < d$ if X is a discrete random variable.

If the *cdf* for X remains constant over any interval, say $F_X(c) = F_X(d)$ where $c < d$, then the probability the interval $(c,d]$ contains the observed value for X is necessarily 0; no observed values for X will ever lie in such an interval $(c,d]$.

 EXERCISES 2.2

1. In Example 2.10, we assumed a *cdf* for Y, the number of winning tickets contained in three lottery tickets. Find the probability function for Y from this *cdf*.

2. Let X be continuous uniform with parameters 0, 10, and, instead of rounding the value for X up to the next larger integer, let the value for Y be the largest integer in X; that is, round X *down* to the next smallest integer. Formally this is written $Y = \lfloor X$. What is the probability law for Y?

3. A hat contains five slips of paper, each of which bears the number 2. One of these slips is selected at random; X is the number on the slip selected. Give the probability function and *cdf* for X.

4. Now suppose a hat contains five slips of paper, two with the number 1 written on them and three with the number 2. One slip is selected at random with X defined to be the number it bears. Find the probability function and *cdf* for X.

5. A hat contains five slips of paper, two with the number 1, and the rest with the number 2. Two slips are selected at random, without replacement; let

Y be the sum of the two numbers drawn. Find the probability function and *cdf* for Y.

6. You drive your automobile from the Atlantic to the Pacific coast. T is the number of flat tires you will suffer during this trip, with *cdf*

$$F_T(t) = \begin{cases} 0, & \text{for } t < 0, \\ .6, & \text{for } 0 \le t < 1, \\ .9, & \text{for } 1 \le t < 2, \\ 1, & \text{for } t \ge 2. \end{cases}$$

Find the probability function for T.

7. Find the *cdf* for a discrete uniform random variable Y with parameter n.

8. A random variable V has probability function $p_V(v) = p(1 - p)^{v-1}$, for $v = 1, 2, 3, \ldots$, and $p_V(v) = 0$, otherwise, where $0 < p < 1$ is a parameter. Find the *cdf* for V.

9. Assume that W is a continuous random variable with *pdf*

$$f_W(w) = \begin{cases} w - 1, & \text{for } 1 < w < 2, \\ 3 - w, & \text{for } 2 \le w < 3, \\ 0, & \text{otherwise,} \end{cases}$$

one type of *triangular density*. Plot this *pdf* and find the *cdf* for W.

10. Modify the *pdf* in Exercise 9 by keeping the density the same, except for the interval from 1.5 to 2.5; over this interval make the density constant. What does this condition imply is the new probability density function, and what is the corresponding *cdf*?

11. Assume the time to diagnose and repair a home dishwasher is a continuous random variable R with *pdf*

$$f_R(r) = \begin{cases} 2e^{-2r}, & \text{for } r > 0, \\ 0, & \text{otherwise.} \end{cases}$$

Evaluate $F_R(t)$, the *cdf* for R.

12. On a trip to a supermarket, let X be the amount of time you must wait in line to check out, in minutes. The *cdf* for X is given as

$$F_X(t) = \begin{cases} 0 & \text{for } t \le 0, \\ \dfrac{t^2}{6}, & \text{for } 0 \le t \le 2, \\ 1 - \dfrac{(t-3)^2}{3}, & \text{for } 2 < t \le 3, \\ 1 & \text{for } t > 3. \end{cases}$$

a. Find the *pdf* for X, and sketch its shape.

b. Evaluate the probability that you must wait more than 1 minute to check out.

13. Let Y represent the amount of time (in minutes) you must wait for a friend whom you agree to meet at 12 noon; assume the *cdf* for Y is

$$F_Y(t) = \begin{cases} 1 - \dfrac{1}{t^3}, & \text{for } t > 1 \\ 0, & \text{otherwise.} \end{cases}$$

    a. Evaluate the probability that you must wait at least 2 minutes for your friend.

    b. Find the *pdf* for Y.

14. A random variable U has *pdf*

$$f_U(u) = \begin{cases} \frac{1}{6} + (u - 1)^2, & \text{for } 0 \leq u \leq 2, \\ 0, & \text{otherwise.} \end{cases}$$

This is called a *U-shaped density*. (Can you see why?) Evaluate the *cdf* for U.

15. It is not necessary that the range of a continuous random variable be a single interval on the real line. Assume W has *pdf*

$$f_W(w) = \begin{cases} w, & \text{for } 0 < w < 1, \\ 3 - w, & \text{for } 2 < w < 3, \\ 0, & \text{otherwise.} \end{cases}$$

    a. What is the range for W?   b. Evaluate the *cdf* for W.

16. Only one of the following functions is a legitimate *cdf* for a random variable X; identify it.

$$F_X(t) = \begin{cases} 0, & \text{for } t < 0 \\ 1 - 3e^{-(t+1)}, & \text{for } t \geq 0 \end{cases} \qquad F_X(t) = \begin{cases} 0, & \text{for } t < 1 \\ 1 - \dfrac{1}{t - 1}, & \text{for } t \geq 1 \end{cases}$$

$$F_X(t) = \begin{cases} 0, & \text{for } t < 0 \\ \frac{1}{5}(1 + t)^2, & \text{for } 0 \leq t < 1 \\ 1 - \dfrac{1}{4t}, & \text{for } t \geq 1 \end{cases}$$

$$F_X(t) = \begin{cases} 0, & \text{for } t < -1 \\ 1 - t^2, & \text{for } -1 \leq t < 0 \\ 1, & \text{for } t \geq 0 \end{cases}$$

17. Let U be a random variable with *pdf*

$$f_U(u) = \begin{cases} \dfrac{1}{\sqrt{u}}, & \text{for } 0 < u < a, \\ 0, & \text{otherwise.} \end{cases}$$

    a. Find the value for a.

    b. What are the values for u such that $f_U(u) > 1$?

2.3 | Summary Measures

Let us briefly summarize the main points discussed so far in this chapter. The probability distribution for a random variable X can be defined by its *cdf*, whether X is discrete or continuous; if X is discrete, its distribution can also be defined by its probability function and if X is continuous, the *pdf* provides a full description of the distribution. This probability distribution can be used to evaluate probability statements about X and provides all the possible information about the random variable.

For many purposes, interest may center only on certain aspects of the distribution:

What is a typical or average value for X?

How much variability does X exhibit?

By how much might an observed value differ from its typical value?

In this section we shall discuss measuring the average value (also called the expected value) for X and computing a number (called the standard deviation) which indicates the variability that occurs over repeated observed values for X. These are both called *summary measures*, descriptors of the distribution for X and of X itself.

Before giving the formal definition of the expected value of a random variable X, let us consider an example. Suppose a computer generates a "random" digit, one of the numbers $0, 1, 2, \ldots, 9$. We can model the digit to occur as a random variable X with the range $R_X = \{0, 1, 2, \ldots, 9\}$; the randomness means that each of these elements could occur as the observed value for X, with equal probability. That is, $X + 1$ is discrete uniform with parameter $n = 10$, and so the probability function for X is

$$p_X(x) = \tfrac{1}{10}, \qquad \text{for } x = 0, 1, 2, \ldots, 9.$$

Recall that probability measures relative frequency; thus, if this computer were to generate a random digit 50 times, the randomness would lead us to expect each digit to occur $50 \times \tfrac{1}{10} = 5$ times; granted this expectation were met, the total of the 50 digits would be $5(1 + 2 + \cdots + 9) = 225$, giving an average of $\mu = \tfrac{225}{50} = 4.5$. This number μ, called the average or expected value for the random variable X, is used as a measure of the "middle" of the probability law for X. This expected value is given by the sum

$$\mu = \sum_{x \in R_X} x\, p_X(x) = \sum_{x=0}^{9} x\, \tfrac{1}{10} = 4.5,$$

weighting the values in the range R_X by their probabilities of occurrence. This average of 4.5 is exactly in the middle of the values in R_X for this case (since the probability function is constant); however, this average value is *not* an element of R_X, since none of the generated digits will be equal to 4.5. The probability function for X, with bars representing the heights $p_X(x)$, is plotted in Fig. 2.10; the balance point μ is also indicated there.

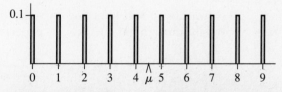

Figure 2.10 Probability function and balance point μ

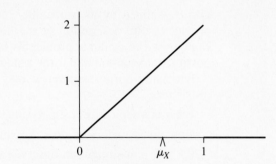

Figure 2.11   Triangular *pdf* and balance point

For a continuous X, the balance point is given by the analogous operation, using the *pdf* instead of the probability function:

$$\mu = \int_{-\infty}^{\infty} x f_X(x)\, dx = \int_{R_X} x f_X(x)\, dx.$$

For example, if a random variable X has the triangular *pdf* $f_X(x) = 2x$ for $0 < x < 1$ (and $f_X(x) = 0$, otherwise), the expected value then is $\mu = \int_0^1 x\, 2x\, dx = \frac{2}{3}$; this *pdf* will balance at the point $\frac{2}{3}$, as illustrated in Fig. 2.11.

The balance point for the probability law (the probability function for discrete X and the *pdf* for continuous X) is called the expected value for the random variable X. (This same number is also frequently called the mean or average value for X.) This is summarized in the following definition.

DEFINITION 2.6

The *expected value* (*mean* or *average value*) *for a random variable X is*

$$E[X] = \mu_X = \begin{cases} \displaystyle\sum_{x \in R_X} x p_X(x), & \text{if } X \text{ is discrete,} \\[2ex] \displaystyle\int_{-\infty}^{\infty} x f_X(x)\, dx, & \text{if } X \text{ is continuous,} \end{cases}$$

as long as the integral or sum is absolutely convergent.

The sum of the first n integers is given by

$$\sum_{x=1}^{n} x = \frac{n(n+1)}{2}$$

(which you will be asked to verify in Exercises 2.3). Thus if X is discrete uniform with parameter n, its expected value is

$$E[X] = \mu_X = \sum_{x=1}^{n} x \frac{1}{n} = \frac{1}{n} \frac{n(n+1)}{2} = \frac{n+1}{2},$$

the midpoint between 1 and n. If X is discrete uniform with parameter $n = 11$, then its expected value is $E[X] = (11 + 1)/2 = 6$, while if $n = 40$ $E[X] = (40 + 1)/2 = 20.5$. The expected value of a discrete uniform random variable increases linearly with the parameter n.

Example 2.15

A small firm has three available positions with equivalent requirements and responsibilities; eight persons, of whom five are female, apply for these positions. If the three persons to get the jobs are chosen at random from the eight and M is the number of males selected, then

$$p_M(m) = \begin{cases} \frac{10}{56}, & \text{for } m = 0, \\ \frac{30}{56}, & \text{for } m = 1, \\ \frac{15}{56}, & \text{for } m = 2, \\ \frac{1}{56}, & \text{for } m = 3, \\ 0, & \text{otherwise.} \end{cases}$$

The expected number of males to be hired is $E[M] = \frac{63}{56} = \frac{9}{8}$.

The expected value of a random variable X is the average of the possible observed values of the variable. To develop a measure of the variability of X it is useful to consider the average of a *function* of the value of the variable. Let X be a random variable, and let $g(X)$ be a function of X; for example $g(X) = X^2$ is one particularly simple function. The average value of $g(X)$ is called its expected value, as given in the following definition.

DEFINITION 2.7

Let $g(\cdot)$ be any real-valued function whose domain includes R_X, the range for a random variable X. Then the *expected value of $g(X)$* is defined to be

$$E[g(X)] = \begin{cases} \displaystyle\sum_{x \in R_X} g(x)p_X(x), & \text{if } X \text{ is discrete,} \\ \displaystyle\int_{-\infty}^{\infty} g(x)f_X(x)\,dx, & \text{if } X \text{ is continuous,} \end{cases}$$

so long as the integral or sum is absolutely convergent.

Example 2.16

The game of roulette (as played in the United States) employs a wheel with 38 spots, of which 18 are red, 18 are black, and the 2 remaining are green. If you bet \$1 on the occurrence of a red number, you will receive \$2 if a red number occurs (the \$1 you bet plus \$1 more you win) and will receive nothing (losing the \$1 you bet) if a red number does not occur. Granted the roulette wheel is fair, the probability of a red number occurring, on any spin, then is $\frac{18}{38}$, and the probability of not getting a red number is $\frac{20}{38}$. Let us define X to equal 1 if a red number occurs and define X to equal 0 if not. Then $R_X = \{0,1\}$, and the probability function for X is simply $p_X(1) = \frac{18}{38}$, $p_X(0) = \frac{20}{38}$, and $p_X(x) = 0$ for any other x. The amount you would win in betting \$1 on the occurrence of a red number then is a random variable W that is a function of X: $W = g(X)$, where $g(1) = 1$ (i.e., you win \$1 if $X = 1$), and $g(0) = -1$ (i.e., you *lose* \$1) if $X = 0$. Our definition of $E[W] = E[g(X)]$ then gives

$$E[g(X)] = g(0)p_X(0) + g(1)p_X(1) = \frac{(-1)20 + (1)18}{38}$$

$$= -\frac{2}{38} = -.0526.$$

You can expect to win \$−.0526 (i.e., lose a little more than a nickel) each time you bet \$1 on the occurrence of a red number.

Example 2.17

A female insect produces a chemical (pheromone) to attract males; at a given moment, in calm windless conditions, assume this chemical will radiate out x feet from the female in all directions. The area of attraction for this female then is $g(x) = \pi x^2$ (the area of a circle of radius x). We assume that the distance this chemical radiates is a random variable X that is continuous uniform with parameters $a = 0$ and $b = 2$. Then the assumed *pdf* for X is $f_X(x) = \frac{1}{2}$, for $0 < x < 2$. The expected area of attraction then is the expected value of $g(X) = \pi X^2$, which gives $E[\pi X^2] = \int_0^2 \pi x^2 \frac{1}{2} \, dx = \frac{4}{3}\pi$ square feet.

Expected values have several simple properties that are easy to establish; these are stated in the following theorem.

THEOREM 2.1 If X is any random variable, then

1. $E[c] = c$, where c is any constant.
2. $E[bg(X)] = b\,E[g(X)]$, where b is any constant.
3. $E\left[\sum_{i=1}^{n} g_i(X)\right] = \sum_{i=1}^{n} E[g_i(X)].$

We shall establish these results for a continuous X. (The proofs for a discrete X are the same, except that summation replaces integration.) Looking at the first result, we see that if $g(X) = c$ (a constant), then $E[g(X)] = E[c] = \int_{-\infty}^{\infty} cf_X(x) \, dx = c \int_{-\infty}^{\infty} f_X(x) \, dx = c$, since the integral of the *pdf* over the full line gives 1. Thus the expected value for any constant is itself. The second result of Theorem 2.1 says that constants factor through the expectation; to see this, let b be a constant; and then

also pictured in Fig. 2.12, $\mu_Y = 0$ and $\sigma_Y = \sqrt{\frac{1}{2}}$, so $2\sigma_Y = \sqrt{2}$ and $P(-\sqrt{2} < Y < \sqrt{2}) = 1$. Y has a larger standard deviation than X, and the interval $(\mu_Y - 2\sigma_Y, \mu_Y + 2\sigma_Y)$ includes R_Y.

Expected values may not exist. The nonexistence of such expectations is not a major topic for our study, simply a technical point of which you should be aware. The following example discusses a probability law whose mean does not exist (the sum defining the mean is not absolutely convergent).

Example 2.21

An urn contains two balls, of which one is red and the other is blue. We remove one ball at random from the urn, observe its color, and then put two balls of that color back into the urn. If the ball we remove is blue, for example, we replace it and add one more blue ball; now there are three balls in the urn, of which two are blue. We continue this process of drawing a ball at random and then, at each turn, we replace it together with one more ball of the same color. The number of balls in the urn increases by 1 with each draw.

Let X be the draw number on which we first get a red ball; the range for X is clearly $R_X = \{1, 2, 3, \dots\}$. Granted the ball selected is chosen at random each time, then $P(X = 1) = \frac{1}{2}$ and $P(X = 2) = \left(\frac{1}{2}\right)\left(\frac{1}{3}\right) = \frac{1}{6}$, since we get the first red ball on the second draw if and only if the first ball drawn is blue and the second ball drawn is red. This scheme generalizes very easily.

The probability of drawing $x - 1$ blue balls in a row is

$$\left(\frac{1}{2}\right)\left(\frac{2}{3}\right) \cdots \left(\frac{x-1}{x}\right) = \frac{1}{x}$$

and the conditional probability that the xth ball is red, given the first $x - 1$ balls were blue, is simply $1/(x + 1)$; thus the probability that we will find $X = x$ is the product of these two terms:

$$p_X(x) = \frac{1}{x(x+1)} = \frac{1}{x} - \frac{1}{x+1}, \quad \text{for } x \in R_X.$$

Clearly $p_X(x) > 0$ for $x \in R_X$, and for any integer $n \geq 1$,

$$\sum_{x=1}^{n} p_X(x) = \left(1 - \frac{1}{2}\right) + \left(\frac{1}{2} - \frac{1}{3}\right) + \left(\frac{1}{3} - \frac{1}{4}\right) + \cdots + \left(\frac{1}{n} - \frac{1}{n+1}\right)$$

$$= 1 - \frac{1}{n+1};$$

as $n \to \infty$ the limit of this sum is 1, so $p_X(x)$ is in fact a legitimate probability function. The infinite series defining the expected value for X, though, is

$$E[X] = \sum_{x=1}^{\infty} x \frac{1}{x(x+1)} = \sum_{x=1}^{\infty} \frac{1}{x+1} = \frac{1}{2} + \frac{1}{3} + \frac{1}{4} + \cdots,$$

which diverges, as is shown in most calculus texts. Thus $E[X]$ does not exist; this probability function has too much probability too far out on the integers for the balance point to exist.

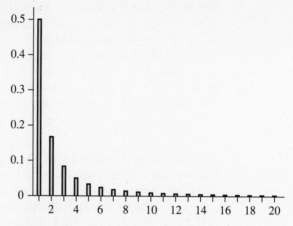

Figure 2.14 $p_X(x) = 1/x(x+1)$ for $x = 1, 2, \ldots, 20$

Figure 2.14 graphs this probability function for $x = 1, 2, \ldots, 20$; while the values for the probability function are decreasing with x, they decrease so slowly that there is no finite location for a balance point. The sum of the 20 probabilities graphed in Fig. 2.14 is $F_X(20) = P(X \leq 20) = \frac{20}{21} = .9524$; if we were to extend this picture out to $x = 200$ and sum the values we would find that $P(X \leq 200) = \frac{200}{201} = .9950$, still not terribly close to 1. Indeed, $P(X \leq 1000) = .9990$ and $P(X \leq 10,000) = .99990$. This "long tail" causes the sum defining the mean to diverge.

 **EXERCISES 2.3**

1. The random variable Z has the probability function given by the following chart:

z	2	3	6
$p_Z(z)$	$\frac{1}{2}$	$\frac{1}{3}$	$\frac{1}{6}$

Evaluate the mean and variance for Z, and graph $p_Z(z)$.

2. A random variable X is equally likely to take on any of the values in $R_X = \{-2, -1, 1, 2\}$.
   a. What is the probability function for X?
   b. Evaluate the mean and variance of X.

3. a. Show that the sum of the first n integers is $n(n+1)/2$.
   b. Show that the sum of the squares of the first n integers is

$$\frac{n(n+1)(2n+1)}{6}.$$

4. Imagine X is discrete uniform with parameter 6 (e.g., X is the number of spots showing on one roll of a fair die), and define $Y = (X - 2)/2$. Find the mean and variance of Y.

5. Show that for any random variable for which $E[X^2]$ exists, the minimum value of $E[(X-a)^2]$ occurs with $a = \mu = E[X]$.

6. a. Show that, for any random variable X, $E[X^2] \geq \mu_X^2$. (*Hint:* Consider the computational formula for σ_X^2.)
   b. Show that the only distribution for which $E[X^2] = \mu_X^2$ is one for which $P(X = \mu_X) = 1$. This is called a *degenerate distribution* (and X is a *degenerate random variable*) since it does not vary.

7. Chuck-a-luck is one of the games offered in Nevada casinos. To play this game you, as the bettor, select one of the integers 1 through 6; suppose you choose to bet $1 on the number 2. The house then rolls 3 fair dice, and the amount you will win depends on the number of 2's to occur (or the number of x's to occur, where you bet on x); if one 2 occurs, you win $1, if two 2's occur you win $2, and if three 2's occur you win $3; if no 2's occur, you lose your $1 bet. Evaluate the expected value of the amount you win on one play of this game.

8. Two fair dice, one red and one green, are rolled one time; let X be the ratio r/g of the red number divided by the green number.
   a. What is the range for X?   b. Find the probability function for X.
   c. Evaluate $E[X]$, the mean of X.

9. Three men and two women apply for the same job; the interviewer will rank these five applicants (assigning each a unique number from 1 for best to 5 for worst) and then choose the one ranked best. Assume that all possible rankings are equally likely, and let M be the best rank attained by a man. Evaluate μ_M and σ_M.

10. Assume the same situation given in Exercise 9, except that now 5 indicates the best and 1 the worst. Again let M be the best rank attained by a man. What are the values for μ_M and σ_M?

11. Suppose the length of time you must wait for a traffic light to turn from red to green, given it was red on your arrival, is a random variable Y whose *pdf* is

$$f_Y(y) = \begin{cases} \frac{1}{44}, & \text{for } 1 < y < 45, \\ 0, & \text{otherwise,} \end{cases}$$

where time is measured in seconds. What is the expected length of time you must wait for the light to turn green? What is the standard deviation of this time?

12. Let U be a continuous uniform random variable with parameters a, b, and evaluate the mean and standard deviation of U.

13. For various reasons the length of time that an aircraft remains at its loading point, after all passengers are on board, will vary. Suppose you are one of those on board an aircraft and that X, the length of time between closing the aircraft door and the instant at which the plane begins to depart, is uniform with $a = 0$, and $b = 10$ (in minutes). What is the expected length of time between the door closing and departure? What is the standard deviation of this time?

14. The continuous uniform random variable with $a = 1$, and $b = n$ should

somewhat mimic the discrete uniform random variable with parameter n. Compare the means and variances of these two probability laws.

15. The length of time (in minutes) you must wait at the checkout counter of a supermarket, before the clerk processes your items, is a random variable W with the following triangular *pdf*:

$$f_W(w) = \begin{cases} w - 1, & \text{for } 1 < w < 2, \\ 3 - w, & \text{for } 2 \le w < 3, \\ 0, & \text{otherwise.} \end{cases}$$

Evaluate $E[W]$ and the variance of W.

16. Suppose your waiting time at the supermarket counter, instead of being described by the *pdf* $f_W(w)$ given in Exercise 15, has the following triangular density with parameter b:

$$f_X(x) = \begin{cases} \dfrac{2x}{b^2}, & \text{for } 0 < x < b, \\ 0, & \text{otherwise.} \end{cases}$$

Evaluate the mean and standard deviation of X. If W and X describe the time you would wait for service at two different markets, for what values of b would you prefer one or the other of the markets?

17. An example of a continuous random variable whose mean does not exist is given by the *pdf*

$$f_X(x) = \begin{cases} \dfrac{1}{x^2}, & \text{for } x \ge 1, \\ 0, & \text{otherwise.} \end{cases}$$

Show that the integral defining $E[X]$ diverges.

18. U-shaped densities do not occur frequently; one observed situation where this type of density does seem to apply is given by the intensities of cloudiness in Greenwich, England, in the month of July. Data observed there during this month, over many years, give high relative frequencies for both 0 (clear) and 2 (quite obscured) days and low frequencies for intermediate values. The observed data have been scaled (and distorted somewhat) to give the following density:

$$f_U(u) = \begin{cases} \frac{1}{6} + (u - 1)^2, & \text{for } 0 < u < 2, \\ 0, & \text{otherwise.} \end{cases}$$

Find the mean and variance of U.

19. The charge on a capacitor does not build up until a critical voltage is reached; then the charge builds linearly to a certain level, at which point it cannot increase any more. Let W represent the charge on such a capacitor, with *pdf*

$$f_W(w) = \begin{cases} \frac{32}{3}(w - .75), & \text{for } .75 < w < 1, \\ \frac{8}{3}, & \text{for } 1 < w < 1.25, \\ 0, & \text{otherwise.} \end{cases}$$

Find the mean and variance of W.

20. A meteorological model includes the maximum daily temperature, at a given location, as a random variable X, where the units for X are degrees Fahrenheit. If this model were changed to employ this temperature in degrees Celsius (call this a new random variable Y), what is the relationship between the mean and standard deviations of X and Y?

21. If X is a discrete random variable with range $R_X = \{0, 1, 2, \ldots\}$, show that its mean can be written

$$E[X] = \sum_{x \in R_X} P(X > x) = \sum_{x \in R_X} [1 - P(X \le x)];$$

this equation gives a second way of visualizing the fact that $E[X]$ moves to the right on the real line as the probability law for X has more weight on larger x.

22. The mean of a continuous random variable locates the balance point for its *pdf*; a second geometric interpretation is given by the fact that $\mu_X = \int_0^\infty (1 - F_X(t))\, dt$, for any positive continuous random variable (meaning $F_X(0) = 0$). Show this result to be true.

23. Let X be the number to occur on one roll of a fair die. Within what interval does the Chebyshev inequality say that X must lie within a probability at least $\frac{3}{4}$? What is the exact probability of finding X in this interval?

24. The Chebyshev inequality must hold for all probability distributions. It is possible to exhibit a number of distributions where the bound is in fact attained, illustrating that the bound could not be made any tighter, if it is to hold universally. Let $k > 1$ be a constant, and take the probability function for X to be

$$p_X(x) = \begin{cases} 1 - 1/k^2, & \text{for } x = 0, \\ 1/2k^2, & \text{for } x = -k, k, \\ 0, & \text{otherwise.} \end{cases}$$

Show that $\mu_X = 0$, and $\sigma_X = 1$ for any value of k; thus we can always find a probability distribution such that the Chebyshev bound is actually attained (gives the actual value for the probability), for any specified value of k.

2.4 | Quantiles for Continuous Random Variables

To summarize what we learned in the previous section, expected values of functions of random variables can be used to summarize various aspects of a probability distribution. The mean (balance point) $\mu_X = E[X]$ is a commonly used measure of the middle of the probability law, for both discrete and continuous random variables. The commonly used measure of spread or variability of a probability law (or random variable) is the variance $\sigma_X^2 = E[X^2] - \mu_X^2$ or its positive square root, the standard deviation.

For continuous random variables, *quantiles* are also useful for measuring the middle and the spread of the probability law. These quantiles (sometimes called *percentiles*) are frequently used in reporting scores achieved on standardized tests. If a person's score is reported to be at the 85th quantile (or percentile), for example, this means that (roughly) 85% of the scores made on the exam are below this person's score; in like manner, 50% of the scores fall below the 50th quantile.

Any continuous *cdf* varies continuously from 0 to 1 (at least in the limit); thus there will always exist some number for which the *cdf* equals k, where $0 < k < 1$. This number is denoted t_k and is called the 100kth quantile for the random variable (or for the distribution). The formal definition follows.

DEFINITION 2.9

If X is a continuous random variable whose *cdf* is $F_X(t)$, the 100kth *quantile* for X is the smallest number t_k such that $F_X(t_k) = k$, where $0 < k < 1$.

Thus the 50th quantile will be denoted by t_5 and the 85th quantile by $t_{.85}$, and so on. The 100kth quantile t_k is the value that makes the *cdf* equal to k; since the value of the *cdf* at t_k is also the area under the *pdf* to the left of t_k, the quantile can be located from either the *cdf* or the *pdf*, as illustrated in the following example.

Example 2.22

The *cdf* pictured in Fig. 2.15 is $F_X(t) = 1 - 1/t$, for $t \geq 1$. Thus the 100kth quantile for this probability law is defined by

$$F_X(t_k) = 1 - \frac{1}{t_k} = k;$$

solving this equation for t_k gives $t_k = 1/(1 - k)$, so the 30th quantile for this probability law is $t_{.3} = 1/.7 = 1.428$, while the 60th quantile is $t_{.6} = 1/.4 = 2.5$.

Assume Y has the *cdf* $F_Y(t) = 1 - e^{-t}$ for $t \geq 0$; then the 100kth quantile for Y is defined by

$$F_Y(t_k) = 1 - e^{-t_k} = k;$$

this gives

$$t_k = -\ln(1 - k) = \ln \frac{1}{1 - k}.$$

Thus the 30th quantile for Y is $t_{.3} = -\ln.7 = .357$, while the 60th quantile is $t_{.6} = -\ln.4 = .916$. Since X and Y have different probability laws (*cdf*s), their quantiles are also different.

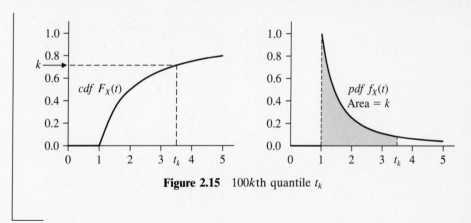

Figure 2.15   100kth quantile t_k

The quantiles of a distribution can be used in quite transparent ways to summarize the probability law. The 50th quantile t_5 locates the value that makes the *cdf* equal to .5; that is, this is the value on the real line that splits the area under the *pdf* into two equal parts. It is frequently used as an alternative measure of the middle of the probability law, because of this property, and is called the median of the distribution (or of the random variable); it may be equal to, less than, or greater than the mean μ_X of the random variable. The mean locates the balance point and measures the middle in a center-of-gravity sense; the median measures the middle in an area sense, giving the number that has equal areas under the *pdf* on its two sides. If the *pdf* is symmetric about some point a, so that $f_X(a - x) = f_X(a + x)$ for all x, then the median and the mean will generally coincide. If the shape of the *pdf* to the left of the median is held constant, while the right-hand tail is stretched out, the median remains constant but the balance point moves to the right.

The quantiles can also be used to measure the variability or spread of a continuous probability law. The most commonly employed measure of variability (using quantiles) is the interquartile range of the distribution. $t_{.25}, t_{.5}$, and $t_{.75}$ are called the *quartiles* of the distribution since they cut the area under the *pdf* into four equal parts. The difference, $t_{.75} - t_{.25}$ is the *interquartile range* r_{iq} of the distribution. Since the area under the *pdf* to the left of $t_{.75}$ is .75, and the area under the *pdf* to the left of $t_{.25}$ is .25, it follows that the area between the two is .50. It is easy to picture the interquartile range, given a graph of the *pdf*, since it is the length of the centrally located interval that includes 50% of the probability (see Fig. 2.16). The longer this interval (the larger the value for r_{iq}), the more dispersed or spread out the probability law is in its "middle," while a shorter interval indicates less dispersion.

Just as the median may differ from the mean, r_{iq} bears no special numerical value relative to the standard deviation σ_X for the same probability law; however, as the probability distribution becomes more dispersed or spread out, both r_{iq} and σ_X must increase.

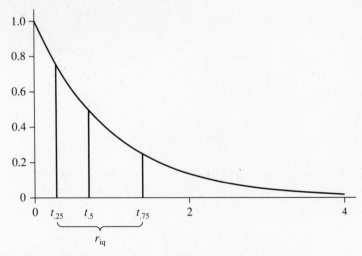

Figure 2.16   Quartiles and r_{iq}. Each section has area $1/4$.

Example 2.23

We have seen that the quantiles of $F_X(t) = 1 - 1/t$ are given by $t_k = 1/(1 - k)$. Thus the median for this probability law is $t_5 = 1/(1 - .5) = 2$; the probability an observed value for this random variable will be less than 2 is $\frac{1}{2}$ (as is the probability that it would exceed 2). The 75th quantile is 4 and the 25th is $\frac{4}{3}$, so the interquartile range is $r_{iq} = \frac{8}{3}$.

The random variable Y whose *cdf* is $F_Y(t) = 1 - e^{-t}$ for $t \geq 0$ has 100kth quantile $t_k = -\ln(1 - k)$; the median for this probability law then is $t_5 = -\ln .5 = .693$, the 75th quantile is $t_{75} = -\ln .25 = 1.386$, and the 25th quantile is $t_{25} = -\ln \frac{3}{4} = .288$. Figure 2.16 pictures these three quantiles. The interquartile range for this random variable Y thus is $r_{iq} = 1.098$ (also pictured in the figure), less than half the interquartile range for X in the previous paragraph.

We saw in Section 2.3 that if Y is a linear function of X, then both the mean and the standard deviation of Y are simply related to the mean and standard deviation of X. This is also true of the quantiles. Suppose X is a continuous random variable with 100kth quantile x_k, and let $Y = a + bX$ where $b > 0$; y_k will be used to denote the 100kth quantile of Y. Then from the definition of the 100kth quantile, we get

$$k = P(Y \leq y_k) = P(a + bX \leq y_k)$$

$$= P\left(X \leq \frac{y_k - a}{b}\right) = F_X\left(\frac{y_k - a}{b}\right),$$

since the events $A = \{Y \leq y_k\}$ and $B = \{a + bX \leq y_k\}$ are equivalent (and so their probabilities must be equal). That is, we have found that $F_X((y_k - a)/b) = k$, but x_k is the value for which $F_X(x_k) = k$. That is,

$$\frac{y_k - a}{b} = x_k.$$

Thus, solving for y_k, we get $y_k = a + bx_k$; the $100k$th quantile of $Y = a + bX$ is the same linear function of the $100k$th quantile for X as long as $b > 0$. This relation becomes $y_k = a + bx_{1-k}$ if $b < 0$ (which you will be asked to show in Exercises 2.4). This establishes the following theorem.

> **THEOREM 2.4**   If X is a continuous random variable, with $100k$th quantile x_k, and $Y = a + bX$ where a and $b \neq 0$ are constants, then the $100k$th quantile y_k for Y is
>
> $$y_k = \begin{cases} a + bx_k, & \text{for } b > 0, \\ a + bx_{1-k}, & \text{for } b < 0. \end{cases}$$

Since $1 - k = k$ for $k = .5$, $y_{.5} = a + bx_{.5}$, regardless of whether b is positive or negative; that is, the median for Y is the same linear function of $x_{.5}$, the median for X. The interquartile range for Y is $y_{.75} - y_{.25} = (a + bx_{.75}) - (a + bx_{.25}) = b(x_{.75} - x_{.25})$ if $b > 0$, and is $y_{.75} - y_{.25} = (a + bx_{.25}) - (a + bx_{.75}) = b(x_{.25} - x_{.75}) = -b(x_{.75} - x_{.25})$ if $b < 0$; thus the interquartile range for Y is $|b|$ times the interquartile range for X, the same relationship that exists for the standard deviations. Suppose X has median $x_{.5} = 3$ and interquartile range $r_{iq} = 2$; then $Y = 100 + 10X$ has median $100 + 10(3) = 130$ and interquartile range $10(2) = 20$, while $Z = 35 - 40X$ has median $35 - 40(3) = -85$ and interquartile range $|-40|(3) = 120$.

Granted x_k is the $100k$th quantile for X, as k varies from 0 to 1, x_k varies over R_X, the range for X. Since $F_X(x_k) = k$, we can use the inverse function to write $x_k = F_X^{-1}(k)$; if we are able to express x_k as a simple function of k, this function must in fact be the inverse of the *cdf* $F_X(\cdot)$. For example, if $x_k = 8 + \sqrt{k}$ for $0 < k < 1$, the range for X is necessarily $R_X = \{x : 8 < x < 9\}$, since these are the values for $8 + \sqrt{k}$ with $0 < k < 1$. The *cdf* for X is given by solving $x = 8 + \sqrt{k}$ for k; that is, since this gives $k = (x - 8)^2$, the *cdf* is $F_X(x) = (x - 8)^2$ for $8 < x < 9$. The *pdf* for X is $f_X(x) = 2(x - 8)$ for $8 < x < 9$, a triangular function.

The *cdf* for a random variable may shift, in functional form, over the range of the random variable. If we want t_k for such a case, it is then necessary to isolate the form of the *cdf* for the desired k. This is illustrated in the following example.

Example 2.24

Suppose a computer is programmed to play a simple game. You must guess a number between 0 and 1 (but not .5); the computer then generates a number between 0 and 1. If both your number and the computer's number are on the same side of .5, then you win; otherwise you lose. The computer actually chooses a value of X where the *pdf* for X is

$$f_X(x) = \begin{cases} 1.2, & \text{for } 0 < x < .5, \\ .8, & \text{for } .5 < x < 1, \\ 0 & \text{otherwise.} \end{cases}$$

If you had this knowledge of how the computer generates its number, you should certainly always guess a number below .5 to maximize your chance of winning.

The *cdf* for X is then

$$F_X(t) = \begin{cases} 1.2t, & \text{for } 0 \le t \le .5, \\ .6 + .8(t - .5), & \text{for } .5 < t \le 1. \end{cases}$$

Note that $0 \le F_X(t) \le .6$ for $0 \le t \le .5$ and $.6 < F_X(t) \le 1$ for $.5 < t \le 1$. Thus the 25th quantile is given by solving $1.2t_{.25} = .25$, giving $t_{.25} = .25/1.2 = .2083$; the 75th quantile is the solution to $.6 + .8(t_{.75} - .5) = .75$, giving $t_{.75} = .5 + .15/.8 = .6875$. The interquartile range for X is $r_{iq} = .6875 - .2083 = .4792$.

Since the 100kth quantile for a continuous random variable simply identifies $t_k \in R_X$, the element in the range that makes $F_X(t_k) = k$, there is no question of its existence, as there is for expected values; that is, for any $0 < k < 1$, granted $F_X(t)$ is continuous, there will be some value in R_X for which the *cdf* is equal to k. In fact, there may be more than one such value, if it happens that the *cdf* remains constant at k for all the values in some interval. To eliminate this possible ambiguity, Definition 2.9 specifies that the 100kth quantile is the smallest value $t_k \in R_X$ that makes $F_X(t) = k$. The following example illustrates a case in which the *cdf* is constant at .5 for an interval of values.

Example 2.25

Let Z be a continuous random variable with *pdf*

$$f_Z(z) = \begin{cases} z, & \text{for } 0 \le z \le 1, \\ 4 - z, & \text{for } 3 \le z \le 4, \\ 0, & \text{otherwise} \end{cases}$$

as pictured in Fig. 2.17. Also pictured in Fig. 2.17 is the *cdf* for Z:

$$F_Z(t) = \begin{cases} 0, & \text{for } t < 0, \\ \int_0^t z \, dz = \dfrac{t^2}{2}, & \text{for } 0 \le t \le 1, \\ \dfrac{1}{2}, & \text{for } 1 < t < 3, \\ \dfrac{1}{2} + \int_3^t (4 - z) \, dz = 1 - \dfrac{(4 - t)^2}{2}, & \text{for } 3 \le t \le 4, \\ 1, & \text{for } t > 4. \end{cases}$$

Since $F_Z(t) = \frac{1}{2}$, for $1 \le t \le 3$, *any* number between 1 and 3 could serve as the median $t_{.5}$ for Z; with our definition the median is the smallest value in this interval; that is, $t_{.5} = 1$. Note that with this definition the median is not equal to μ_Z for this probability law, although $f_Z(z)$ is symmetric about $a = 2$. This type of behavior could occur for any value of $F_Z(t)$; that is, any of the quantiles might occur for all values in an interval.

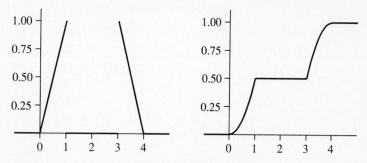

Figure 2.17 Probability density function, cumulative *pdf* (left) and *cdf* (right) distribution function

Quantiles can equally well be defined for discrete random variables, although they are not as frequently used for such random variables. Recall that the *cdf* for a discrete random variable is a step function; it jumps in value at each element of the range of the random variable. This in turn implies that there may be no value t_k such that $F_X(t_k) = k$, since the accumulation may have jumped from a value smaller than k to a value greater than k. (Indeed, this must in fact happen for almost all values of k where $0 < k < 1$.) The usual modification to extend the definition of quantiles to cover the discrete case is to let t_k be the *smallest* value such that $F_X(t_k) \geq k$ for $0 < k < 1$. For example, assume that X is discrete with *cdf*

$$F_X(t) = \begin{cases} 0, & \text{for } t < 1, \\ .2, & \text{for } 1 \leq t < 2, \\ .9, & \text{for } 2 \leq t < 3, \\ 1, & \text{for } t \geq 3, \end{cases}$$

Then $R_X = \{1, 2, 3\}$, and the probability function for X has values .2, .7, .1, respectively, at these points. Our extended discrete quantile definition then gives $t_{.5} = 2$, since the smallest observed value such that $F_X(t) \geq .5$ is $t = 2$. Thus the median for X would be 2. Note that we then also have $t_{.75} = 2 = t_{.25}$, so $r_{iq} = 0$ for this random variable. This is caused by the fact that the *cdf* jumps at one point ($t = 2$) from having value .2 to having value .9 (since $P(X = 2) = .7$, as noted).

With this definition for the discrete case, the same number t_k occurs as the 100kth quantile for a range of k values; the length of this range is in fact given by $p_X(t_k)$, the value of the probability function at that point. That is, for the random variable X whose *cdf* was given in the previous paragraph, $t_k = 1$ for $0 < k \leq .2$, $t_k = 2$ for $.2 < k \leq .9$, and $t_k = 3$ for $.9 < k \leq 1$, so the quantiles vary discretely (rather than continuously) for a discrete random variable.

This behavior in turn allows us to easily recover the *cdf* (or probability function) for a discrete X, given the values for the quantiles. That is, if we are given a random variable X whose quantiles are $t_k = 3$ for $0 < k \leq .4$, $t_k = 5$ for $.4 < k \leq .7$, $t_k = 7$ for $.7 < k \leq .95$, and $t_k = 9$ for $.95 < k \leq 1$, then the *cdf* for X is

$$F_X(t) = \begin{cases} 0, & \text{for } t < 3, \\ .4, & \text{for } 3 \le t < 5, \\ .7, & \text{for } 5 \le t < 7, \\ .95, & \text{for } 7 \le t < 9, \\ 1, & \text{for } 9 \le t. \end{cases}$$

The probability function for X can then be gotten in the usual way from $F_X(t)$.

EXERCISES 2.4

1. The length of time you must wait at a supermarket checkout point is a random variable W with a triangular *pdf*:

$$f_W(w) = \begin{cases} w - 1, & \text{for } 1 < w < 2, \\ 3 - w, & \text{for } 2 \le w < 3, \\ 0, & \text{otherwise.} \end{cases}$$

Find $t_{.5}$ and r_{iq} for W.

2. Instead of your waiting time at the supermarket counter being described by the *pdf* $f_W(w)$ given in Exercise 1, suppose it has the triangular density with parameter b given by

$$f_X(x) = \begin{cases} \dfrac{2x}{b^2}, & \text{for } 0 < x < b, \\ 0, & \text{otherwise.} \end{cases}$$

Find the median and interquartile range of X. If W and X describe the time you would wait for service at two different markets, for what values of b would you prefer one or the other of the two?

3. The mean of X with *pdf*

$$f_X(x) = \begin{cases} \dfrac{1}{x^2}, & \text{for } x \ge 1, \\ 0, & \text{otherwise,} \end{cases}$$

does not exist. Evaluate the median and interquartile range of X.

4. Evaluate the $100k$th quantile for the uniform probability law with parameters a, b.

5. If X is a continuous random variable, the *deciles* of the distribution for X are $t_{.1}, t_{.2}, \ldots, t_{.9}$, the values that cut the total probability mass of 1 into 10 equal parts. The difference, $r_{id} = t_{.9} - t_{.1}$, called the *interdecile range*, is sometimes used as a measure of variability for a probability law. Evaluate the deciles and r_{id} for the triangular density

$$f_X(x) = \begin{cases} \dfrac{2}{b^2}(b - x), & \text{for } 0 < x < b, \\ 0, & \text{otherwise.} \end{cases}$$

6. Let W be a random variable whose *pdf* is

$$f_W(w) = \begin{cases} \frac{32}{3}(w - .75), & \text{for } .75 < w < 1, \\ \frac{8}{3}, & \text{for } 1 < w < 1.25, \\ 0, & \text{otherwise.} \end{cases}$$

Find the 100kth quantile for W for $0 < k < 1$.

7. Let $Y = a + bX$, where X is a continuous random variable, and $b < 0$ and a are constants. Show that $y_k = a + bx_{1-k}$, where y_k, x_k are the 100kth quantiles of Y, and X, respectively.

8. A meteorological model includes the maximum daily temperature, at a given location, as a random variable X, where the units for X are degrees Fahrenheit. Assume this model is changed to employ this temperature in degrees Celsius, and call this a new random variable Y. What is the relationship between the median and interquartile range for X and Y?

9. The 100kth quantile for a continuous random variable T is k^3 for $0 \leq k \leq 1$; find the *pdf* for T.

10. If the *pdf* for a continuous random variable Y is symmetric about a point a, (so that $f_Y(a - y) = f_Y(a + y)$ for all possible values of y), then it is easy to see that $\mu_Y = t_{.5}$ (i.e., the mean for Y and the median for Y are equal) as long as $f_Y(a) > 0$ (the *cdf* is increasing over any neighborhood including a). Their common value, of course, is the point of symmetry a. It is possible for μ_Y and $t_{.5}$ to be equal, even if $f_Y(y)$ has no point of symmetry. Show that the mean and median are equal if

$$f_Y(y) = \begin{cases} y, & \text{for } 0 < y < 1, \\ \frac{3}{4}, & \text{for } 1 < y < \frac{5}{3}, \\ 0, & \text{otherwise.} \end{cases}$$

2.5 | Summary

Random variables are numeric quantities whose values depend on "chance." A discrete random variable X has a discrete range R_X, while a continuous random variable has a continuous range R_X. The probability law for a discrete random variable may be specified by its probability function $p_X(x)$, while the probability law for a continuous random variable is specified by its probability density function (*pdf*). The requirements for these functions are as follows:

PROBABILITY FUNCTION REQUIREMENTS

1. $0 \leq p_X(x) \leq 1$ for $x \in R_X$.

2. $\displaystyle\sum_{x \in R_X} p_X(x) = 1$.

***pdf* REQUIREMENTS**

1. $f_X(x) \geq 0$ for all real x.

2. $\displaystyle\int_{-\infty}^{\infty} f_X(x)\, dx = 1$.

The discrete uniform and continuous uniform probability laws are described by the following characteristics:

DISCRETE UNIFORM PROBABILITY LAW

Probability function: $p_X(x) = \dfrac{1}{n}$

Range: $R_X = \{1, 2, \ldots n\}$

Parameter: $n = 1, 2, 3, \ldots$

Mean: $\dfrac{n+1}{2}$

Variance: $\dfrac{n^2 - 1}{12}$

CONTINUOUS UNIFORM PROBABILITY LAW

pdf: $f_X(x) = \dfrac{1}{b-a}$

Range: $R_X = \{a < x < b\}$

Parameters: $-\infty < a < b < \infty$

Mean: $\dfrac{a+b}{2}$

Variance: $\dfrac{(b-a)^2}{12}$

The cumulative distribution function (cdf) for a random variable X is defined for any real t by the following:

CUMULATIVE DISTRIBUTION FUNCTION

$$F_X(t) = P(X \leq t)$$

$$= \sum_{x \leq t} p_X(x), \quad \text{for } x \in R_X \text{ for discrete } X$$

$$= \int_{-\infty}^{t} f_X(x)\, dx, \quad \text{for continuous } X.$$

If X is a discrete random variable, then $F_X(t)$ is a step function, jumping in value at each $t \in R_X$; the size of the jump is given by $p_X(t)$, the value for the probability function at that number. If X is a continuous random variable, then $F_X(t)$ is a continuous function of t.

REQUIREMENTS FOR $F_X(t)$

1. $0 \le F_X(t) \le 1$ for all real t.
2. $\lim_{t \to -\infty} F_X(t) = 0, \quad \lim_{t \to \infty} F_X(t) = 1$.
3. If $c < d$, then $F_X(c) \le F_X(d)$.
4. $F_X(t)$ must be continuous from the right.

The expected value of a function $g(X)$ of a random variable X is given by the following:

EXPECTED VALUE OF $g(X)$

$$E[g(X)] = \begin{cases} \sum_{x \in R_X} g(x)p_X(x), & \text{if } X \text{ is discrete,} \\ \int_{-\infty}^{\infty} g(x)f_X(x)\,dx, & \text{if } X \text{ is continuous.} \end{cases}$$

Some important consequences of this definition are

CONSEQUENCES

$E[a + bX] = a + b\,E[X]$, for any constants a, b.
$E[g_1(X) + g_2(X)] = E[g_1(X)] + E[g_2(X)]$, for any functions $g_1(\cdot)$ and $g_2(\cdot)$.

The mean μ_X for a random variable is $E[X]$, its expected value; this measures the middle of the probability law in a center-of-gravity sense. The variance of a random variable is $\sigma_X^2 = E[X^2] - \mu_X^2$; the positive square root of the variance is called the standard deviation for the random variable. It provides a measure of the variability of the probability law (or of the random variable). The standard deviation provides a natural scale factor for measuring the amount of probability contained in intervals centered at the mean, as shown by the Chebyshev inequality:

CHEBYSHEV INEQUALITY

Let X be a random variable with expected value μ_X and standard deviation σ_X. Then

$$P(|X - \mu_X| < k\sigma_X) \ge 1 - \frac{1}{k^2},$$

no matter what the distribution for X.

Quantiles can also be used to easily summarize continuous probability laws; the 100*k*th quantile t_k is defined by $F_X(t_k) = k$, for $0 < k < 1$. The 50th quantile $t_{.5}$, called the median for the random variable, cuts the *pdf* into two equal areas and provides an alternative measure of the middle of the probability law. The difference $t_{.75} - t_{.25}$, called the interquartile range, provides an alternative measure of the variability of a continuous probability law.

Linear functions of random variables play an important role in many applications; if *a* and *b* are any constants, the mean and variance of $Y = a + bX$ are both simply related to μ_X and σ_X^2. If *X* is continuous, the quantiles are also simply related.

LINEAR FUNCTION RELATIONSHIPS

Let *X* be a random variable with mean μ_X, variance σ_X^2, and (if *X* is continuous) 100*k*th quantile x_k. Let $Y = a + bX$, where *a* and *b* are constants. Then

$$\mu_Y = a + b\mu_X, \qquad \sigma_Y^2 = b^2\sigma_X^2$$

$$y_k = \begin{cases} a + bx_k, & \text{for } b > 0, \\ a + bx_{1-k}, & \text{for } b < 0. \end{cases}$$

 **EXERCISES 2.5**

1. The profit *P* to be made on an investment, as a percentage of the amount invested, is modeled with *pdf*

$$f_P(p) = 12(p + \tfrac{1}{4})^2(\tfrac{3}{4} - p), \quad \text{for } -\tfrac{1}{4} < p < \tfrac{3}{4}.$$

   a. What is the probability of suffering a loss with this investment (i.e., what is the probability that $P < 0$)?

   b. Evaluate E[*P*], the expected profit for this investment.

2. A fair die is rolled until the face with six spots first occurs. Let *X* be the number of rolls required, and evaluate the probability that
   a. exactly six rolls are required.   b. fewer than six rolls are required.
   c. an odd number of rolls are required.

3. Person $\mathcal{A}$ selects a number at random from $\{1, 2\}$ and person $\mathcal{B}$ independently selects a number at random from $\{1, 2, 3\}$. What is the probability that $\mathcal{A}$'s number is (strictly) larger than $\mathcal{B}$'s? That $\mathcal{B}$'s is (strictly) larger? (*Hint*: Consider a sample space of 2-tuples, with the first entry being $\mathcal{A}$'s number.)

4. A computer algorithm will generate one of the digits in $R_X = \{1, 2, 3, 4\}$; let *X* represent the digit generated, and assume the probability function for *X* is $p_X(x) = x/10$ for $x \in R_X$. You win \$1 if the digit generated is odd and lose \$1 if the digit generated is even.
   a. What is the probability that you win \$1?
   b. What are your expected winnings?

5. Assume that a stoplight you pass by each day has a 4-minute cycle: red for

1.5 minutes, green for 2.5 minutes, then red again for the same period, and so on. Let X represent the time at which you arrive at the light, relative to this cycle, where 0 indicates the start of the red cycle and 1.5 indicates the start of the green cycle. Assume X is continuous uniform with $a = 0$ and $b = 4$. What is the probability that you have to stop for the red light?

6. Define the function

$$g(t) = \begin{cases} 0, & \text{for } t < 0, \\ \dfrac{t^2}{2}, & \text{for } 0 \le t < 1, \\ 1 - \dfrac{(t-2)^2}{2}, & \text{for } 1 \le t < 2, \\ 1, & \text{for } t \ge 2. \end{cases}$$

Does this function satisfy the requirements to be the *cdf* for some random variable X? If so, is X discrete or continuous?

7. A 12-inch string is cut into two pieces at a random point along its length. Let X represent the point at which the cut occurs and assume X is continuous uniform, with $a = 0$ and $b = 12$. What is the probability that the length of the longer piece is at least twice that of the shorter, after the cut is made?

8. Define the function

$$g(t) = \begin{cases} 0, & \text{for } t < 1, \\ \frac{1}{3}, & \text{for } 1 \le t < 3, \\ \frac{5}{6}, & \text{for } 3 \le t < 6, \\ 1, & \text{for } 6 \le t. \end{cases}$$

Show that this function is a *cdf* for a random variable; call the random variable Y, and find R_Y and $p_Y(y)$.

9. The proportional gain in weight by a newborn puppy, in its first 2 weeks of life, is modeled as a random variable W whose *pdf* is $f_W(w) = 3w^2$ for $0 < w < 1$.
   a. Evaluate $P(W > .5)$.
   b. Find the mean and variance of W.

10. Let U be a continuous uniform random variable with $a = 0$ and $b = 1$.
   a. We know that $P(0 < U < 1) = 1$; this interval $(0,1)$ can be written as

$$\{u : |U - \mu_u| < k\sigma_U\}$$

   for some value of k. What is this value of k?
   b. What is the Chebyshev lower bound for the probability in part *a*?

11. A computer game selects a value for V whose *pdf* is

$$f_V(v) = \begin{cases} \dfrac{1}{2\sqrt{v}}, & \text{for } 0 < v < 1, \\ 0, & \text{otherwise.} \end{cases}$$

   a. Suppose you pay \$1 to play the game; if the number generated by the computer exceeds $\frac{1}{3}$ (i.e., if $V > \frac{1}{3}$), you receive \$2 (meaning you win $1), and if the number generated does not exceed $\frac{1}{3}$, you receive nothing (i.e., you lose $1). What is the probability that you win $1?

b. What is the expected amount you will win on one play of this game?

c. Evaluate the $100k$th quantile for V.

12. The *cdf* for Y is

$$F_Y(t) = \begin{cases} 0, & \text{for } t < -1, \\ \frac{1}{2}(t^3 + 1), & \text{for } -1 \le t \le 1, \\ 1, & \text{for } t > 1. \end{cases}$$

a. Evaluate the median and interquartile range of Y.

b. Evaluate the mean and standard deviation of Y.

13. Assume Y is defined as in Exercise 12, and let $X = 3 - 2Y$. Find the mean, median, standard deviation, and interquartile range of X.

14. Let X be a discrete random variable with probabilities $\frac{1}{6}, \frac{1}{3}, \frac{1}{6}$, and $\frac{1}{3}$ of equaling 1, 2, 3, and 4, respectively. Evaluate the mean and variance of X.

15. A commuter drives his car to the train station every weekday morning, arriving between 6:45 AM and 7:05 AM each time. Let X represent the instant at which he arrives, and assume X is continuous uniform with $a = 0$ and $b = 20$ (where 0 corresponds to 6:45 AM, and 20 to 7:05 AM). If trains he can use leave this station (promptly) at 6:48 AM, 6:55 AM, 7:03 AM, and 7:10 AM, evaluate the probability that he will have to wait no more than 4 minutes for one of his trains to leave.

16. A continuous random variable X has *cdf* $(3 + t)^2/16$ for $t \in R_X$. What is the range for X? Find its *pdf*, and evaluate x_k, its $100k$th quantile.

CHAPTER 3

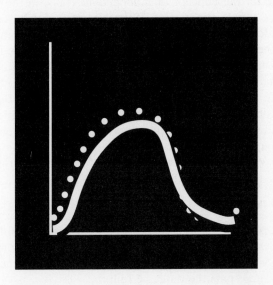

DISCRETE PROBABILITY
DISTRIBUTIONS

Probability theory has been found useful across a broad spectrum of disciplines in the modern world, almost universally through the use of random variables and their probability distributions (or probability laws). In this chapter we shall investigate several of the most frequently employed discrete distributions, seeing the assumptions that lead to their use, examples of such uses, as well as their means and standard deviations.

One of the simplest, and most useful, concepts in probability theory is the Bernoulli trial, a simple experiment that has only two distinct outcomes; such trials can be used to define many different discrete random variables, which find applications in diverse circumstances. In Sections 3.1 and 3.2 we will define, and explore the distributions of, several different random variables defined for sequences of such trials. The Poisson process is discussed in Section 3.3, along with a deriviation of the Poisson probability law as the limit of the binomial. Section 3.4 covers sampling without replacement and the resulting hypergeometric distribution.

3.1 | Bernoulli and Binomial Distributions

Recall the concept of independent trials, introduced in Chapter 1. A trial is a simple experiment. If a series of such trials is performed, the trials are independent if the outcome of *any* trial has no effect on the outcome of any other trial. The probability measure appropriate for a sequence of independent trials is given by the product of the measures used for the individual trials. In this section we shall build on these ideas in some simple ways. We shall discuss random variables defined on Bernoulli trials and find the probability law for the Bernoulli random variable. We shall also derive the binomial probability law, one of the most frequently employed distributions in applications.

We begin with the definition of a Bernoulli trial.

DEFINITION 3.1

A *Bernoulli trial* is an experiment with two different possible outcomes, labeled *success* and *failure*. The sample space for a single Bernoulli trial is defined as $T = \{s, f\}$, where s represents the outcome success and f represents the outcome failure.

Examples of Bernoulli trials include the flip of a coin (with heads called success), a play of any game (with a win called success), treating a patient with a new medicine (with a cure called success), investing in a business venture (with profit called success), and so forth. Indeed, given any experiment whatsoever, and any event $A \subset S$, we can call the occurrence of A a success and the occurrence of $\overline{A}$ a failure, thus converting the experiment into a Bernoulli trial.

Bernoulli trials are quite general in nature and quite simple to treat from a probabilistic point of view. With the sample space $T = \{s, f\}$ it is clear that there is a single free parameter involved in the probability measure for the trial: the probability of the event $\{s\}$, observing a success when the trial is performed. Once the value for this probability is known, the probability of the only other single-element event, $\{f\}$, is necessarily $P(\{f\}) = 1 - P(\{s\})$. We shall let p represent the probability of success, that is, $p = P(\{s\})$, and let $q = 1 - p$ denote the probability of observing $\{f\}$. If an experiment consists of a *single* Bernoulli trial with parameter p (so that $P(\{s\}) = p$) and we let X be the number of successes to occur, then X is called a *Bernoulli random variable with parameter p*. The range for X is then $\{0, 1\}$, and its probability function is especially simple:

$$P(X = 1) = p = p^1(1 - p)^0, \qquad P(X = 0) = 1 - p = p^0(1 - p)^1,$$

which we can write succinctly as

$$p_X(x) = \begin{cases} p^x(1 - p)^{1-x}, & \text{for } x = 0, 1, \\ 0, & \text{otherwise.} \end{cases}$$

This Bernoulli distribution has a the single parameter p, which must lie between 0 and 1.

Let us assume a computer is asked to generate a random digit, and we let $X = 1$ if the digit generated is an element of $\{0, 2, 4, 6, 8\}$ and $X = 0$ otherwise. Then X is Bernoulli with $p = \frac{1}{2}$ (assuming the computer generating scheme works as it should). If we let $Y = 1$ when the computer-generated digit is an element of $\{0, 1, 2\}$ and let $Y = 0$ otherwise, then Y is Bernoulli with parameter $p = .3$. In many senses the Bernoulli random variable is the simplest possible random variable, since it has only two different observed values.

The mean and variance for a Bernoulli random variable are easily evaluated: $E[X] = \mu_X = 0\, p_X(0) + 1\, p_X(1) = p$, so the parameter p is actually the expected value or mean of X. Similarly, $E[X^2] = 0^2\, p_X(0) + 1^2\, p_X(1) = p$, so the Bernoulli variance is $\sigma_X^2 = E[X^2] - \mu_X^2 = p - p^2 = p(1 - p) = pq$, and the standard deviation is $\sigma_X = \sqrt{pq}$. If X is Bernoulli with parameter $p = \frac{1}{2}$, then its mean is $\frac{1}{2}$ and its variance is $\frac{1}{4}$; if Y is Bernoulli with parameter $p = .3$, then its mean is .3 and its variance is .21.

This parameter p must lie between 0 and 1 since it represents the probability of a success occurring. The graph of the Bernoulli variance, $p(1 - p) = p - p^2$ versus p, is a segment of a parabola with maximum value $\frac{1}{4}$ at $p = \frac{1}{2}$ and minimum value 0 when $p = 0$ or $p = 1$. (See Fig 3.1.) If X is Bernoulli with $p = 1$, then every observed value is 1, there is no variation, and the variance of X is 0. (This also occurs for $p = 0$, when all observed values are 0 rather than 1.) The biggest variance of X occurs with $p = \frac{1}{2}$; this case involves the most "bouncing around" of the observed 0's and 1's, over repeated observed values.

Bernoulli trials are extremely simple, since only two possible outcomes can occur on each trial. Many experiments can be modeled as a sequence of independent Bernoulli trials, for example:

Ten scratch-off lottery tickets are purchased; each ticket either will (success) or will not (failure) win some prize, where p is the probability of a success occurring for each.

A sample of 30 items is selected at random from a production line; each is either defective (failure) or not (success), with probability of success equal to p.

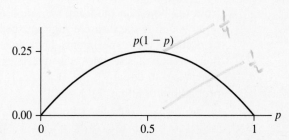

Figure 3.1 $p(1 - p)$ versus p

Each of 100 patients with the same affliction is given medication $\mathcal{A}$; each patient will either be cured (success) or not (failure), with the same success probability p.

A nursery worker plants 120 seeds in the same flat, each of which will (success) or will not (failure) germinate.

Every person applying for consumer credit either is a good risk (success) or not (failure), with successes occurring independently from person to person, with the same probability.

We have, in fact, encountered and used this type of structure several times already. Let us build a model for a *sequence* of repeated, independent Bernoulli trials, each with parameter p. The experiment consists of observing the results of n independent trials. The sample space S for the experiment is $S = T \times T \times \cdots \times T$, with $T = \{s, f\}$, a collection of n-tuples. Each position in each n-tuple is either s (success) or f (failure), with the ith position reporting the result observed on the ith trial. Because the experiment consists of independent trials, the probability of occurrence for each single-element event in S is the product of n terms, and each term in the product is either p or q, depending on whether the corresponding entry in the n-tuple is s or f. That is, for each n-tuple $(t_1, t_2, \ldots, t_n) \in S$,

$$P(\{(t_1, t_2, \ldots, t_n)\}) = p^y q^{n-y},$$

where $y = 0, 1, 2, \ldots, n$ is the number of s's in $(t_1, t_2, \ldots, t_n)$. There are 2^n elements in S, giving a total of 2^n different single-element events; however, only $n + 1$ different values occur in specifying the single-element event probabilities: $p^0 q^n, p^1 q^{n-1}, p^2 q^{n-2}, \ldots, p^n q^0$.

This sample space S contains all 2^n possible n-tuples that can be constructed by placing s or f in each position; if y is any integer between 0 and n, inclusive, the number of n-tuples in S that contain exactly y s's (and $n - y$ f's, so that it is an n-tuple) is then $\binom{n}{y}$. Thus the probability of observing exactly y successes is $\binom{n}{y} p^y q^{n-y}$ since each n-tuple with y successes has the same probability. The sum of all the single-element probabilities is

$$\sum_{y=0}^{n} \binom{n}{y} p^y q^{n-y} = (p + q)^n = (p + (1 - p))^n = 1 = P(S),$$

from the binomial theorem, as required.

Now define the discrete random variable Y to be the number of successes to occur in this experiment. The range for Y is $R_Y = \{0, 1, 2, \ldots, n\}$, since the number of successes will necessarily be one of these integers. We will observe Y to equal $y \in R_Y$ if and only if exactly y successes occur. Thus the probability function for Y is

$$p_Y(y) = P(Y = y) = \binom{n}{y} p^y q^{n-y}, \qquad \text{for } y \in R_Y.$$

These results are summarized in the following theorem.

> **THEOREM 3.1**   If Y is the number of successes to occur in n repeated, independent Bernoulli trials, each with probability of success p, then Y is a binomial random variable with parameters n and p. The range for Y is $R_Y = \{0, 1, 2, \ldots, n\}$, and its probability function is
>
> $$p_Y(y) = \begin{cases} \binom{n}{y} p^y q^{n-y}, & \text{for } y \in R_Y, \\ 0, & \text{otherwise.} \end{cases}$$

Note that Y is called a binomial random variable because the values of its probability function are the terms in the binomial expansion of $(p + q)^n$. This binomial probability law occurs frequently across a broad spectrum of applications. The next three examples illustrate some of its uses.

Example 3.1

An automobile dealer has 15 new cars in stock on a given day. He assumes that each of these cars represents a Bernoulli trial, for this given day, in the sense that each car will be sold (success) or not. He assumes that the probability of success (i.e., that it will be sold on this day) is .06 for each car, and that these sales are independent. If Y is the number of new cars he will sell this day, then Y is binomial with parameters $n = 15$ and $p = .06$. The probability that he makes no sales on this day is then $p_Y(0) = .94^{15} = .3953$, and the probability that he makes at least one sale is $P(Y \geq 1) = 1 - p_Y(0) = .6047$.

Example 3.2

Suppose you buy 10 scratch-off lottery tickets. Assume the probability of winning some prize is $\frac{1}{9}$ for each ticket and that these tickets are independent (i.e., the fact that any one ticket may bring a prize has no effect on the probability that any other will, or will not, also win a prize). Thus each ticket defines a Bernoulli trial, and if X is the number of tickets to win some prize, the probability law for X is binomial, with $n = 10$ and $p = \frac{1}{9}$:

$$p_X(x) = \binom{10}{x} \left(\frac{1}{9}\right)^x \left(\frac{8}{9}\right)^{10-x}, \qquad \text{for } x = 0, 1, 2, \ldots, 10.$$

The probability that none of your tickets wins a prize is then $P(X = 0) = p_X(0) = \left(\frac{8}{9}\right)^{10} = .3079$, while the probability that three or more tickets win some prize is given by the sum

$$P(X \geq 3) = \sum_{x=3}^{10} p_X(x) = \sum_{x=3}^{10} \binom{10}{x} \left(\frac{1}{9}\right)^x \left(\frac{8}{9}\right)^{10-x} = .0906.$$

Example 3.3

A camera flashbulb manufacturer sells a bulb of a certain type in packages of six. Assume that individual flashbulbs are independent Bernoulli trials with $p = .9$, where success means the bulb flashes correctly and failure that it is a dud. If Y is the number of good bulbs in a package of six, then Y is binomial with parameters $n = 6$ and $p = .9$. The probability a given package contains at least five good bulbs is

$$P(Y \geq 5) = \sum_{i=5}^{6} \binom{6}{i}(.9)^i(.1)^{6-i} = .8857.$$

Indeed, each package of six bulbs could itself be modeled as a Bernoulli trial, defining success to mean at least five good bulbs are in the package; with this definition for success, $p = .8857$ (to four decimal places). Now suppose these packages of six bulbs are sold in cartons of 24; that is, each carton contains 24 packages of six, for a total of 144 bulbs. Let X be the number of successful packages (containing at least five good bulbs) in a carton. It is easy to see that X is binomial with parameters $n = 24$ and $p = .8857$ and, for example,

$$P(X \geq 20) = \sum_{i=20}^{24} \binom{24}{i}(.8857)^i(.1143)^{24-i} = .8685.$$

The probability law for Z, the number of good bulbs in a carton, is binomial with parameters $n = 144$ and $p = .9$.

The shape of the binomial probability function is determined by the values of its two parameters n and p. As with the discrete uniform probability law, the parameter n delimits the range of Y, while the parameter p affects the probabilities with which these values in the range will be observed. Figure 3.2 graphs the probability functions for a binomial random variable X where $n = 5, 10$, and $p = .2, .5$. Note that the probability function is symmetric with $p = .5$ for both values of n. This will be true for any value of n: with $p = .5$,

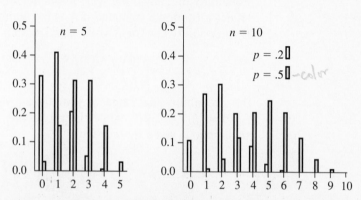

Figure 3.2 Two binomial probability laws

then $q = 1 - p = .5$ as well, and

$$P(Y = y) = \binom{n}{y}(.5)^y(1 - .5)^{n-y} = \binom{n}{y}(.5)^n$$

$$= \binom{n}{n-y}(.5)^n = P(Y = n - y).$$

When $p \neq .5$, the binomial probability function goes through a maximum that is shifted toward $y = 0$ if $p < .5$ and shifted toward $y = n$ if $p > .5$. To investigate this behavior consider the ratio of two succeeding values of the probability function:

$$\frac{p_Y(y)}{p_Y(y+1)} = \frac{\binom{n}{y}p^y(1-p)^{n-y}}{\binom{n}{y+1}p^{y+1}(1-p)^{n-y-1}} = \frac{(y+1)(1-p)}{(n-y)p}.$$

This ratio then is less than 1 (meaning $p_Y(y) < p_Y(y+1)$) whenever $(y+1)(1-p) < (n-y)p$, which in turn gives $y < (n+1)p - 1$. Thus, the amount of probability at any value of y is less than the probability at $y+1$ for any $y < (n+1)p - 1 = np - q$; if it is true that $(n+1)p$ is an integer, then $y = (n+1)p - 1$ and $y + 1 = (n+1)p$ have the same probability of occurrence. This common value also maximizes the probability function, with the values for $p_Y(y)$ trailing off toward 0 on both sides. In all other cases the unique maximum value for $p_Y(y)$ occurs at $y = \lfloor (n+1)p$, the largest integer in $(n+1)p$.

If, for example, Y is binomial with $n = 10$, and $p = \frac{4}{11}$, then the most likely values for Y are $y = 3, 4 : p_Y(3) = p_Y(4) = .2438$. If Y is binomial with $n = 11$ and $p = \frac{4}{11}$, then the most likely value for Y (the highest value for $p_Y(y)$) is $y = \lfloor (11+1)\frac{4}{11} = 4$, with $p_Y(4) = .2438$, and this is the unique maximum.

Note as well that the occurrence of Y successes in n trials means that $X = n - Y$ failures must also occur, so the probabilities of these two events are equal; that is,

$$p_Y(y) \equiv p_X(n - y),$$

where Y is binomial with parameters n, p, and X is binomial with parameters n and $q = 1 - p$. This means that the binomial probability function, with p replaced by $1 - p$, takes on equal values except that the values are reversed in order, since y is replaced by $n - y$. For example, the probability function for the binomial probability law with $n = 5$ and $p = .8$ looks just like the probability function in Fig. 3.2 for $n = 5$ and $p = .2$, except the figure is reversed; $x = 5$ replaces $y = 0$, $x = 4$ replaces $y = 1$, and so forth. The binomial *cdf* is

$$F_X(t) = P(X \le t) = \begin{cases} 0, & \text{for } t < 0, \\ \sum_{x=0}^{\lfloor t} \binom{n}{x}p^x q^{n-x}, & \text{for } 0 \le t \le n, \\ 1, & \text{for } t > n, \end{cases}$$

where $\lfloor t$ is the greatest integer in t. As with all discrete random variables, this *cdf* is a step function with jumps in value at the possible observed values; the size of the jump is given by the amount of probability at the observed value. Table A.1 in Appendix A gives the values for this binomial *cdf* for $n = 5, 10, 15, 20, 25, 30$ and $p = .05, .10, \ldots, .50$. Since $F_X(n) = 1$, this table presents the *cdf* values only for the integers $0, 1, 2, \ldots, n-1$ for which $.00005 < F_X(x) < .99995$; for example, for $n = 30$, the table entries stop with $x = 24$. This is done since $F_X(x) \geq .99995$ for $x \geq 25$ for each of $p = .05, .10, \ldots, .50$.

As just mentioned, if Y is binomial with parameters n, p, then $X = n - Y$ is also binomial with parameters $n, 1 - p$, so $p_Y(y) \equiv p_X(n - y)$. This relationship allows Table A.1 to be used to evaluate the binomial *cdf* for the same values of n, with $p = .55, .60, \ldots, .95$. The following examples illustrate some uses of Table A.1.

Example 3.4

A telephone salesperson is given a list of 30 prospects who may have interest in the purchase of a vacation home in Florida. Each prospect either will or will not respond favorably to her offer to send a prospectus. Assume that the probability of getting a favorable response is .1 and that these responses are independent. Let Y be the number of favorable responses she generates from the 30 prospects; then Y has the binomial distribution with $n = 30$ and $p = .1$. The probability that she gets three or fewer favorable responses is $P(Y \leq 3) = F_Y(3) = .6474$ (taken directly from Table A.1). The probability that she gets at least five favorable responses is $P(Y \geq 5) = 1 - F_Y(4) = 1 - .8245 = .1755$, and the probability that she gets from one to five favorable responses (inclusive) is $P(1 \leq Y \leq 5) = P(0 < Y \leq 5) = F_Y(5) - F_Y(0) = 0.9268 - 0.0424 = .8844$.

Example 3.5

A random sample of 20 items is selected from a production line; each item either is defective or not, and defectives occur independently. The probability each item selected is nondefective is .95. What is the probability no more than 18 of the 20 are nondefective? that at least 17 are nondefective? X, the number of nondefectives in the sample, is binomial with $n = 20$ and $p = .95$; Y, the number of defectives in the sample of 20, is binomial with $n = 20$ and $p = .05$. Then $P(X \leq 18) = F_X(18) \equiv P(Y \geq 2) = 1 - F_Y(1) = 1 - 0.7358 = .2642$, while $P(X \geq 17) \equiv P(Y \leq 3) = .9841$, from Table A.1.

The balance point for the binomial probability function is given by its mean or expected value:

$$\mu_X = \mathrm{E}[X] = \sum_{x=0}^{n} x \binom{n}{x} p^x q^{n-x}.$$

In evaluating this sum, notice first of all that the beginning term $(x = 0)$ contributes nothing, since $0 \cdot p_X(0) = 0$, so we have the same value, if the range of summation is from $x = 1$ to $x = n$. Note as well that

$$x\binom{n}{x} = x\frac{n!}{x!(n-x)!} = \frac{n(n-1)!}{(x-1)!(n-x)!}$$

$$= n\binom{n-1}{x-1}, \qquad \text{for } x = 1, 2, \ldots, n,$$

$p^x = p \cdot p^{x-1}$, and $q^{n-x} = q^{(n-1)-(x-1)}$. Thus

$$\mu_X = \mathrm{E}[X] = np \sum_{x=1}^{n} \binom{n-1}{x-1} p^{x-1} q^{(n-1)-(x-1)}$$

$$= np(p+q)^{n-1} = np,$$

where we have recognized the value for this final sum from the binomial theorem. The expected value for the binomial random variable is simply the product of its two parameters. If X is binomial with $n = 20$ and $p = .4$, say, the expected value for X is then $20(.4) = 8$, a result that is probably not surprising.

We saw in Chapter 2 that the variance of a random variable can be evaluated from the computational formula $\sigma_X^2 = \mathrm{E}[X^2] - \mu_X^2$. In evaluating $\mathrm{E}[X^2]$ for discrete random variables whose probability function involves $x!$ in the denominator (like the binomial), it is useful to realize that $x(x-1) = x^2 - x$, for any x, or equivalently, $x^2 = x(x-1) + x$, so

$$\mathrm{E}[X^2] = \mathrm{E}[X(X-1)] + \mathrm{E}[X] = \mathrm{E}[X(X-1)] + \mu_X.$$

A second computational formula for σ_X^2 is then

$$\sigma_X^2 = \mathrm{E}[X(X-1)] + \mu_X - \mu_X^2 = \mathrm{E}[X(X-1)] - \mu_X(\mu_X - 1).$$

For many of the standard discrete probability laws it is more straightforward to find the value for the sum defining $\mathrm{E}[X(X-1)]$ than to find the value for the sum defining $\mathrm{E}[X^2]$. This alternate computational formula for σ_X^2 is useful in such cases. Let us illustrate this approach in evaluating the variance for the binomial probability law with parameters n, p.

If X is binomial with parameters n, p, we know $\mathrm{E}[X] = np$; using our earlier discussion, then, the only additional quantity needed to evaluate the variance and standard deviation for X is

$$\mathrm{E}[X(X-1)] = \sum_{x=0}^{n} x(x-1)\binom{n}{x} p^x q^{n-x}$$

$$= \sum_{x=2}^{n} x(x-1)\frac{n!}{x!(n-x)!} p^x q^{n-x}$$

$$= n(n-1)p^2 \sum_{x=2}^{n} \frac{(n-2)!}{(x-2)!(n-x)!} p^{x-2} q^{n-x}$$

$$= n(n-1)p^2 (p+q)^{n-2} = n(n-1)p^2.$$

Note that $x(x-1) = 0$ for both $x = 0$ and $x = 1$, so the range of summation can be taken as $x = 2, 3, \ldots, n$; after factoring $n(n-1)p^2$ through the summation, the remaining sum is just $(p+q)^{n-2}$ from the binomial theorem. Thus the variance for the binomial probability law is

$$\sigma_X^2 = n(n-1)p^2 - np(np-1) = np(1-p) = npq;$$

the standard deviation is

$$\sigma_X = \sqrt{npq}.$$

Recall that the Bernoulli mean is p and the Bernoulli variance is pq. Suppose we perform n independent Bernoulli trials, each with parameter p. Let

$$y_1 = \begin{cases} 1, & \text{if a success occurs on trial 1} \\ 0, & \text{otherwise} \end{cases}$$

$$y_2 = \begin{cases} 1, & \text{if a success occurs on trial 2} \\ 0, & \text{otherwise} \end{cases}$$

and so forth, with $Y_n = 1$ only if a success occurs on trial n, $Y_n = 0$ otherwise. Then $Y_1, Y_2 \ldots Y_n$ are Bernoulli random variables, and the sum of these Bernoulli variables gives the total number of successes (number of 1's) to occur in the n trials. That is, $X = \sum_{i=1}^{n} Y_i$ is binomial with parameters n, p; the mean for X is np, the sum of the n Bernoulli means; the variance of X is npq, the sum of the Bernoulli variances. Later we shall see why this must be true. Because of this, the binomial variance (or its standard deviation) is largest for $p = q = \frac{1}{2}$ for any given value of n, as with the Bernoulli variance; the binomial variance decreases as p moves away from $\frac{1}{2}$ and becomes 0 for either $p = 1$ or $p = 0$.

Example 3.6

If six fair dice are rolled one time, then the number of these to show the face with six spots is a binomial random variable with $n = 6, p = \frac{1}{6}$. The expected number of dice to show six spots is $np = 6\left(\frac{1}{6}\right) = 1$, while the variance of this distribution is $npq = 6(\frac{1}{6})(\frac{5}{6}) = \frac{5}{6}$. The standard deviation is $\sqrt{\frac{5}{6}} = .913$.

If n fair dice are rolled, and X is the number of these to show six spots, then X is binomial with parameters n and $p = \frac{1}{6}$. The ratio $Y = X/n$ gives the *proportion* of these dice to show six spots when rolled; the expected value for Y is $\frac{1}{6}$ ($1/n$ times E[X]), and the variance of Y is $5/36n$ ($1/n^2$ times the variance of X). Note that the variance of the proportion gets smaller as n increases; this fact has important consequences for statistical methods, as we shall see.

EXERCISES 3.1

1. A nurseryworker plants 105 flower seeds in a flat to germinate for sale to customers. Assume that the probability each seed will successfully germinate is .95 and that the individual seeds constitute independent Bernoulli trials. If Y is the number of these seeds to germinate, what

following. Let X be the number of times she is late to these seven classes. Granted she is late a total of two times during these seven classes, what is the probability that both late occurrences were in the second week? (See Exercise 15.)

18. Ten independent Bernoulli trials are performed, each with probability of success $p = \frac{1}{2}$. Evaluate the probability that successes and failures alternate (i.e., the results are given by (s,f,s,f,s,f,s,f,s,f) or (f,s,f,s,f,s,f,s,f,s)). Thus show that this lock-step alternating behavior is not very likely to occur.

19. Ten independent Bernoulli trials are performed, each with probability of success p. Given that five successes occur, evaluate the probability that successes and failures alternate (i.e., the results are given by (s,f,s,f,s,f,s,f,s,f) or (f,s,f,s,f,s,f,s,f,s)).

3.2 | Geometric and Negative Binomial Probability Laws

If n is some fixed positive integer, and n independent Bernoulli trials are performed with probability p of success for each, then the number of successes to occur is binomial with parameters n, p. In this section we shall again consider performing independent Bernoulli trials, where the number of trials is not fixed in advance but determined by the outcomes of the trials themselves.

First, assume independent Bernoulli trials are performed, each with probability of success equal to $p > 0$, and let N be the (random) trial number of the *first* success to occur. The probability function for N is easily derived.

The range for N is $R_N = \{1, 2, 3, \dots\}$. The observed value for N will be 1 if and only if the first trial performed is a success, so that $P(N = 1) = p_N(1) = p$. The observed value for N will be 2 if and only if the first trial produces a failure and the second trial produces a success (so that the first success occurs on the second trial), giving $P(N = 2) = p_N(2) = qp$. In general, the observed value for N will be the positive integer n if and only if the first $n - 1$ trials result in failures and the nth trial is a success. This establishes the following theorem.

THEOREM 3.2 Let N be the trial number of the first success in a sequence of independent Bernoulli trials, each with parameter p. The probability function for N is

$$p_N(n) = \begin{cases} pq^{n-1}, & \text{for } n \in R_N = \{1, 2, 3, \dots\} \\ 0, & \text{otherwise.} \end{cases}$$

N is called a geometric random variable with parameter p.

Note that $\sum_{n=1}^{\infty} pq^{n-1} = p\sum_{i=0}^{\infty} q^i = p/(1-q) = 1$, as required, since $q = 1 - p$; the values for $p_N(n)$ form a geometric progression, which is the reason that N is called a geometric random variable.

Example 3.7

A medical research team wants to locate one person with a given malady in order to attempt a new method of treatment. Suppose that 10% of the population has this malady, and that the research team will examine persons at random until they locate the first one with the disease. We assume the persons examined constitute independent Bernoulli trials, with success defined to mean the person examined has the disease ($p = .1$ for each); then the number of persons the researchers must examine to locate the first person with the disease is a geometric random variable N with $p = .1$. The probability that they will have to examine no more than three persons is

$$P(N \le 3) = \sum_{n=1}^{3} (.1)(.9)^{n-1} = 1 - (.9)^3 = .271.$$

The probability that they will have to examine at least six persons is

$$P(N \ge 6) = \sum_{n=6}^{\infty} (.1)(.9)^{n-1} = (.9)^5 = .590.$$

Figure 3.3 presents graphs of two different geometric probability laws, one with $p = .8$ and the other with $p = .5$. In each case the range is $R_N = \{1, 2, 3, \ldots\}$, the set of positive integers. Because the values of the probability function for N form a geometric progression, the ratio of the heights of any two successive bars is q, giving the smooth decreasing picture. The larger the value of q, the more spread out the probability distribution is, or, more exactly, the longer (larger value of n) it takes for the probability function to become smaller than .001 (or any other value).

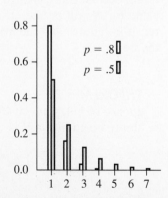

Figure 3.3 Geometric probability laws

The probability that $N > n$, where n is any positive integer, is

$$P(N > n) = pq^n + pq^{n+1} + pq^{n+2} + \cdots$$

$$= pq^n \sum_{i=0}^{\infty} q^i = \frac{pq^n}{1-q} = q^n. \tag{1}$$

We used Eq. (1) in Example 3.7 with $n = 5$. (The two events defined by $N > 5$ and $N \geq 6$ are the same since N is discrete with range $\{1, 2, 3, \ldots\}$.) Equivalently, the number of Bernoulli trials required to observe the first success exceeds n ($N > n$) if and only if the first n trials are all failures ($Y = 0$, where Y is binomial with parameters, n, p), so $P(N > n) = P(Y = 0) = \binom{n}{0} p^0 q^n = q^n$. Since $P(N \leq n) = 1 - P(N > n) = 1 - q^n$, the geometric *cdf* is simply expressed as

$$F_N(t) = P(N \leq t) = \begin{cases} 0, & \text{for } t < 1, \\ 1 - q^{\lfloor t}, & \text{for } t \geq 1, \end{cases}$$

where $\lfloor t$ is the greatest integer in t.

Example 3.8

Evaluating the probability of occurrence of certain events, defined in terms of geometric random variables, may involve evaluating other geometric series. For example, let N be a geometric random variable with parameter p and consider the probability that N is an odd number; thus

$$P(N \in \{1, 3, 5, \ldots\}) = p + pq^2 + pq^4 + \cdots = p\frac{1}{1-q^2}.$$

It is easy to see (as you will be asked to do in Exercises 3.2) that the probability N is a multiple of $k = 2, 3, 4, \ldots$ is simply $pq^{k-1}/(1-q^k)$.

The structure of the geometric probability function gives it the memoryless property, which follows directly from the form of its *cdf*: The geometric random variable records the trial number of the first success to be observed in repeated, independent Bernoulli trials. Now suppose we start performing such trials, find that we have observed, say, five failures in a row, and ask ourselves the question: What is the probability that it will take at least three more trials before we get the first success? Probably not too surprisingly, the answer is, the same probability as when you started observing the trials. This is referred to as the *memoryless property*; in terms of the probability of what may yet be observed in the future, independent Bernoulli trials "forget" that several trials have been performed without the occurrence of a success.

To make this discussion of the memoryless property more precise, let us assume that N has the geometric probability distribution with parameter p. Then $P(N > n) = q^n$, where $q = 1 - p$ is the probability of a failure occurring on each trial. If a and b are any two positive integers, then $q^{a+b} = q^a q^b$; that is,

$$P(N > a + b) = P(N > a)P(N > b);$$

The probability N exceeds $a + b$ is the product of the probability that $N > a$ times the probability that $N > b$. From this it follows that the ratio

$$\frac{P(N > a + b)}{P(N > b)} = P(N > a).$$

Now formally consider the events $B = \{N > b\}$, and $C = \{N > a + b\}$, where both a and b are positive integers. If $N > a + b$, then $N > b$. Therefore, $C \subset B$, so $B \cap C = C$ and

$$P(C \mid B) = \frac{P(C \cap B)}{P(B)} = \frac{P(C)}{P(B)}.$$

That is,

$$P(N > a + b \mid N > b) = \frac{P(N > a + b)}{P(N > b)}$$

$$= \frac{q^{a+b}}{q^b} = q^a = P(N > a). \qquad (2)$$

Equation (2) expresses more exactly what was said earlier, that the geometric random variable "forgets" that b trials have occurred with no successes and sets the probability that it will take more than a additional trials to get the first success equal to $P(N > a)$. Although we shall not attempt to do so, it can be shown that this is the only discrete probability law which has this memoryless property, summarized in the following theorem.

THEOREM 3.3 If N is a geometric random variable with parameter p, then

$$P(N > a + b \mid N > b) = P(N > a),$$

where a and b are any positive integers. This is the only discrete probability law to have this memoryless property.

Example 3.9

Stated mathematically, the memoryless property may require a little time to absorb. If you think of a particular case, though, it is quite easy to see that it must hold. Think again of examining patients until the first one having a certain malady is found, where individuals are independent Bernoulli trials with $p = .1$. If eight patients have already been examined, with no success, the probability the next one may have the malady is unchanged (by assumption), as is the probability that it will take four more examinations to find one with the malady. If we consider any other discrete probability law, this property does not hold for all possible positive integers a and b.

For example, consider the discrete uniform law on the integers from 1 to 10, that is, a random variable X with probability function $p_X = \frac{1}{10}$ for $x = 1, 2, \ldots, 10$. Let $a = 2$ and $b = 3$, so $a + b = 5$. Then $P(X > a) = P(X > 2) = .8$, $P(X > b) = P(X > 3) = .7$, and $P(X > a + b) = P(X > 5) = .5$. This gives

$$\frac{P(X > a + b)}{P(X > b)} = \frac{.5}{.7} \neq .8 = P(X > a).$$

The uniform probability law does not have the memoryless property, nor does any other discrete probability law, except the geometric.

The expected value of the geometric random variable with parameter p is

$$E[N] = \sum_{n=1}^{\infty} n \, p_N(n) = \sum_{n=1}^{\infty} n \, p q^{n-1}$$

$$= p(1 + 2q + 3q^2 + \cdots) = p \frac{1}{(1-q)^2} = \frac{1}{p}. \tag{3}$$

Equation (3) is the first derivative of a geometric series. (See Appendix A for a discussion of various geometric series.) If N is geometric with $p = \frac{1}{6}$, its expected value is 6, while if $p = \frac{1}{10}$, its expected value is 10.

The expected value for $N(N - 1)$ is

$$E[N(N - 1)] = \sum_{n=1}^{\infty} n(n - 1) \, p_N(n) = \sum_{n=2}^{\infty} n(n - 1) \, p q^{n-1}$$

$$= pq(2 + 6q + 12q^2 + \cdots) = pq \frac{2}{(1-q)^3} = \frac{2q}{p^2}. \tag{4}$$

This expected value is the second derivative of a geometric series. Then the variance for N (using the discrete computational formula) is

$$\text{Var}[N] = E[N(N - 1)] - \mu_N(\mu_N - 1) = \frac{2q}{p^2} - \frac{1}{p}\left(\frac{1}{p} - 1\right) = \frac{q}{p^2}.$$

If N is geometric with $p = \frac{1}{6}$, its variance is $\frac{5}{6}(6)^2 = 30$, while if $p = \frac{1}{10}$, its variance is $\frac{9}{10}(10)^2 = 90$; the *smaller* the parameter p the larger both μ_N and σ_N^2 become.

The probability law describing the occurrence of the first success in a sequence of repeated, independent Bernoulli trials is geometric. What is the probability law for the occurrence of the second success in a sequence of repeated, independent Bernoulli trials? Let N_2 represent the trial number of the second success to occur in this sequence. Clearly, the range for N_2 must be $R_{N_2} = \{2, 3, 4, \ldots\}$ since it will require at least two trials for the second success to occur, and $P(N_2 = 2) = p^2$. We will find $N_2 = 3$ only if the second success occurs on the third trial; this entails two requirements: that the third trial be a success *and* that we have exactly one success in the first two trials, i.e., $P(N_2 = 3) = p(2pq) = 2p^2q$. Since N_2 is the trial number of the second success, the last trial to be observed is necessarily a success, and there *must* have been exactly one other success in the preceding trials. Thus, for $n \in R_{N_2}$,

$$p_{N_2}(n) = p\binom{n - 1}{1} p q^{n-2} = (n - 1)p^2 q^{n-2}. \tag{5}$$

Equation (5) is called the *negative binomial probability law* with parameters 2 and p.

Example 3.10

On a good day, a trout fisherman assumes the probability that he will catch a trout each time he casts his line into the water is .4; he also assumes that these casts behave like independent Bernoulli trials. If we let N_2 be the number of casts he must make to catch his second fish on such a day, then N_2 follows the negative binomial probability law with parameters 2 and .4. The probability function for N_2 is

$$p_{N_2}(n) = (n-1)(.4)^2(.6)^{n-2}, \qquad n = 2, 3, \ldots.$$

From this we see that the probability he catches his second trout on his third cast is $2(.16)(.6){=}.192$. If his limit is two fish per day, he is virtually certain to reach his limit by his 13th cast (with these assumptions) since

$$\sum_{n=2}^{13}(n-1)(.4)^2(.6)^{n-2} = .9919.$$

It is easy to generalize this discussion. Again assume a sequence of independent Bernoulli trials, each with probability of success equal to p. Define N_r to be the trial number of the rth success, where r is a positive integer; the range of N_r is then $R_{N_r} = \{r, r+1, r+2, \ldots\}$, since it will require r or more trials to get the rth success. If n trials are required to get the rth success, then (as with $r = 2$) two things must be true: the final (nth) trial must have been a success *and* exactly $r - 1$ successes must have occurred in the $n - 1$ trials preceding the nth. This establishes the following theorem.

THEOREM 3.4   Independent Bernoulli trials, each with probability of success equal to p, are performed until the rth success occurs. The number of trials required, N_r, is called a negative binomial random variable with parameters r, p; its probability function is

$$p_{N_r}(n) = \begin{cases} \dbinom{n-1}{r-1}p^r q^{n-r}, & \text{for } n \in R_{N_r} = \{r, r+1, r+2, \ldots\}, \\ 0, & \text{otherwise.} \end{cases}$$

The kth derivative of a geometric progression gives the result that

$$\sum_{j=k}^{\infty}\binom{j}{k}q^{j-k} = \frac{1}{(1-q)^{k+1}} \tag{6}$$

for any $|q| < 1$. Setting $j = n - 1$ and $k = r - 1$ in Eq. (6) gives

$$\sum_{n-1=r-1}^{\infty}\binom{n-1}{r-1}q^{(n-1)-(r-1)} = \frac{1}{(1-q)^{(r-1)+1}} = \frac{1}{p^r},$$

showing that $\sum_{n \in R_{N_r}} p_{N_r}(n) = 1$, as of course it must.

Example 3.11

A consumer action group is interested in gathering evidence about the shoddiness of the product put out by a major corporation. The group assumes that fully $\frac{1}{4}$ of all the products sold by this corporation is substandard; to bolster its argument, the group would like to obtain $r = 10$ examples of the corporation's substandard work. Assume that each item sold is a Bernoulli trial, where success indicates the item is substandard, that these trials are independent, and that $p = \frac{1}{4}$. Then N_{10} is the number of items the group would have to purchase in order to find 10 substandard items (i.e., N_{10} = the number of trials to get the 10th success). N_{10} then has the negative binomial distribution with parameters $r = 10$ and $p = \frac{1}{4}$. Its probability function is

$$p_{N_{10}}(n) = \binom{n-1}{9}\left(\frac{1}{4}\right)^{10}\left(\frac{3}{4}\right)^{n-10}, \qquad \text{for } n = 10, 11, 12, \ldots.$$

The value for this probability function at $n = 10$ is $p_{N_{10}}(10) = (\frac{1}{4})^{10} = 9.5 \times 10^{-7}$, so it is quite unlikely that this group will get 10 substandard items by purchasing only 10 items, if in fact $p = \frac{1}{4}$. The values for $p_{N_{10}}(n)$ increase slowly, as you can see numerically, reaching a maximum of .0380 at both $n = 36$ and $n = 37$ before decreasing again, approaching 0 as $n \to \infty$. The most likely number of items the group will have to purchase, to get 10 that are substandard, is 36 or 37 (but neither of these numbers is terribly likely since their probabilities of occurrence are so small). The probability that the group will have to purchase 40 or fewer is $P(N_{10} \leq 40) = .5605$, while the probability that it will have to purchase 60 or fewer is .9548.

Figure 3.4 presents two graphs, each plotting the negative binomial probability function for two sets of parameters. The graph on the left shows the negative binomial with $r = 2$ for $p = .8$ and $p = .5$, while the one on the right plots the negative binomial with $r = 4$ for the same two p values. The parameter r sets the smallest value for the range of N_r, while p and r together control the shape. The range is always unbounded and the ratio $p_{N_r}(n)/p_{N_r}(n+1)$

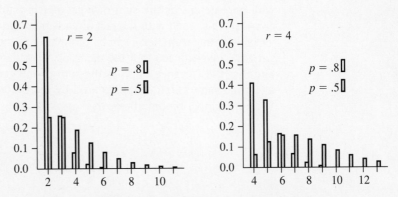

Figure 3.4 Negative binomial distributions

is no larger than 1 for $n \le (r-1)/p$. Thus if $(r-1)/p$ is an integer, the probability function is largest for $n = (r-1)/p$ and $n = 1 + (r-1)/p$, while if $(r-1)/p$ is not an integer, the largest value occurs at $n = \lceil (r-1)/p$, the next integer above $(r-1)/p$. This could in fact be at r itself, as shown in Fig. 3.4 for two of the cases given.

The expected value for the negative binomial random variable is

$$E[N_r] = \sum_{n=r}^{\infty} n p^r \binom{n-1}{r-1} q^{n-r}$$

$$= rp^r \sum_{n=r}^{\infty} \binom{n}{r} q^{n-r} = rp^r \frac{1}{(1-q)^{r+1}} = \frac{r}{p},$$

where the second sum is the $(r+1)$st derivative of a geometric progression. For example, if N_{10} has parameter $r = .25$, then $E[N_{10}] = 10/.25 = 40$, and the mean of the negative binomial random variable with $r = 2$ and $p = .4$ is $2/.4 = 5$. You will be asked in Exercises 3.2 to show that the variance for the negative binomial probability law is $\sigma_{N_r}^2 = rq/p^2$. With $r = 10$ and $p = .25$, the negative binomial variance is $10(.75)/(.25)^2 = 120$, while $r = 2$ and $p = .4$ gives a variance of $2(.6)/(.4)^2 = 7.5$.

With $r = 1$, the negative binomial random variable N_1 actually has the geometric distribution; that is, the geometric probability law is a special case of the negative binomial probability law, in much the same way that the Bernoulli is a special case of the binomial. Let us shift our notation slightly to illustrate one further fact. Suppose we perform independent Bernoulli trials, with parameter p, and let Y_1 be the number of trials to the first success; let Y_2 be the number of *additional* trials (after the first success) to get the second success. Y_1 and Y_2 are then geometric, and $Y_1 + Y_2 = N_2$, the number of trials (from the beginning) to get the second success; that is, the negative binomial random variable with $r = 2$ can be thought of as the sum of two geometric random variables. This relationship continues: If Y_j is the number of additional trials after the $(j-1)$st success needed to get the jth success, we have $N_r = Y_1 + Y_2 + \cdots + Y_r$; that is, the negative binomial random variable with parameter r is the sum of r geometric random variables. Compare (or contrast) this result with the fact already mentioned that the binomial random variable with parameter n can be thought of as the sum of n Bernoulli random variables. We can see that the mean of N_r is the sum of the geometric means and that the variance of N_r is the sum of the geometric variances; these are again the same relationships shared by the binomial and Bernoulli.

The negative binomial *cdf* is

$$F_{N_r}(t) = P(N_r \le t) = \begin{cases} 0, & \text{for } t < r, \\ p^r \sum_{n=r}^{\lfloor t \rfloor} \binom{n-1}{r-1} q^{n-r}, & \text{for } t \ge r, \end{cases}$$

where again $\lfloor t$ is the greatest integer in t. This *cdf* and the negative binomial probability function are both simply related to the binomial *cdf* and probability function. If N_r is the (random) number of trials to the rth success, and

X_n is the (random) number of successes to occur in n trials, then

$$p_{N_r}(n) = \binom{n-1}{r-1}p^r q^{n-r} = pP(X_{n-1} = r-1). \qquad (7)$$

Equation (7) simply restates the fact that $N_r = n$ only if there are exactly $r-1$ successes in the first $n-1$ trials, followed by a success on the nth trial. Additionally, the event $\{N_r > n\}$ is equivalent to $\{X_n \leq r-1\}$, since it will require *more than* n trials to get the rth success if and only if there are $r-1$ or fewer successes in the first n trials; thus we also have $P(N_r > n) = P(X_n \leq r-1)$. That is,

$$1 - F_{N_r}(n) \equiv F_{X_n}(r-1),$$

or

$$F_{N_r}(n) \equiv 1 - F_{X_n}(r-1). \qquad (8)$$

Equation (8) allows us to evaluate certain values for the negative binomial *cdf* from tables of the binomial *cdf*, as illustrated in the following example.

Example 3.12

Assume that the probability a scratch-off lottery ticket wins some prize is $p = .1$; you buy tickets until you win $r = 3$ prizes. The number of tickets you must purchase then is N_3, which is negative binomial with $r = 3$ and $p = .1$. The probability you must purchase no more than five tickets is $P(N_3 \leq 5) = F_{N_3}(5) = 1 - F_{X_5}(2) = 1 - .991 = .009$ from Table A.1. The probability that the number of tickets you must purchase lies between 6 and 10 inclusive is $P(5 < N_3 \leq 10) = F_{N_3}(10) - F_{N_3}(5) = F_{X_5}(2) - F_{X_{10}}(2) = .991 - .930 = .061$, again from Table A.1.

This section and the previous one have been concerned with random variables defined on independent Bernoulli trials. It is important to note the distinctions between the binomial and negative binomial probability laws. The terminology and notation have been chosen to try to simplify this task. If we perform n Bernoulli trials, where n is a fixed positive integer, the random number of successes to occur, Y, has the binomial distribution. On the other hand, if we want to observe r successes, where r is a fixed positive integer, the random number of trials required, N_r, has the negative binomial distribution. In moving from the binomial to the negative binomial distribution, the two quantities—number of trials and number of successes—switch roles in terms of what is fixed and what is random.

EXERCISES 3.2

1. A nationally sold candy bar advertises that the wrapper in which it is contained may give the purchaser a ticket for a free large bottle of soda. On the inside of this wrapper is the statement "The odds of a winning ticket are 1:229." The word *odds* is a gambling term meant to express a probability; if the *odds* of a winning event are quoted as $a:b$, then the probability of the winning event is the ratio $a/(a+b)$. Thus the candy bar wrapper statement means the probability of a given wrapper having

the free ticket is $\frac{1}{230}$. Suppose you repeatedly buy this brand of candy bar until you get your first winning ticket; let N be the number of bars that you must buy.

    a. What is the probability law for N?   b. Evaluate $P(N \le 230)$.

2. A man arrives at his door, late at night, with a key chain containing five keys, one of which opens his door. Not being too well organized, he tries the keys one at a time at random, with replacement, until he gets the door open. Let N be the number of keys he tries.

    a. What is the probability law for N?

    b. What is the probability that $N \le 5$?

3. Assume that Jessica is a well-practiced dart thrower with probability .8 of hitting the bull's-eye on each throw; also assume that her tosses are independent, that the result of any toss has no effect on the result of any other. Evaluate the mean and variance of the number of tosses she requires to get her first bull's-eye.

4. Assume the same situation described in Exercise 3. What is the expected value and variance of the number of tosses this player requires to get her 10th bull's-eye?

5. A fair coin is flipped until the first head occurs. What is the probability an even number of flips is required?

6. Some commercially sold products require the purchaser to collect a certain seal a specified number of times to win a prize; the seal to be collected is hidden in the package, so the package must be purchased before it can be determined whether it contains the seal or not. For example, suppose a manufacturer includes such a seal in one box out of five produced (independently and at random) and that you must collect three seals to win a prize.

    a. Let N_3 be the number of boxes you must buy to collect three seals. What is the probability law for N_3?

    b. Evaluate $P(N_3 = n)$ for $n = 3, 4, 5, 6, 7$.

7. Let N_r be negative binomial with parameter p, and evaluate the ratio

$$\frac{p_{N_r}(n)}{p_{N_r}(n+1)}.$$

Show that this ratio is 1 or less for $n \le (r-1)/p$, as mentioned in the text.

8. Some presentations define the geometric random variable X to be the number of failures *before* the first success, thus yielding the range $R_X = \{0, 1, 2, \ldots\}$. Let N be geometric as we have defined it, here, thus $X = N - 1$. What are the mean and variance for X?

9. Some presentations define the negative binomial random variable X to be the number of failures *before* the rth success, thus giving $R_X = \{0, 1, 2, \ldots\}$. Let N_r be geometric as we have defined it, so $X = N_r - r$. What are the mean and variance for X?

10. Suppose the probability of success occurring in a sequence of independent Bernoulli trials is .6 for each trial; a prize is awarded on the occurrence of the fifth success. What is the most likely trial number when the prize is awarded? (That is, what value of n maximizes $p_{N_5}(n)$?)

11. Show that the probability the observed value for a geometric random variable equals a multiple of k, where $k = 2, 3, 4, \ldots$, is $pq^{k-1}/(1 - q^k)$.

12. Independent Bernoulli trials, each with probability p of success, are performed. Assume that the second success occurred on the sixth trial, and let X be the trial on which the first success occurred. Find the probability law for X.

13. A woman arrives at her door, late at night, with a key chain containing five keys, one of which opens her door. She tries the keys one at a time, *without* replacement, until she gets the door open. Let N be the number of keys she tries.
    a. What is the probability law for N? (*Hint*: Consider the use of conditional probability.)
    b. What is the probability that $N \leq 5$?

14. Generalize the situation described in Exercise 13. Assume an urn contains M balls of which r are red. Balls are removed *without replacement* until the first red ball is drawn; what is the probability law for X, the number of balls drawn?

15. Assume that the probability is $\frac{1}{3}$ that any telephone call you receive is an unsolicited call for an opinion or a contribution; let N be the call number of the first such call you receive on a given day. (Idealize the situation by assuming an unlimited number of calls are possible.)
    a. What is the probability law for N?
    b. What is the probability that your first call of the day is for one of these two purposes?
    c. What is the probability that the first such call is among the first four calls you receive on a given day?

16. Reconsider Exercise 15: Recognize the fact that you will not receive an indefinitely large number of telephone calls on any day. More specifically, assume the total number of calls you will receive is some integer M; also assume that these calls are independent Bernoulli trials, each with parameter $\frac{1}{3}$, and let N be the trial number of the first success (if a success occurs). What is the probability law for N? Would the answers for parts b or c be any different?

17. Show that the variance for the negative binomial random variable, N_r, with parameters r, p, is rq/p^2.

18. If N is geometric with $p = .3$, evaluate $P(|N - \mu_N| < 3\sigma_N)$ and contrast the result with the Chebyshev bound for this probability.

3.3 | The Poisson Distribution

The binomial probability distribution occurs frequently across many different disciplines. Another discrete distribution that has found a great many applications is the Poisson.

DEFINITION 3.2

X is called a *Poisson random variable* with parameter μ if its probability function is

$$p_X(x) = \frac{\mu^x}{x!} e^{-\mu}, \qquad \text{for } x = 0, 1, 2, \dots.$$

Note that a Poisson random variable is discrete with range $R_X = \{0, 1, 2, \dots\}$, the positive integers. From the power-series expansion of e^z, namely,

$$e^z = 1 + z + \frac{z^2}{2!} + \frac{z^3}{3!} + \cdots = \sum_{x=0}^{\infty} \frac{z^x}{x!}$$

for any real z, we have

$$\sum_{x=0}^{\infty} p_X(x) = \sum_{x=0}^{\infty} \frac{\mu^x}{x!} e^{-\mu} = e^{-\mu} \sum_{x=0}^{\infty} \frac{\mu^x}{x!} = e^{-\mu} e^{\mu} = 1.$$

Thus the Poisson probability function sums to 1 as required, for any parameter value $\mu > 0$. (Why can't μ be negative?)

The Poisson probability distribution is employed in cases where occurrences are counted over time; for example, the number of telephone calls to arrive at a business phone in a 1-hour period, the number of customers to request service in a 10-minute period, the number of fatal auto accidents to occur on a given stretch of road over a 1-month period. The basic assumption (which we shall examine shortly in more detail) is that the phenomena being counted occur independently, at random, and at a constant rate over the period of observation.

Example 3.13

Assume that the number of injury-causing accidents to occur in a 4-week period, at a large industrial plant, is a Poisson random variable X with parameter $\mu = 2$. Then the probability function for X is

$$p_X(x) = \frac{2^x}{x!} e^{-2}, \qquad \text{for } x = 0, 1, 2, \dots.$$

The probability of there being no accidents in this period is $p_X(0) = .1353$, and the probability of at least three accidents occurring is

$$P(X \geq 3) = \sum_{x=3}^{\infty} \frac{2^x}{x!} e^{-2} = 1 - 5e^{-2} = .3233,$$

while the probability of there being one or two accidents is $p_X(1) + p_X(2) = 4e^{-2} = .5414$.

The mean value for the Poisson random variable X is

$$\mathrm{E}[X] = \mu_X = \sum_{x=0}^{\infty} x \frac{\mu^x}{x!} e^{-\mu} = \mu \sum_{x=1}^{\infty} \frac{\mu^{x-1}}{(x-1)!} e^{-\mu} = \mu,$$

since the value for the sum is $e^\mu e^{-\mu} = 1$; that is, the mean for the Poisson random variable is its parameter μ. You will be asked in Exercises 3.3 to show that the variance of the Poisson probability law is also μ; if Y is Poisson with parameter $\mu = 6.5$, say, then both its mean and its variance are equal to 6.5. The Poisson *cdf* is

$$F_X(x) = \sum_{j \le \lfloor x} \frac{\mu^j}{j!} e^{-\mu}, \qquad \text{for any } x \ge 0.$$

Table A.2 in Appendix A presents this *cdf* for the parameter μ varying from .1 to 1, in increments of .1, then in increments of .25 from 1.25 up to 6, in increments of .5 from 6.5 up to 11, and finally in increments of 1 from 12 up to 21. The table is constructed so that, in a given section, the range for x is from the smallest value (typically 0) for which $F_X(x) > .00005$ up to the largest value for which $F_X(x) < .99995$.

Example 3.14

Based on recent statistics, fatal automobile accidents occur in the state of California over a holiday period at a rate of six per (24-hour) day. That is, the expected number in a given day is 6; if we assume the actual number to occur in such a 24-hour period is Poisson with $\mu = 6$, we can see from Table A.2 that the probability of there being six or fewer is $P(X \le 6) = F_X(6) = .6063$. The probability of there being between four and eight (inclusive) is $P(4 \le X \le 8) = F_X(8) - F_X(3) = .8472 - .1512 = .6962$. Since the mean for X is $\mu_X = 6$ and its standard deviation is $\sigma_X = \sqrt{6} = 2.449$, the interval ranging from $2\sigma_X$ below the mean to $2\sigma_X$ above the mean is $(6 \pm 4.898) = (1.102, 10.898)$. Thus $P(1.102 < X < 10.898) = F_X(10) - F_X(1) = .9574 - .0174 = .94$, considerably above the Chebyshev bound of .75 for this probability.

Figure 3.5 presents graphs of the Poisson probability function. The graph on the left presents the probability functions for $\mu = .5, 1$, while the graph

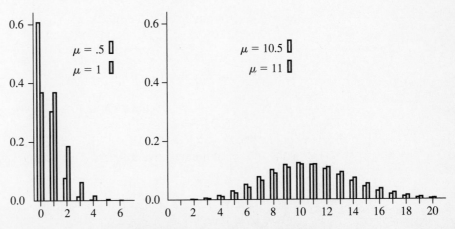

Figure 3.5 Poisson probability laws

on the right presents the probability functions for $\mu = 10.5, 11$. The range of X for any μ is $R_X = \{0, 1, 2, \ldots\}$, the set of nonnegative integers; for small values of μ ($\mu < 1$) the probability function decreases with increasing x, while for larger values of μ the probability function increases to a maximum (possibly not unique) and then decreases.

To formally explore this behavior, suppose X is a Poisson random variable with parameter μ. By considering the ratio $p_X(x)/p_X(x+1)$ we can easily investigate the relative magnitudes of succeeding values of the probability function. Thus

$$\frac{p_X(x)}{p_X(x+1)} = \frac{\frac{\mu^x}{x!}e^{-\mu}}{\frac{\mu^{x+1}}{(x+1)!}e^{-\mu}} = \frac{x+1}{\mu}.$$

Whenever $(x+1)/\mu \le 1$, $P(X = x) \le P(X = x+1)$; that is, the probability function increases as x increases. This requirement is equivalent to $x \le \mu - 1$; as long as $x \le \mu - 1$, the probability function is increasing. If μ is an integer, then $P(X = \mu - 1) = P(X = \mu)$; for $x > \mu$ the probability function decreases and goes to 0 as $x \to \infty$. As the parameter μ gets bigger, the Poisson probability function gets flatter (the individual probabilities decrease) and the location of the largest probability moves farther to the right.

This Poisson distribution occurs as the limit of $P(X = x)$, where X is binomial with parameters, n, p if $n \to \infty$, and $p \to 0$ such that $np = \mu$ remains constant. To see that this is true, suppose X is binomial with parameters n and $p = \mu/n$; thus

$$P(X = x) = \binom{n}{x}\left(\frac{\mu}{n}\right)^x\left(1 - \frac{\mu}{n}\right)^{n-x}, \quad \text{for } x = 0, 1, 2, \ldots, n.$$

Now

$$\binom{n}{x}\left(\frac{\mu}{n}\right)^x = \frac{n!}{x!(n-x)!}\left(\frac{\mu}{n}\right)^x$$

$$= \frac{\mu^x}{x!}\left(\frac{n}{n}\right)\left(\frac{n-1}{n}\right)\cdots\left(\frac{n-x+1}{n}\right) \to \frac{\mu^x}{x!},$$

as $n \to \infty$ (for any fixed value of $x \le n$). The remaining term in $P(X = x)$ is

$$\left(1 - \frac{\mu}{n}\right)^{n-x} = \left(1 - \frac{\mu}{n}\right)^n\left(1 - \frac{\mu}{n}\right)^{-x} \to e^{-\mu}$$

as $n \to \infty$ (for any fixed x). Thus we have

$$\lim_{n \to \infty} P(X = x) = \frac{\mu^x}{x!}e^{-\mu},$$

for any fixed $x \in R_X = \{0, 1, 2, \ldots, n\} \to \{0, 1, 2, \ldots\}$; that is, the Poisson range is the full set of nonnegative integers. This establishes the following theorem.

THEOREM 3.5 If X is a binomial random variable with parameters n and μ/n, then

$$\lim_{n \to \infty} P(X = x) = \frac{\mu^x}{x!} e^{-\mu}, \qquad \text{for } x \in R_X = \{0, 1, 2, \dots\}.$$

This is called the Poisson probability law with parameter μ.

If X is binomial with "large" n and "small" p, Theorem 3.5 suggests that the distribution for X should be well approximated by the Poisson probability law, where $\mu = np$, the product of the two binomial parameters. This relationship is illustrated in the following example.

Example 3.15

Assume that the probability is .0001 that a baby will be born with a rare malady and that a large city hospital has 5000 births in 1 year. Let us assume that these births behave like independent Bernoulli trials with $p = .0001$. Then the number of such births, X, to occur at this hospital in 1 year is binomial with parameters $n = 5000$ and $p = .0001$ and with probability function

$$p_X(x) = \binom{5000}{x} (.0001)^x (.9999)^{5000-x}, \qquad \text{for } x = 0, 1, 2, \dots, 5000.$$

Since n is quite large and p is quite small, the discussion preceding this example suggests that these exact binomial probabilities should be well approximated by the probability function for a Poisson random variable Y with parameter $\mu = np = \frac{1}{2}$. You can easily verify from Table A.2 that the probabilities of observing zero, one, two, or three such births are .6065, .3033, .0758, and .0126, respectively, for this Poisson distribution; they agree with the exact binomial values to a full four decimal places.

This Poisson approximation to the binomial distribution is generally quite good as long as $n \geq 100$ and $np \leq 10$, and is very good if $n \geq 100$ with $p \leq .05$. Figure 3.6 pictures the binomial probability function with $n = 100$ and $p = .1$, together with the Poisson probability function with $\mu = 10$, the same value as np. The agreement of the two functions is quite good for all observed values of x. The binomial variance is $npq < np = \mu$, the Poisson variance. This

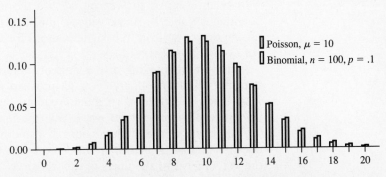

Figure 3.6 Poisson and binomial probability laws

phenomenon is evident in Fig. 3.6 by observing that the Poisson probabilities are smaller than their binomial counterparts in the neighborhood of the mean (10), and larger at both extremes (values for $x \leq 7$ or $x \geq 14$).

A *Poisson process* is a simple mechanism that may govern the time instants at which occurrences are observed as time passes. The occurrences could be such things as fatal auto accidents, arrivals of customers at a service center, coverage claims received by an insurance company, emergency calls placed to a 911 number, and so on. In any one of these cases, we could in theory observe the times of the occurrences, and for any given period of time t, we could then count the number of occurrences, giving the observed value for a discrete counting variable. The probability distribution for this count would clearly be dependent on the way in which the occurrences were generated.

In a Poisson process with parameter λ, the occurrences are assumed to be independent and to happen at random at a constant rate λ. More exactly, the independence assumption means the following: If we take *any k* nonoverlapping periods of time and A_{i_k} is the event that exactly i_k occurrences will be observed in the kth period, then $A_{i_1}, A_{i_2}, \ldots, A_{i_k}$ are independent events for all possible values of $i_1, i_2, \ldots, i_k$, where $k = 2, 3, \ldots$. The "at random with constant rate λ" assumption means we can convert any fixed period of time (of length $t > 0$) into n nonoverlapping equal-length increments, each of length $\Delta t = t/n$, that, for sufficiently large values of n, are independent Bernoulli trials (i.e., each time period of length Δt will contain either zero or one occurrence). That is, if n is sufficiently large, it is impossible for two or more occurrences to be observed in the same increment of time of length Δt. Furthermore, the probability of one occurrence in each increment (a success) is $p = \lambda \, \Delta t = \lambda t/n$, where p is inversely proportional to n.

We assume occurrences (also called events) are generated by a Poisson process with parameter λ (the *rate* at which occurrences are observed per unit of time) and let X be the number of occurrences to be observed in the time interval $(0, t]$, where $t > 0$ is some fixed constant. From these assumptions, X is approximately binomial with parameters n and $p = \lambda t/n$; as $n \to \infty$, the probability law for X becomes Poisson with parameter $\mu = np = \lambda t$ from Theorem 3.5. That is, if occurrences are generated in accord with the Poisson process assumptions and if X is the number to be observed in some interval of fixed length t, then X has the Poisson distribution with $\mu = \lambda t$. This relationship is the reason we use the name Poisson for this discrete counting variable X. As we will see in Chapter 4, these same Poisson process assumptions also lead to easy derivations for the *times* at which occurrences (or events) are observed.

This way of thinking of the Poisson random variable X gives a natural decomposition of its parameter μ into two parts: the product of the rate λ at which occurrences are generated and the length of time t over which the process is observed. If we double the length of time t of observation, the parameter of the random count X also doubles. (However, this doubling does not generally lead to an observed count that is twice as big because of the randomness of the times of occurrence.)

Example 3.16

The occurrences of accidents are frequently assumed to behave like events generated by a Poisson process. Suppose that injury-causing accidents at a large industrial plant occur this way, at a rate of $\lambda = \frac{1}{2}$ per week, and let X represent the number of such accidents to occur in this plant during the first 4 weeks of the next year. Then X is a Poisson random variable with parameter $\mu = \frac{1}{2}(4) = 2$, and the probability function for X is

$$p_X(x) = \frac{2^x}{x!}e^{-2}, \qquad \text{for } x = 0, 1, 2, \ldots.$$

The probability of exactly two accidents occurring during this period is $p_X(z) = 2^2 e^{-2}/2! = .2707$, while the probability that there will be no such accidents during this period is $p_X(0) = (2^0/0!)e^{-2} = .1353$. The latter probability could be computed in several other ways.

Clearly there will be no accidents in the 4-week period if and only if there are no accidents in the first week, in the second week, and so forth. Furthermore, counts from nonoverlapping periods in a Poisson process are independent. Thus if Z is the number of accidents to occur in a 1-week period, Z is Poisson with parameter $\frac{1}{2}$, so the probability of zero accidents in any week is $p_Z(0) = e^{-1/2} = .6065$. Because of the independence of occurrences from week to week, then, the probability of four 1-week periods in a row (equivalent to saying there are no accidents in 4 weeks) must be $(.6065)^4 = .1353 = p_X(0)$, the same value derived in the previous paragraph.

Example 3.17

The earth occasionally passes through "meteor belts" in space, periods when shooting stars can be seen frequently. Suppose the earth is passing through one of these belts when you are outside after dark, and that visible shooting stars are occurring at a rate of 10 per hour (like events in a Poisson process). If you scan the sky, what length of time t should be sufficient for you to have probability .95 of sighting at least one shooting star? If you measure time in minutes, starting from time 0 when you start looking, and consider any fixed interval $(0, t]$, the number of visible shooting stars is Poisson with parameter $\mu = t/6$. The probability that you see no shooting stars in this time period is then $P(X = 0) = e^{-t/6}$, and the probability that you see at least one is $P(X \geq 1) = 1 - e^{-t/6}$. Setting $1 - e^{-t/6} = .95$ gives $t = -6 \ln .05 = 17.97$. Thus you have probability .95 of seeing one or more shooting stars if you plan on watching the sky for about 18 minutes.

Our Poisson process assumptions say the events being observed occur independently and at random through time. If we say X is Poisson with parameter μ (for some period $(0, t]$), and condition on the fact that $X = k$, say, where k is some positive integer, we should expect these k observed events to be uniformly distributed over $(0, t]$ in some sense. That this is the case is illustrated in the following example (referred to as *Poisson sampling*).

Example 3.18

Assume that particles from a (low-level) radioactive source are emitted at a rate of 1 per second, like events in a Poisson process. Suppose we have a Geiger counter that records these emissions and find that $k = 3$ emissions occurred in the first 3 seconds (in the period $(0,3]$). What is the conditional probability that exactly one particle was registered in each of the 1-second segments observed?

Subdivide $(0,3]$ into the three nonoverlapping 1-second intervals (from 0 to 1, from 1 to 2, from 2 to 3); let A_i be the event that exactly one event occurs in interval i, where $i = 1,2,3$. Then $P(A_i) = e^{-1}$, the probability of observing $x = 1$ for a Poisson random variable X with parameter $\mu = 1$. Since nonoverlapping intervals are independent, $P(A_1 \cap A_2 \cap A_3) = (e^{-1})^3$, the product of the three individual probabilities. Now let B be the event that exactly three events occur in $(0,3]$, the full 3-second period. Then

$$P(B) = \frac{3^3}{3!}e^{-3},$$

the probability of observing $x = 3$ for a Poisson random variable X with $\mu = 3$. It is clear that $(A_1 \cap A_2 \cap A_3) \subset B$, so

$$P(A_1 \cap A_2 \cap A_3 \,|\, B) = \frac{P(A_1 \cap A_2 \cap A_3)}{P(B)} = \frac{2}{9}.$$

This probability is exactly the same as the one we would get from the following "random" construct: Suppose we have three balls and three urns. We will take each ball in turn and randomly select one of the three urns to put it into. What is the probability that each urn gets exactly one ball? For this event to occur, the first ball can be put into any of the urns, the second then must be put into one of the empty urns, and the remaining ball must be placed into the single remaining empty urn. That is, the required probability is $(1)(\frac{2}{3})(\frac{1}{3}) = \frac{2}{9}$, the same as the conditional probability just evaluated for the Poisson process.

EXERCISES 3.3

1. Assume that automobiles pass a given point on a rural road at a rate of three per hour, like events in a Poisson process. Let X be the number of automobiles to pass this point in a 10-minute period, and let Y be the number of automobiles to pass this point in a 3-hour period.
   a. Evaluate $P(X \geq 2)$ and $P(X = 0)$.
   b. Evaluate $P(Y = y)$ for $y = 6, 7, \ldots, 12$.
2. Assume that fatal traffic accidents in the state of California over a 72-hour holiday weekend occur like events in a Poisson process with parameter $\lambda = \frac{1}{2}$ per hour. Let T be the number of such accidents to occur on the first day (24 hours) of the weekend. Evaluate $P(T = 12)$.
3. Telephone calls arrive at the toll-free 800 number of a major airline like events in a Poisson process with parameter $\lambda = 100$ per hour, between the hours of 9 AM and noon each day.

    a. What is the probability there will be no calls arriving during the first minute of this period?

    b. What is the probability of no calls arriving in each of the first 3 minutes of this period?

4. A used car salesperson assumes that her sales of automobiles occur like events in a Poisson process at a rate of $\lambda = 2$ per (5-day) week.

    a. Evaluate the probability that she makes no sales on a given day.

    b. What is the probability that she makes two sales on a given day?

    c. What is the probability that she will make at least six sales in a 4-week period?

5. Assume that solar flares of a given magnitude occur on the surface of the sun like events in a Poisson process at the rate of two per day. What proportion of days would then pass with no such flares occurring?

6. Emergency calls for help from small-boat owners are received at a Coast Guard station like events in a Poisson process at a rate of $\frac{1}{3}$ per day. Let C be the number of such calls this station receives in a (7-day) week. What is the most likely number of calls this station will receive in a week?

7. Assume that a Poisson process with parameter λ is observed and that X is the number of events to occur in the fixed interval $(0, t]$. Let Y_1 be the number of events to occur in the first quarter of this period [the interval $(0, t/4)]$, while Y_i is the number to occur in the ith quarter of this same period, where $i = 2, 3, 4$. Thus $X = \sum_{i=1}^{4} Y_i$.

    a. What is the probability law for Y_i?

    b. Show that $P(X = 0) = [P(Y_1 = 0)]^4$.

    c. Show that $P(X = 1) = \binom{4}{1} P(Y_1 = 1) [P(Y_1 = 0)]^3$.

8. Assume that rocks of a given size are scattered at random and independently on the surface of the moon like events in a Poisson process with parameter $\lambda = .2$ per square meter. An unmanned space shuttle will land on the surface of the moon and send out a vehicle to systematically investigate 9 square meters of the surface. What is the probability that this vehicle will find at least one of these rocks in its search?

9. Molecules of a rare gas occur at a rate of 5 per cubic foot of air at sea level, like events in a Poisson process. How many cubic feet of air must be analyzed to have probability .95 of finding at least one molecule of this rare gas?

10. Let X be Poisson with parameter $\mu = 3$, and compare $P(|X - \mu| \geq 2\sigma_X)$ with the Chebyshev bound for this same event occurring.

11. It might seem reasonable to make the Poisson process assumptions in terms of fatal automobile *accidents* over a given period of time, for a given geographic area. This is not necessarily true for the number of automobile accident *fatalities*. Why?

12. Assume that automobiles pass a given point on a rural road at a rate of 3 per hour, like events in a Poisson process. What is the probability, if we start looking for automobiles passing this point at time $t = 0$, that the first automobile to pass comes in the 15th minute? (*Hint:* This will occur only if no automobiles come past in the interval $(0, 14]$ and then at least

one comes in the interval (14,15]; recall that nonoverlapping intervals are independent.)

13. Assume that injury-causing accidents at a large industrial plant occur at a rate of $\lambda = 2$ per 5-day week, like events in a Poisson process.
    a. Evaluate the probability of there being (exactly) two accidents in a given 5-day week.
    b. Granted that exactly two accidents occurred in a given 5-day week, evaluate the (conditional) probability that one occurred on Monday and one occurred on Tuesday.

14. A pet store owner claims that a "talking" parrot for sale audibly utters the particular phrase "Come here!" four times per hour. Assume that these utterances occur like events in a Poisson process (independently and randomly at the rate $\lambda = 4$ per hour). You will observe this bird for t minutes; let X be the number of times the parrot utters "Come here!" in this period. How large must t be to have $P(X \geq 1) = .9$?

15. Assume that each time you drive your car the probability you are stopped by a police officer, for some reason, is .001; each such time you drive generates a Bernoulli trial, and these trials are independent.
    a. Let X be the number of times you are stopped by a police officer in $n = 1000$ such times in your car. Evaluate $P(X = x)$ for $x = 0, 1, 2$.
    b. Let Y be the appropriate Poisson variable to approximate the distribution of X in part a, and evaluate $P(Y = x)$ for $x = 0, 1, 2$.

16. *For the mathematically inclined*   If X is Poisson with parameter μ, show that the probability that X is an even number is given by $\frac{1}{2}(1 + e^{-2\mu})$, where 0 is considered even. (*Hint:* Consider the power-series expansion of $\cosh \mu = (e^{\mu} + e^{-\mu})/2$.)

17. If X is binomial with $n = 40, p = .1$, and Y is binomial with $n = 100, p = .04$, then the probability laws of each of these random variables could be approximated by a Poisson random variable Z with $\mu = np = 4$. Evaluate $P(X = 3)$ and $P(X = 4)$; compare these numbers with the probabilities that Y and Z equal these same values.

18. Show that the variance for the Poisson random variable with parameter μ is $\sigma_X^2 = \mu$.

19. An unloading point for oil tankers has three berths. Assume that tankers arrive at this unloading point like events in a Poisson process with parameter $\lambda = 2$ per day. Any tanker to arrive at this unloading point is serviced (unloaded) at this point as long as a berth is available. (If no berth is available, the tanker proceeds to the next unloading point.) Each tanker serviced at this unloading point is finished the same day it arrives, so there are always three free berths at the start of each day.
    a. What is the probability that this unloading point can service all tankers to arrive on a given day?
    b. How many berths should this unloading point provide, if it wants to have roughly probability .95 of being able to service all tankers to arrive on a given day?

20. Events are generated in accord with a Poisson process with parameter λ. Let X be the number of events to occur in $(0, t]$, while Y is the number

of events to occur in $(0, 2t]$. It was mentioned in the text that doubling the length of the period of observation is not the same as doubling the observed count.

a. What are the probability distributions for X and Y?

b. Evaluate $P(2X = 2)$ and $P(Y = 2)$.

c. What is the probability that we will find $Y = 2X$?

3.4 | The Hypergeometric Distribution

Many common and important practical problems can be phrased in terms of sampling at random from a given finite population, where each population member falls into one of two classes. Table 3.1 lists some examples of populations categorized into two classes. These are only a few of the many applications of sampling without replacement from a population of items, a topic which naturally leads to the hypergeometric probability law, the subject for this section.

As a prototype for these types of situation, let us consider sampling balls from an urn. Suppose we have an urn that contains m balls (the population contains m members), of which r are red (the class A population members) and the remaining $m - r$ are not red (those not in class A). We select $n \leq m$ of the balls from the urn, at random and without replacement; that is, we can conceive of reaching into the urn one time and selecting n balls, which then constitute the sample. Or, equivalently, we may select one ball, then a second ball (not having replaced the first), then a third ball, and so on, until we have selected a total of n balls (without replacing any already drawn). Clearly, it is necessary that $n \leq m$ with this procedure, since we could not take more balls from the urn than are present initially. Let X be the number of red balls that occur in the sample. What is the probability law for X?

Since X counts the number of red balls in the sample, the range for X is a set of nonnegative integers; because this is a discrete set, X is a discrete random variable. For the moment, let us simply use $R_X = \{0, 1, 2, \ldots, n\}$ as

Table 3.1

Populations with Two Types of Elements

Population	Type A	Not Type A
Voters	Prefers candidate 1	Prefers someone else
Cats in a given area	Has leukemia	Doesn't have leukemia
Consumers	Prefers cereal 1	Does not prefer cereal 1
Television viewers	Watches program A	Doesn't watch program A
Manufactured items	Defective	Not defective
Trees in a forest	Oak	Other genus

the range for X. (It may be that not all of these values could occur. Consider the case in which the urn has $m = 100$ balls, $r = 95$ of which are red, and we select a sample of $n = 10$; or consider the case where $m = 100$, $r = 5$, and $n = 10$. We shall discuss this point later.) There are $\binom{m}{n}$ different selections of n balls that could be drawn from the urn; if the drawing is done "at random," each of these then has probability $1 / \binom{m}{n}$ of being selected.

It may aid the reasoning process to assume that the balls in the urn are numbered from 1 to m, with the red balls bearing the numbers 1 to r; any selection of n balls from the urn can then be thought of as being a subset, of size n, from the integers $\{1, 2, \ldots, m\}$. Recall that the number of subsets, each of size n, that contain *exactly* x of the numbers between 1 and r (and thus $n - x$ numbers that are not in this range) is given by the product $\binom{r}{x}\binom{m-r}{n-x}$. The random variable X will equal x if and only if *any* one of these subsets with exactly x red balls occurs when the selection is made. Thus, the probability function for X is given by the following theorem:

THEOREM 3.6 If an urn contains m balls, of which r are red, and X is the number of red balls to occur in a random sample of n balls removed from the urn without replacement, the probability function for X is

$$p_X(x) = \begin{cases} \dfrac{\dbinom{r}{x}\dbinom{m-r}{n-x}}{\dbinom{m}{n}}, & \text{for } x = 0, 1, 2, \ldots, n, \\ \\ 0, & \text{otherwise.} \end{cases} \tag{1}$$

X is called a hypergeometric random variable with parameters m, n, r, and $p_X(x)$ is the hypergeometric probability function.

The set $\{1, 2, \ldots, m\}$ has $\binom{m}{n}$ subsets of size n, for $n = 0, 1, \ldots, m$; the number of these subsets that contain exactly x of the integers from 1 to r, inclusive, as well as $n - x$ integers not in this range, is $\binom{r}{x}\binom{m-r}{n-x}$. If we sum this product over the possible values for x, we must get the total number of subsets of size n; that is,

$$\sum_{x=0}^{n} \binom{r}{x}\binom{m-r}{n-x} = \binom{m}{n},$$

which, dividing both sides of this equation by $\binom{m}{n}$, shows that the sum of the hypergeometric probability function values totals 1.

Example 3.19

A statistics class contains 25 students, of whom nine are female. Four of these students will be selected to work on a class project. If the four are selected at random, what is the probability that all four are female? all four are male? Here we have a "population" of $m = 25$ members, of whom $r = 9$ are class A (female). If a random sample of $n = 4$ population members is selected and we let X be the number of females in the sample, then X is hypergeometric with parameters $m = 25, r = 9$, and $n = 4$. The probability that those selected are all female is

$$P(X = 4) = \frac{\binom{9}{4}\binom{16}{0}}{\binom{25}{4}} = .00996,$$

while the probability that they are all male is

$$P(X = 0) = \frac{\binom{9}{0}\binom{16}{4}}{\binom{25}{4}} = .1439.$$

If we let Y be the number of males in this same sample of 4, how is Y related to X? What is the probability distribution for Y?

The hypergeometric probability distribution has three integer parameters m, n, and r. Figure 3.7 plots the probability functions for four different sets of values of m, n, and r. The left-most graph is a plot of the probability function $p_X(x)$ for $m = 20$ balls in the urn, where $n = 5$ is the size of the sample; the unfilled bars are for $r = 4$ red balls in the urn, while the solid bars give the distribution for X with $r = 10$ red balls in the urn. (The *proportions* of red balls are .2 and .5.) The right-most graph in Fig. 3.7 is a plot of $p_X(x)$ for $m = 100$ balls in the urn, where the sample size is $n = 10$; the *proportions* of red balls are the same as for the left-hand graph, namely the unfilled bars are for $r = 20$ (a proportion of .2) while the filled bars correspond to $r = 50$.

Notice that, if the proportion of red balls is .5, the probability function for X is symmetric for both graphs. If the proportion of red balls is .2, most of the probability mass is shifted toward 0. This is true in general; if the proportion of red balls is .5, regardless of m or n, the probability function for X will be symmetric. A proportion of red balls less than .5 (like the

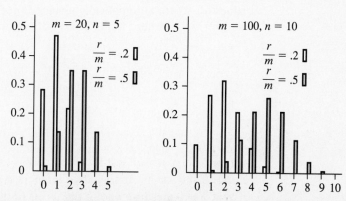

Figure 3.7 Hypergeometric probability laws

graph for $r/m = .2$) places more probability on the lower end of R_X, while a proportion of red balls greater than $.5$ places more probability on the upper end of R_X. As the sample size n increases, the probability function is spread over a wider range (since $R_X = \{0, 1, 2, \ldots, n\}$), like the discrete uniform, which in turn implies that the individual probability masses at particular x-values get smaller. (We can think of the whole probability function as being "pushed down" to accommodate the wider range, since the sum of all the probabilities must still equal 1.)

It is straightforward to see that the inequality $p_X(x)/p_X(x + 1) < 1$ reduces to

$$x + 1 < \frac{(r + 1)(n + 1)}{m + 2};$$

the probability function is increasing as long as this inequality holds, since $p_X(x)$ is then smaller than $p_X(x + 1)$. If $(r + 1)(n + 1)/(m + 2)$ is an integer, then the maximum value for $p_X(x)$ occurs at the two values

$$x = -1 + \frac{(r + 1)(n + 1)}{m + 2} \quad \text{and} \quad x = \frac{(r + 1)(n + 1)}{m + 2}.$$

If this quantity is not an integer, the unique maximum occurs at $\lfloor (r + 1)(n + 1)/(m + 2) \rfloor$, the largest integer in the ratio.

Example 3.20

Suppose your local newspaper publishes 12 letters to the editor each weekday, giving $m = 60$ published letters in 1 week. If $r = 25$ letters are concerned with actions taken by the city council, and you randomly select 10 of the letters published in 1 week, the most likely number of those selected that deal with council actions is $\lfloor 26(11)/62 = 4$. The values of the probability function for X, the number of the letters selected that deal with council actions, are 0.2051, 0.2723, and 0.2288, respectively, for $x = 3, 4,$ and 5.

Some of the values in $\{0, 1, 2, \ldots, n\}$ may not be possible observed values for X, as already mentioned; if we redefine our meaning of the combinatorial coefficient $\binom{u}{v}$, those elements in this range that cannot serve as observed values for X are automatically assigned probability 0. Recall from Chapter 1 that, for $v \leq u$ where u and v are both nonnegative integers,

$$\binom{u}{v} = \frac{u!}{v!(u - v)!}$$
$$= \frac{u(u - 1)(u - 2) \cdots (u - v + 1)(u - v)!}{v!(u - v)!}$$
$$= \frac{u(u - 1)(u - 2) \cdots (u - v + 1)}{v!}.$$

We could just as well use this last ratio as the definition for $\binom{u}{v}$, because it is equal to $\binom{u}{v}$ for all nonnegative integers $v \leq u$. When $v > u$, this definition gives $\binom{u}{v} = 0$, because one of the terms in the product in the numerator then equals 0. We shall employ this extended definition for $\binom{u}{v}$ from this point on.

The numerator for the hypergeometric probability law is given by the product $\binom{r}{x}\binom{m-r}{n-x}$, which, with this extended definition, will equal 0 if either $x > r$ or $n - x > m - r$. This is exactly what is desired; we want to assign probability 0 to those values in $\{0, 1, 2, \ldots, n\}$ that correspond either to drawing more red balls than possible ($x > r$) or to drawing more nonred balls than is possible ($n - x > m - r$). Thus we can always use $R_X = \{0, 1, 2, \ldots, n\}$ as the range for the hypergeometric random variable X with parameters m, n, r, no matter what their values may be. Only those elements of R_X that can in fact occur will be given positive probability with this extended definition for combinatorial coefficients.

Example 3.21

Acceptance sampling is frequently employed in deciding whether or not a large lot of items is of acceptable quality. Assume that a manufacturer of electronic calculators purchases the central computing chip for his product in lots of size 100; assume further that he has decided that a defective chip rate of 5% (5 out of 100) is tolerable. The complete testing of such a chip is time-consuming and expensive, so he employs a sampling plan to decide whether or not to accept a lot of 100 chips when it arrives; that is, he will select a sample (or subset) of the chips in the lot. Those selected are thoroughly tested, and the decision to accept or reject the whole lot rests totally on the number of bad chips found in the sample. Assume arbitrarily that the plan he employs calls for a random sample of eight chips to be selected from the lot of 100; each of these eight is thoroughly tested, and the number of these eight that are defective is then noted.

Now let X be the number of defective chips found in the sample, and assume that his acceptance sampling plan calls for the lot to be accepted if $X \leq 1$. That is, he will accept the lot of 100 if he finds no bad chips in the sample of 8 *or* if he finds exactly one bad chip in the sample of 8; if two or more of the eight chips in the sample are bad, he will reject the whole lot and return it to the supplier. Granted the sample is selected at random from the lot, X is then a hypergeometric random variable with parameters $m = 100$, $n = 8$, and r, where the defective chips in the lot play the role of the red balls in the previous discussion. Of course, r is unknown or there would be no need to employ sampling and testing of the chips in the sample. The probability function for X is

$$p_X(x) = \begin{cases} \dfrac{\binom{r}{x}\binom{100-r}{8-x}}{\binom{100}{8}}, & \text{for } x = 0, 1, 2, \ldots, 8, \\ 0, & \text{otherwise.} \end{cases}$$

The probability that the lot is accepted, with this plan, is then

$$P(X \leq 1) = p_X(0) + p_X(1)$$
$$= \frac{\binom{r}{0}\binom{100-r}{8} + \binom{r}{1}\binom{100-r}{7}}{\binom{100}{8}},$$

which is easily evaluated for any given value for r. If in fact exactly 5% of the chips in the lot are defective, then $r = 5$, and you can verify that $P(X \leq 1) = .950$, the lot is accepted with probability .950 using this rule. If 10 of the chips in the lot are defective ($r = 10$), the lot is accepted with probability .818, and if $r = 20$ the lot is accepted with probability .497. By varying the size n of the sample employed and the number of defectives tolerated in the sample before deciding to reject the lot, one can choose from a wide variety of acceptance sampling plans.

Example 3.22

Assume that the stock in a family-owned corporation, in the second generation of the family, is split evenly among 15 persons (siblings and cousins). Also assume that a bare majority of these stockholders (eight of the 15) favor a public stock issue for the corporation. If we poll six of these stockholders at random, what is the probability that a majority *in our poll* will favor the public stock issue? Since a majority in the group favors the stock issue, it seems intuitive that the probability a majority in the poll will favor it should be at least $\frac{1}{2}$; this is not the case for these values of m, n, and r.

Again the hypergeometric probability law is appropriate for answering this question. Here we have $m = 15$, $n = 6$, and $r = 8$, and if we let X represent the number in the sample who favor the public issue, the probability function for X is

$$p_X(x) = \begin{cases} \dfrac{\dbinom{8}{x}\dbinom{7}{6-x}}{\dbinom{15}{6}}, & \text{for } x = 0, 1, 2, \ldots, 6, \\ 0, & \text{otherwise.} \end{cases}$$

The value for x that maximizes $p_X(x)$ is $\lfloor (9)(7)/(17) = 3$, which is not surprising. A majority in the sample will favor the issue if the observed value for X is 4 or more. Thus the probability that a majority in the sample favor the issue is

$$P(4 \leq X \leq 6) = \frac{\dbinom{8}{4}\dbinom{7}{2} + \dbinom{8}{5}\dbinom{7}{1} + \dbinom{8}{6}\dbinom{7}{0}}{\dbinom{15}{6}} = .378.$$

The probability that exactly three persons in the sample favor the stock issue is $p_X(3) = \binom{8}{3}\binom{7}{3}/\binom{15}{6} = .392$, larger than the probability of getting a majority in the sample.

The hypergeometric probability law provides a prototype for any situation in which a random sample is selected from a finite population, without replacement, and each element of the population either does or does not have a specific attribute (red ball, defective item, favors item, likes product, and so forth). The random variable counts the number in the sample having the attribute, and the probability function gives the probability of occurrence for the possible observed values of this variable.

The mean value for the hypergeometric random variable with parameters m, n, r is

$$E[X] = \sum_{x=0}^{n} x \frac{\binom{r}{x}\binom{m-r}{n-x}}{\binom{m}{n}}.$$

Since $x\binom{r}{x} = r\binom{r-1}{x-1}$, and the term with $x = 0$ contributes 0 to the sum, we have

$$E[X] = \frac{r}{\binom{m}{n}} \sum_{x=1}^{n} \binom{r-1}{x-1}\binom{m-r}{n-x}$$

$$= \frac{r}{\binom{m}{n}} \binom{m-1}{n-1} = n\frac{r}{m}. \tag{2}$$

The value of the sum in Eq. (2) is $\binom{m-1}{n-1}$ because it gives the total number of subsets of size $n-1$ that can be constructed from a set with $m-1$ elements (by summing the number of subsets of this size that have $j = x - 1$ of the first $r - 1$ elements of the set, over the possible x-values). In a very similar manner, we can show that the variance for X is

$$\sigma_X^2 = n\frac{r}{m}\left(1 - \frac{r}{m}\right)\frac{m-n}{m-1}.$$

Suppose an urn contains m balls of which r are red. Instead of selecting balls from the urn without replacement, now assume n balls are selected *with* replacement. By sampling with replacement we mean that the first ball is removed from the urn, its color is noted, and then it is replaced in the urn *before* the next ball is drawn; this procedure is repeated for each draw. If the sampling is done in this way, the composition of the urn does not change from draw to draw. When each draw is made, there are m balls in the urn, of which r are red; granted each ball is selected at random, the individual draws are independent trials. Letting the occurrence of a red ball be a success, the probability of success is $p = r/m$ for each draw. The number of red balls to occur in a sample of size n is then a binomial random variable with parameters n and $p = r/m$.

When a sample of size n is drawn *without* replacement, the number of red balls to occur is hypergeometric with parameters m, n, and r. If m is quite large compared with n (i.e., the number of balls in the urn is much bigger than the sample size) and if r/m is not too close to 0 or 1, it should not matter much whether the sample is selected with or without replacement. After all, if there are $m = 100$ balls in the urn, of which 20 are red, and we select a sample of $n = 5$ without replacement, the fact that the number of balls available is decreasing to 99, 98, etc. should not have a great effect on the probability of drawing (exactly) one red ball in the sample of five. That is, we should expect $P(X = 1)$, say, computed from either the hypergeometric or the binomial probability function, to give about the same value.

Figure 3.8 presents a graphic comparison of the probability functions for the binomial and hypergeometric probability laws. Two separate graphs are given there, each presenting a comparison of a binomial and a hypergeometric probability function. In each plot the hypergeometric probability law has $m = 100$, and $r = 20$, so the *proportion* of red balls is $p = .2$, the same value of p as in the binomial random variable.

The left-most graph compares the two distributions with a sample size of $n = 5$, and the right-most graph compares them with a sample size of $n = 10$; in both cases the two distributions are very similar. They are "closer," over all x, for $n = 5$ than they are for $n = 10$. It matters less whether balls are replaced with the smaller sample size. (If $n = 1$, it doesn't matter at all.) In any case, for the given values of m and n, the two distributions are most similar with $p = r/m = .5$, which corresponds to the situation with half the balls being red; as p moves toward either 0 or 1, the differences between the two distributions increase for any fixed values of m and n. If n and $p = r/m$ are held constant, then for any value of x, the hypergeometric probability function actually converges to the binomial probability function for the same x, as m increases. (Note that r must increase at the same time to hold $p = r/m$ constant.) Thus if you must rely on tables to evaluate probability values, the more readily available binomial probability function values can be used to approximate the hypergeometric values if m is relatively large. Many texts suggest this approximation should be quite good as long as n is no larger than $.2r$ or $.2(m - r)$, that is, as long as $n \leq .2\min(r, m - r)$. This rule suggests that the approximation should be good for $n \leq 4$, with $m = 100$ and $r = 20$, as pictured in Fig. 3.8.

The similarity between the binomial and hypergeometric probability laws is also apparent in comparing their means and variances. If we sample at random with replacement from the urn with m balls, r of which are red, and use a sample of n, then the number of red balls in the sample is binomial with

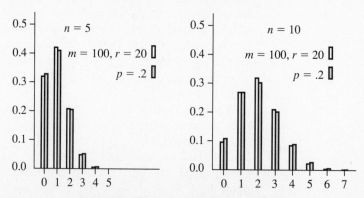

Figure 3.8 Binomial versus hypergeometric

parameters n and $p = r/m$. The expected value for this random variable is nr/m, the same as the hypergeometric mean. The variance for this binomial random variable is

$$npq = n \frac{r}{m} \left(1 - \frac{r}{m}\right),$$

while the hypergeometric variance is

$$n \frac{r}{m} \left(1 - \frac{r}{m}\right) \left(\frac{m - n}{m - 1}\right),$$

smaller by the factor $(m - n)/(m - 1)$, which is called the *finite population correction factor*. If $n = m$, this factor is 0, so the hypergeometric variance is in fact 0; this of course must be the case since using $n = m$ means we have removed all the balls in the urn and the number of red balls we get must be r every time. The hypergeometric probability law is degenerate with $n = m$.

 **EXERCISES 3.4**

1. A regular 52-card deck of playing cards contains cards of 13 different denominations $(A, 2, 3 \ldots, 10, J, Q, K)$ in each of four suits: hearts $(\heartsuit)$, diamonds $(\diamondsuit)$, spades $(\spadesuit)$, and clubs $(\clubsuit)$. The first two suits are red, the last two are black.
   a. Assume that five cards are selected at random from this deck, and let X be the number of red cards to occur. What is the probability function for X?
   b. Assume that five cards are selected at random from this deck, and let Y be the number of hearts to occur. What is the probability function for Y?
   c. Assume that 13 cards are selected at random from this deck, and let Z be the number of red cards to occur. What is the probability function for Z?
   d. Assume that 13 cards are selected at random from this deck, and let U be the number of diamonds to occur. What is the probability function for U?

2. If five cards are selected from the 52-card deck mentioned in Exercise 1, evaluate the probability that the majority of them are red.

3. You are dealt a five-card hand from a regular 52-card deck (as defined in Exercise 1). What is the probability that you get
   a. No aces?   b. Three aces?

4. A lot of 50 items is received, of which seven are defective.
   a. The lot will be accepted if and only if there are no defectives found in a random sample of four selected for inspection. Evaluate the probability that the lot is accepted.
   b. Now suppose the lot is accepted if there is no more than one defective found in the sample of four, and evaluate the probability that the lot is accepted.

5. A lot of 50 items is received, of which seven are defective. A sample of n items is selected from the lot to be inspected. The lot will be accepted

only if there are no defectives found in the sample. How large must n be so that this lot will be accepted with probability no greater than .1?

6. If there are 10 defective items in a lot of 50, and the lot is accepted only if there are no defectives in a sample of n, how large must n be so that the probability is no more than .1 that the lot will be accepted?

7. A hat contains 10 slips of paper, nine of which are blank; the remaining one contains a message. If you remove slips from the hat, one at a time without replacement, and let X be the number of slips you have looked at by the time you get the message, what is the probability law for X?

8. Show that the variance for the hypergeometric random variable X with parameters m, n, r is

$$\sigma_X^2 = n\frac{r}{m}\left(1 - \frac{r}{m}\right)\frac{m-n}{m-1}.$$

9. The lotto games adopted by many states for legalized gambling are variations of the game of Keno, long a favorite at gambling casinos. In Keno the house selects 20 numbers at random from the first 100 integers (once per hour or more frequently, depending on the crowd available for betting). You as a bettor are allowed to select integers from 1 to 100 as well, and the actual number of integers you select—from 5 to 20—is also your choice. You decide to try to match five numbers from the 100 to be selected at a given drawing and let X be the number of matches that you score. The probability function for X can be thought of in two distinct ways:

a. The 20 numbers the house selects are marked red, say, with the remainder white, and you pick five from these 100 colored numbers.

b. The five numbers you select are marked blue, say, with the remainder white, and the house picks 20 from these 100 colored numbers.

Express the probability function for X in the two ways suggested by these statements, and show that the two are identical.

10. By writing out the combinatorial terms involved and rearranging the factorials involved, show that

$$\binom{r}{x}\binom{m-r}{n-x}\binom{m}{r} = \binom{n}{x}\binom{m-n}{r-x}\binom{m}{n},$$

as long as $m \geq \max(r, n)$. This says that $P(X = x) = P(Y = x)$ for all x, where X is hypergeometric with parameters m, n, r, and Y is hypergeometric with parameters m, r, n. In one case a sample of r balls, say, is selected from an urn with n red balls, and in the other a sample of n balls is selected from an urn with r red balls. This is a general statement of the equality of the two ways of thinking of the number of matches scored in the Keno game in Exercise 9.

11. The owner of a small firm is forced, by a business slowdown, to lay off three of her 20 employees. Eleven of the employees are female. If the three to be laid off are chosen at random, what is the probability they are all male? that most of those laid off are male?

12. The owner of a small firm is forced, by a business slowdown, to lay off nine of her 20 employees. Seventeen of the employees are female. If the nine to be laid off are chosen at random, what is the probability that three are male? that two or three of those laid off are male?

13. Box 1 contains 100 balls, of which 40 are red, while box 2 is empty. Fifty balls are selected at random from box 1 and placed in box 2; three balls are then selected at random from box 2. Let X be the number of red balls to occur in this sample of size 3, and evaluate $P(X = 0)$; compare this value with the probability of drawing no red balls in a sample of 3 from the initial contents of box 1. (*Hint*: Use conditional probability.)

14. The Internal Revenue Service "randomly" selects income tax returns for audit. Suppose 1% of the tax forms are audited in a given tax year. Also assume that 10% of the forms submitted have one or more errors, which would increase the amount of tax due. Let X be the number of those audited that include errors that would increase the amount of tax due. What is the approximate probability law for X? Approximate the mean and variance of X.

3.5 | Summary

This chapter has examined several discrete probability laws. The discrete uniform probability law describes an integer-valued random variable whose range is $R_X = \{1, 2, \ldots, n\}$ and which has equal probability at all values $x \in R_X$.

DISCRETE UNIFORM RANDOM VARIABLE

Probability function: $p_X(x) = \dfrac{1}{n}$

Range: $R_X = \{1, 2, \ldots n\}$

Parameter: $n = 1, 2, 3, \ldots$

Mean: $\dfrac{n+1}{2}$

Variance: $\dfrac{n^2 - 1}{12}$

Many random variables can be defined on a sequence of repeated, independent Bernoulli trials with parameter p. The Bernoulli random variable, with range $R_X = \{0, 1\}$, counts the number of successes in a single trial.

BERNOULLI RANDOM VARIABLE

Probability function: $p_X(x) = p^x(1-p)^{1-x}$

Range: $R_X = \{0,1\}$

Parameter: $0 \le p \le 1$

Mean: p

Variance: $p(1-p)$

In a fixed number of trials n, the binomial random variable X counts the random number of successes to occur, and has range $R_X = \{0, 1, 2, \ldots, n\}$.

BINOMIAL RANDOM VARIABLE

Probability function: $p_X(x) = \binom{n}{x} p^x(1-p)^{n-x}$

Range: $R_X = \{0, 1, \ldots, n\}$

Parameters: $n = 1, 2, 3, \ldots, \ 0 \le p \le 1$

Mean: np

Variance: $np(1-p)$

The geometric random variable N counts the (random) number of Bernoulli trials necessary to observe the first success; its range is $R_N = \{1, 2, 3, \ldots\}$.

GEOMETRIC RANDOM VARIABLE

Probability function: $p_N(n) = p(1-p)^{n-1}$

Range: $R_N = \{1, 2, \ldots\}$

Parameter: $0 < p \le 1$

Mean: $\dfrac{1}{p}$

Variance: $\dfrac{1-p}{p^2}$

The negative binomial random variable N_r counts the (random) number of trials to the rth success; its range is $R_X = \{r, r+1, r+2, \ldots\}$.

NEGATIVE BINOMIAL RANDOM VARIABLE

Probability function: $p_{N_r}(n) = \binom{n-1}{r-1} p^r (1-p)^{n-r}$

Range: $R_{N_r} = \{r, r+1, \ldots\}$

Parameters: $r = 1, 2, 3, \ldots,\ 0 < p \leq 1$

Mean: $\dfrac{r}{p}$

Variance: $\dfrac{r(1-p)}{p^2}$

The hypergeometric random variable X occurs when sampling without replacement from a finite population; it counts the number of red balls to occur in a random sample of size n from an urn with m balls, of which r are red. Its range is $R_X = \{0, 1, 2, \ldots, n\}$.

HYPERGEOMETRIC RANDOM VARIABLE

Probability function: $p_X(x) = \dfrac{\binom{r}{x}\binom{m-r}{n-x}}{\binom{m}{n}}$

Range: $R_X = \{0, 1, \ldots n\}$

Parameters: $m = 1, 2, 3, \ldots, \quad r = 0, 1, \ldots, m$

$n = 1, 2, \ldots, m$

Mean: $n\dfrac{r}{m}$

Variance: $n\dfrac{r}{m}\left(1 - \dfrac{r}{m}\right)\left(\dfrac{m-n}{m-1}\right)$

Events that occur independently and at random through time, at a constant rate, are frequently assumed to follow the Poisson process assumptions:

POISSON PROCESS ASSUMPTIONS

1. Any sufficiently small interval of time can be modeled as a Bernoulli trial.
2. The probability of an event occurring in such a small interval is proportional to the length of the interval.
3. The numbers of events to occur in nonoverlapping time segments are independent.

If X counts the number of events to occur in an interval of length t, and these events are generated in a manner satisfying the Poisson process assumptions with rate parameter λ, then X is a Poisson random variable with parameter $\mu = \lambda t$.

POISSON RANDOM VARIABLE

Probability function: $p_X(x) = \dfrac{\mu^x}{x!}e^{-\mu}$

Range: $R_X = \{0, 1, 2, \dots\}$

Parameters: $\mu > 0$

Mean: μ

Variance: μ

EXERCISES 3.5

1. Assume that orders for a particular part arrive at an inventory control point at a rate of two per working day, like events in a Poisson process; let Y be the number of orders to arrive during a (5-day) week. What is the mean for Y? What is the probability of eight or fewer orders arriving in a given week?

2. Each call for assistance received by a municipal fire department either is or is not legitimate. Assume that past records indicate that two calls out of 100 received by this department are not legitimate; assume then that each call received has probability .02 of not being legitimate, and that these occurrences are independent. Also assume that this department receives 50 calls for assistance each 24-hour day. What is the probability of getting at least one illegitimate call for assistance in a 2-day period? in a 7-day period?

3. Requests for charter reservations for a sportfishing boat occur, for a given date, like events in a Poisson process with parameter λ. Granted that the probability of receiving no requests for this date is .25, what is the probability of two or more requests (for this same date)?

4. A machine used to shrink-wrap textbooks has probability .001 of producing a defective wrapping (wrinkled or torn, and so forth), and such defects occur independently. Let X be the number of books wrapped each workday until the first defective occurs. What is the probability law for X (idealizing the situation so the number of books wrapped is unbounded)? Evaluate the mean for X and $P(X > 1500)$.

5. For the situation described in Exercise 4, let Y be the number of books wrapped until the third defective occurs. What is the probability law for Y? Evaluate the mean and standard deviation of Y.

6. A box contains 400 identical-appearing chips, 15 of which are defective. A sample of 20 of these is selected, without replacement, for testing; let X be the number that are found defective in the test. What is the probability law for X? Evaluate μ_X and σ_X.

7. Suppose that the sampling in Exercise 6 is done with replacement and that

Y is the number of defective chips in the sample. What is the probability law for Y? Evaluate μ_Y and σ_Y.

8. The audit of a large firm's accounts can be arduous and time-consuming; frequently audits are based on sampling from the accounts rather than a complete inspection of all of them. Suppose this sampling approach is taken for the accounts payable to a given firm, 60% of which have a positive amount due. If the firm has a total of 10,000 accounts payable and a sample of 200 of these is selected at random without replacement, what is the probability law for Y, the number in the sample with a positive amount due? What is the expected value for Y?

9. Assume that the last four digits in telephone numbers, as listed in a local directory, occur in a random order and that all possibilities occur equally frequently. You glance at the listed numbers, starting with the first one listed on a given page, and count until you find the first number whose last two digits are both 0. What is the expected count at which you find the first such number? At which you find the second such number?

10. If the expected value for a Poisson random variable X is 3, evaluate $P(X = 1)$.

11. The number of independent Bernoulli trials performed until the first success occurs is X. If $E[X] = 20$, evaluate $P(X > 100)$.

CHAPTER 4

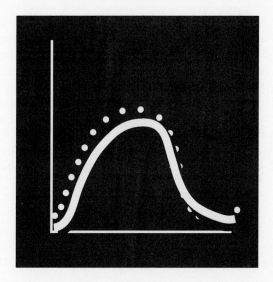

CONTINUOUS DISTRIBUTIONS
AND TRANSFORMATIONS

In Chapter 3, we have seen a number of standard discrete probability laws and some assumed circumstances that led to their use. The *cdf* for a discrete random variable is a step function, which jumps in value at each of the possible observed values; the *cdf* for a continuous random variable is a continuous function of its argument t. The *pdf* for a continuous random variable is given by the derivative of its *cdf*. The *pdf* plays essentially the same role for continuous random variables as the probability function does for discrete random variables; the probability that the observed value for a continuous random variable lies in any interval can be found by evaluating the difference in the values of the *cdf* at the endpoints of the interval, or by integrating the *pdf* over the interval.

The Poisson process assumptions are useful in modeling the occurrences of events in a continuum like time; granted the assumptions of a Poisson process, the count of the number of events to occur in a fixed interval must follow the Poisson probability law. We shall find in Section 4.1 that these assumptions allow a straightforward derivation of the continuous probability laws (exponential and Erlang) for the times of occurrence of events generated by a Poisson process; we shall also discuss the gamma distribution in this section.

One of the most frequently used probability measures in practical applications is the normal probability distribution, discussed in Section 4.2. Many different distributions converge to the normal under certain limiting operations; this behavior allows us to approximate many such distributions by the normal, as will be illustrated.

Section 4.3 introduces the beta probability law, and illustrates how it occurs through modeling uniform random variables. Finally, Section 4.4 discusses transformations of one-dimensional probability laws, including the probability integral transform for both the continuous and discrete cases. A number of additional continuous distributions will also be derived in this section.

4.1 | The Exponential, Erlang, and Gamma Distributions

A continuous random variable has a continuous range; modeling times, weights, distances, or other variables that can vary continuously is most easily accomplished with continuous distributions. Recall that in a Poisson process *events* (e.g., requests for emergency service, telephone calls, or injury-causing accidents) occur at particular time instants, independently and at random at a constant rate λ per unit of time. If a fixed interval of time of length t is observed, and $X_{\lambda t}$ counts the *number* of events to occur in this interval, we have seen that $X_{\lambda t}$ has the Poisson distribution (or follows the Poisson probability law) with parameter $\mu = \lambda t$ (we are appending the value of the Poisson parameter as a subscript to clarify some of the reasoning employed). $X_{\lambda t}$ provides a count and is a discrete variable. Assuming a Poisson process is generating events, there are other random variables whose probability laws may also be deduced. For example, if we begin observing the process at an arbitrary point in time, labeled $t = 0$ for convenience, the probability law for the *time of occurrence* of, say, the first succeeding event is easy to find; this random variable is continuous, granted we want to conceive of time as being "infinitely divisible." Indeed, the probability law for the *time of occurrence* of the rth event, where $r > 1$, is also easily found.

Perhaps it is best to think of a particular example. Suppose we were able to observe the arrivals of telephone calls to a police station, in particular calls requesting the service of a police car. Also, let us assume that these calls arrive at a rate of 2 per hour according to a Poisson process. We shall label $t = 0$ the time at which we begin observing these calls and let T_1 be the time (in hours) of occurrence of the first call. What is the probability law for T_1? Granted that time passes continuously, the observed value for T_1 could be any point in an interval; T_1 is then a continuous random variable. The

Poisson process assumptions allow us to derive the probability distribution for T_1.

Let t represent some fixed number of hours. Since we are assuming $\lambda = 2$, X_{2t} is the number of calls to arrive between 0 and t. With the Poisson process assumptions, X_{2t} is Poisson with parameter $\mu = \lambda t = 2t$. Now let T_1 represent the time at which the first call arrives. This first call will arrive after t hours $(T_1 > t)$ if and only if there are no calls in the interval $(0, t]$, that is, if and only if we find $X_{2t} = 0$. Said another way, the two events $A = \{X_{2t} = 0\}$ and $B = \{T_1 > t\}$ are equivalent; one happens if and only if the other does, so $P(X_{2t} = 0) \equiv P(T_1 > t)$. Thus the *cdf* for T_1 is

$$F_{T_1}(t) = P(T_1 \le t) = 1 - P(T_1 > t)$$
$$= 1 - P(X_{2t} = 0) = 1 - e^{-2t},$$

for any fixed $t > 0$ (and $F_{T_1}(t) = 0$ for any $t \le 0$).

Once we have the *cdf* for a continuous random variable, its *pdf* is given by its derivative; that is,

$$f_{T_1}(t) = \frac{d}{dt} F_{T_1}(t) = 2e^{-2t}, \qquad t > 0, \tag{1}$$

Equation (1) is an example of an *exponential pdf*. The results of this discussion are expressed in the following theorem.

THEOREM 4.1 If T_1 is the time of occurrence of the *first* event in a Poisson process with parameter λ, starting at an arbitrary origin $t = 0$, the *cdf* for T_1 is

$$F_{T_1}(t) = \begin{cases} 1 - e^{-\lambda t}, & \text{for } t \ge 0, \\ 0, & \text{otherwise}, \end{cases} \tag{2}$$

and its *pdf* is

$$f_{T_1}(t) = \begin{cases} \lambda e^{-\lambda t}, & \text{for } t > 0, \\ 0, & \text{otherwise}. \end{cases} \tag{3}$$

These define the exponential probability law with parameter λ.

This probability law is called exponential (sometimes *negative exponential*) because of the shapes of the *cdf* and *pdf*, given in Fig. 4.1. The height of the *pdf* at $t = 0$ is given by λ, trailing asymptotically to 0 as $t \to \infty$. The *cdf* gives the area under the *pdf*, starting at 0 with $t = 0$ and approaching 1 asymptotically as $t \to \infty$.

Example 4.1

Assume again that a police station receives requests for a police car at a rate of 2 per hour like events in a Poisson process. If we begin monitoring this activity at time $t = 0$ and let T_1 be the time of the first such call to arrive, then T_1 is exponential with parameter $\lambda = 2$. The probability the first call arrives within the first hour is then $P(T_1 \le 1) = F_{T_1}(1) = 1 - e^{-2} = .865$; if these calls

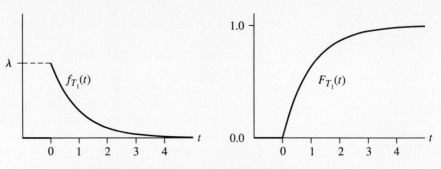

Figure 4.1   Exponential *pdf* (*left*) and *cdf* (*right*)

are arriving like events in a Poisson process with rate 2 per hour, it is quite likely the first call will arrive in the first hour. The probability the first call arrives in the first $\frac{1}{2}$ hour is $F_{T_1}(\frac{1}{2}) = .632$, while the probability it arrives in the second hour is $P(1 < T_1 \le 2) = F_{T_1}(2) - F_{T_1}(1) = e^{-2} - e^{-4} = .117$. The probability that we must wait more than 10 minutes ($\frac{1}{6}$ hour) for the first call is $P(T_1 > \frac{1}{6}) = e^{-1/3} = .717$.

The expected value for T_1 is

$$\mathrm{E}[T_1] = \int_0^\infty t\lambda e^{-\lambda t}\, dt = \frac{1}{\lambda},$$

while the expected value for T_1^2 is

$$\mathrm{E}[T_1^2] = \int_0^\infty t^2 \lambda e^{-\lambda t}\, dt = \frac{2}{\lambda^2};$$

thus the variance of T_1 is $\mathrm{Var}[T_1] = \mathrm{E}[T_1^2] - \left(\mathrm{E}[T_1]\right)^2 = 1/\lambda^2$, so the standard deviation is $\sigma_{T_1} = 1/\lambda = \mu_{T_1}$, the same value as the mean. For Example 4.1, the expected time of the arrival of the first call is $1/\lambda = \frac{1}{2}$ hour; the standard deviation of this time of arrival is also $\frac{1}{2}$ hour.

The mean for an exponential random variable with parameter λ is $1/\lambda$, and $F_{T_1}(1/\lambda) = 1 - e^{-1} = .632$, so the area under the *pdf* to the left of the mean is greater than .5; thus the median value $t_{.5}$ for T_1 lies to the left of $\mu_{T_1} = 1/\lambda$. This is caused by the long right-hand tail of the *pdf*, which shifts the mean (the balance point) in its direction. The $100k$th quantile for the exponential probability law is the solution to

$$1 - e^{-\lambda t_k} = k,$$

which immediately gives

$$t_k = -\frac{1}{\lambda}\ln(1 - k).$$

From this formula with $k = .5$, the median value for T_1 is $t_{.5} = (-1/\lambda)\ln .5 = .693/\lambda$, a number to the left of the mean as expected. The interquartile range is $r_{\mathrm{iq}} = (\ln 3)/\lambda = 1.099/\lambda$, a little larger than the standard deviation.

Recall that, in discussing the Poisson process and deriving the Poisson probability law, we subdivided the continuum of time into small pieces, each of which we treated as a Bernoulli trial. The discrete number of events to occur in a fixed length of time is then analogous to the binomial counting variable; indeed, recall that the Poisson distribution was found as the limit of the binomial. The analog to the random amount of time T_1 to pass before the occurrence of the first event is the (random) number of trials needed to get the first success (which follows the geometric law). The exponential probability law can be considered the continuous analog of the geometric in this sense. Indeed, the exponential is the only continuous probability law to share the memoryless property (see Section 3.2) of the geometric.

As with the geometric, let a and b be any two positive constants; then the conditional probability we will find $T_1 > a + b$, given that $T_1 > b$, is

$$P(T_1 > a + b \mid T_1 > b) = \frac{P(T_1 > a + b)}{P(T_1 > b)}$$

$$= \frac{e^{-\lambda(a+b)}}{e^{-\lambda b}} = e^{-\lambda a} = P(T_1 > a).$$

The exponential probability law "forgets" that b units of time have already passed without the first event occurring; the *conditional* probability that at least a more units of time will be required to observe the first event is identical with the probability that at least a units of time would have been required in the first place. This is a property of the exponential probability law, regardless of whether one considers it to have been embedded in a Poisson process, and establishes the following theorem.

THEOREM 4.2 If X is an exponential random variable with parameter λ, and a and b are any two positive contants, then

$$P(X > a + b \mid X > b) = P(X > a).$$

Example 4.2

Assume that measurable earthquakes (most of which cannot be felt) occur at the rate of 1 per hour in an active area of the San Andreas fault. Suppose you arrive to monitor this activity at a time labeled 0 and let T_1 be the time of occurrence of the first measurable quake after your arrival. With these assumptions T_1 is exponential with parameter $\lambda = 1$ if we measure time in hours. The expected time of the first such quake is 1 (hour), and the median time of the first quake is .693 hours (about 41.6 minutes). If 1 hour has already passed with no such quake occurring, the (conditional) probability that at least 41.6 more minutes will pass before the first quake occurs is still .5. This probability remains constant at .5 as time passes, with these Poisson process assumptions.

Recall from Chapter 3 that finding the probability law for the occurrence of the second, third, or the rth event in repeated independent Bernoulli trials (giving the negative binomial probability law) is a straightforward process. In similar manner we can easily find the probability law of the *time of occurrence* of the second, third, ... event in a Poisson process. Consider the time of occurrence, T_2, of the second event, measured from $t = 0$, the time at which we begin observing the process. Let $t > 0$ be an arbitrary fixed constant and consider the event that $T_2 > t$ (i.e., that this second event occurs *after* time t). This event will happen if and only if there are either no occurrences by time t or there is exactly one event to occur by time t. Said another way, if we let $X_{\lambda t}$ represent the (discrete) number of events to occur between times 0 and t (so $X_{\lambda t}$ is Poisson with parameter $\mu = \lambda t$) then the two events $\{T_2 > t\}$ and $\{X_{\lambda t} \leq 1\}$ are equivalent, so their probabilities of occurrence must be equal. Thus we have, for any value $t > 0$,

$$P(T_2 > t) \equiv P(X_{\lambda t} \leq 1) = (1 + \lambda t)e^{-\lambda t}.$$

The *cdf* for T_2 is then the probability of the complement of this event:

$$F_{T_2}(t) = 1 - P(T_2 > t) = 1 - (1 + \lambda t)e^{-\lambda t}, \qquad \text{for } t > 0.$$

The *pdf* for T_2 again is the derivative of its *cdf*:

$$f_{T_2}(t) = \lambda^2 t e^{-\lambda t}, \qquad \text{for } t > 0.$$

Example 4.3

Return to the assumptions given in Example 4.2, where earthquakes occur at a rate of 1 per hour like events in a Poisson process (and our arbitrary time to start observing is again $t = 0$). If T_2 is the time of the *second* earthquake, the *cdf* for T_2 is $F_{T_2}(t) = 1 - (1+t)e^{-t}$, for $t > 0$ and the *pdf* for T_2 is

$$f_{T_2}(t) = \begin{cases} te^{-t}, & \text{for } t > 0, \\ 0, & \text{for } t \leq 0. \end{cases}$$

Thus, the probability the second earthquake occurs in the first hour is $P(T_2 \leq 1) = F_{T_2}(1) = 1 - 2e^{-1} = 0.2642$, and the probability the second earthquake occurs after $t = 3$ hours have passed is $P(T_2 > 3) = 1 - F_{T_2}(3) = 4e^{-3} = 0.1991$.

This same reasoning is simple to employ for the time of occurrence of the *r*th event in a Poisson process, where $r = 3, 4, \ldots$. Consider a general value for r, letting T_r be the time of occurrence for this *r*th event, and again let $t > 0$ be an arbitrary constant. What does it mean for the time of the *r*th event to exceed t? This will happen if and only if the (discrete) count of the number of events in $(0, t]$ is $r - 1$ or less. That is, the event $\{T_r > t\}$ is equivalent to the event $\{X_{\lambda t} \leq r - 1\}$, where $X_{\lambda t}$ is the count of the number of events to occur between time 0 and time t. Thus the *cdf* for T_r is

$$F_{T_r}(t) = 1 - P(T_r > t) \equiv 1 - P(X_{\lambda t} \leq r - 1)$$

$$= P(X_{\lambda t} \geq r) = \sum_{k=r}^{\infty} \frac{(\lambda t)^k}{k!} e^{-\lambda t}.$$

The *pdf* for T_r is given by the slope (i.e., the derivative) of the *cdf*. Since the *cdf* for T_r is an accumulation of values of the Poisson probability function, we can find this *pdf* by summing the derivatives of the Poisson probability function with respect to t. The derivative of $P(X_{\lambda t} = k)$ with respect to t is (for any integer $k \geq 1$)

$$\frac{d}{dt} P(X_{\lambda t} = k) = \frac{d}{dt} \left(\frac{\lambda^k t^k}{k!} e^{-\lambda t} \right)$$

$$= \frac{\lambda^k t^{k-1}}{(k-1)!} e^{-\lambda t} - \frac{\lambda^{k+1} t^k}{k!} e^{-\lambda t}$$

$$= \lambda (P(X_{\lambda t} = k - 1) - P(X_{\lambda t} = k)).$$

Thus the *pdf* for T_r is

$$f_{T_r}(t) = \frac{d}{dt} F_{T_r}(t) = \sum_{k=r}^{\infty} \frac{d}{dt} P(X_{\lambda t} = k)$$

$$= \lambda \sum_{k=r}^{\infty} (P(X_{\lambda t} = k - 1) - P(X_{\lambda t} = k))$$

$$= \lambda (P(X_{\lambda t} = r - 1) - P(X_{\lambda t} = r) + P(X_{\lambda t} = r)$$

$$- P(X_{\lambda t} = r + 1) + \cdots)$$

$$= \lambda P(X_{\lambda t} = r - 1) = \frac{\lambda^r t^{r-1}}{(r-1)!} e^{-\lambda t},$$

since all terms cancel in this sum, except for the first. This establishes the following theorem.

THEOREM 4.3 If T_r is the time of occurrence of the rth event in a Poisson process with parameter λ, the *cdf* for T_r is

$$F_{T_r}(t) = \sum_{k=r}^{\infty} \frac{(\lambda t)^k}{k!} e^{-\lambda t}, \qquad \text{for } t > 0,$$

and the *pdf* for T_r is

$$f_{T_r}(t) = \frac{\lambda^r t^{r-1}}{(r-1)!} e^{-\lambda t}, \qquad \text{for } t > 0.$$

T_r is called an Erlang random variable with parameters r and λ.

Example 4.4

Calls for a police car are assumed to arrive at a rate of $\lambda = 2$ per hour. Consider the time T_2 at which the *second* call arrives after we start observing the process; T_2 has the Erlang distribution with $\lambda = 2$ and $r = 2$. The probability that this second call arrives within the first hour, then, is

$$P(T_2 \leq 1) = F_{T_2}(1)$$

$$= 1 - P(X_2 = 0) - P(X_2 = 1) = 1 - 3e^{-2}$$

$$= .594,$$

where X_2 is Poisson with parameter $\lambda t = 2(1) = 2$. The probability that the second call arrives in the second hour is

$$P(1 < T_2 \le 2) = F_{T_2}(2) - F_{T_2}(1)$$
$$= 1 - P(X_4 = 0) - P(X_4 = 1) - (1 - P(X_2 = 0) - P(X_2 = 1))$$
$$= .314,$$

where again the subscript on the Poisson variable X identifies the parameter of its probability law.

As the number r of the event changes, so does the corresponding parameter in the Erlang probability law; the probability that the *third* call arrives in the first hour is

$$P(T_3 \le 1) = F_{T_3}(1)$$
$$= 1 - P(X_2 = 0) - P(X_2 = 1) - P(X_2 = 2) = 1 - 5e^{-2}$$
$$= .323,$$

while the probability the third call arrives in the second hour is

$$P(1 < T_3 \le 2) = F_{T_3}(2) - F_{T_3}(1) = .439.$$

Note that the number of Poisson terms to be summed depends only on r, while the parameter of the Poisson distribution employed depends on t, the argument of the Erlang *cdf* (and on λ, the parameter of the process).

Before discussing the mean and variance of the Erlang distribution, let us discuss the gamma function, first described by Euler in 1729 and still employed for many purposes today. Euler showed that the definite integral

$$\int_0^\infty x^{n-1}e^{-x}\,dx = \Gamma(n)$$

converges to a positive number for all $n > 0$ (so the exponent of x in this integral is $n - 1 > -1$); the value of this integral depends only on n and is called the *gamma function* (with argument n). If we apply integration by parts, it is easy to see that $\Gamma(n) = (n-1)\Gamma(n-1)$; that is,

$$\int_0^\infty x^{n-1}e^{-x}\,dx = (n-1)\int_0^\infty x^{n-2}e^{-x}\,dx;$$

indeed, if n is a positive integer, repeated integration by parts yields

$$\int_0^\infty x^{n-1}e^{-x}\,dx = (n-1)(n-2)\cdots 1 \int_0^\infty e^{-x}\,dx$$
$$= (n-1)!\,,$$

since this last integral has value 1. Thus, if n is a positive integer, the value of the gamma function is $\Gamma(n) = (n-1)!$, numbers that are familiar from counting problems; note then, as mentioned in Chapter 1, that $\Gamma(1) = 0! = 1$.

The gamma function, shown in Fig. 4.2, varies continuously with $n > 0$. Note that $\Gamma(1) = 0! = \Gamma(2) = 1! = 1$, while $\Gamma(3) = 2! = 2$ and $\Gamma(4) = 3! = 6$,

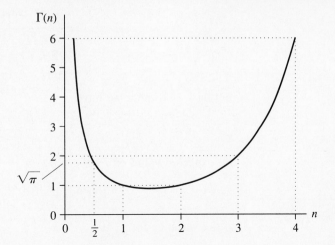

Figure 4.2 The gamma function

as we already know. (Because of the "U shape" of the gamma function, $\Gamma(n)$ also equals 2 for $n \approx .443$ and equals 6 for $n \approx .155$, neither of which is identified in the figure.) The figure also notes that $\Gamma(\frac{1}{2}) = \sqrt{\pi}$, a fact we shall not verify but which is of use in evaluating certain integrals. (See Exercises 4.1 for a sketch of the proof of this fact.)

The gamma function provides the value of the definite integral of x^{n-1} times e^{-x} over the interval $(0, \infty)$ for $n > 0$, or for any integral that can be transformed into such a form. In particular, we saw earlier that the *pdf* for the time T_r of the rth occurrence in a Poisson process is $\lambda^r t^{r-1} e^{-\lambda t}/(r-1)!$ for $t > 0$. Of course the integral of $f_{T_r}(t)\, dt$ from 0 to ∞ must equal 1; the gamma function is useful in establishing this fact, as follows:

$$
\begin{aligned}
\int_0^\infty f_{T_r}(t)\, dt &= \int_0^\infty \frac{\lambda^r t^{r-1}}{(r-1)!} e^{-\lambda t}\, dt \\
&= \frac{\lambda^r}{(r-1)!} \int_0^\infty \left(\frac{u}{\lambda}\right)^{r-1} e^{-u} \frac{du}{\lambda} \\
&= \frac{\lambda^r}{(r-1)!} \frac{1}{\lambda^r} \int_0^\infty u^{r-1} e^{-u}\, du \\
&= \frac{\lambda^r}{(r-1)!} \frac{\Gamma(r)}{\lambda^r} = 1,
\end{aligned}
$$

since $\Gamma(r) = (r-1)!$, where we have made the change of variable $u = \lambda t$.

To evaluate the mean and variance of T_r we require $\mathrm{E}[T_r]$ and $\mathrm{E}[T_r^2]$; these are also easily evaluated by making use of the gamma function. Again letting $u = \lambda_t$, we can easily see that, for any real number $k > -r$,

$$
\begin{aligned}
\mathrm{E}[T_r^k] &= \int_0^\infty t^k f_{T_r}(t)\, dt \\
&= \frac{\lambda^r}{(r-1)!} \int_0^\infty t^{k+r-1} e^{-\lambda t}\, dt
\end{aligned}
$$

$$= \frac{\lambda^r}{(r-1)!} \frac{\Gamma(k+r)}{\lambda^{k+r}}$$

$$= \frac{\Gamma(k+r)}{\lambda^k(r-1)!}.$$

Setting $k = 1$, then, we have

$$\mathrm{E}[T_r] = \frac{\Gamma(1+r)}{\lambda^1(r-1)!} = \frac{r}{\lambda};$$

the expected value of T_r^2 is

$$\mathrm{E}[T_r^2] = \frac{\Gamma(2+r)}{\lambda^2(r-1)!} = \frac{(r+1)r}{\lambda^2},$$

so the variance for T_r is

$$\mathrm{Var}[T_r] = \frac{(r+1)r}{\lambda^2} - \left(\frac{r}{\lambda}\right)^2 = \frac{r}{\lambda^2}.$$

Figure 4.3 pictures the Erlang *pdf*s for $r = 1, 2, 3$; notice that the Erlang distribution with $r = 1$ is the same as the exponential probability law. As r increases, the bulk of the probability shifts further to the right and becomes more spread out as indicated by the facts that $\mu_{T_r} = r/\lambda$, and $\sigma_{T_r} = \sqrt{r}/\lambda$. For $r > 1$ the *pdf* equals 0 at $t = 0$, passes through a maximum value at $t = (r-1)/\lambda$, and then decreases back to 0 as $t \to \infty$.

Recall the relationship between the negative binomial and the geometric random variables: If N_r is negative binomial with parameters r, p, it can be thought of as the sum of r independent geometric random variables, and both the mean and variance of N_r are r times the corresponding geometric values. This same relationship holds for Erlang and exponential random variables: T_r can be written as the sum of r independent exponential random variables, each with parameter λ. To see that this is the case, suppose we start observing the occurrences of events in a Poisson process with parameter λ; and let Y_1 be the time of occurrence of the first event, Y_2 be the *additional* time (after y_1) until the second event, Y_3 be the *additional* time (after $y_1 + y_2$) until

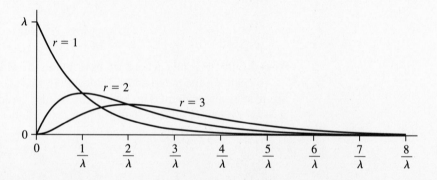

Figure 4.3 Erlang *pdf*s

the third event, and so on. These individual Y_i values are independent and exponential with parameter λ; the time of occurrence of the rth event T_r is the sum of these values: $T_r = Y_1 + Y_2 + \cdots + Y_r$. The expected value for each of the Y_i values is $1/\lambda$, and their variances are all equal to $1/\lambda^2$; the expected value for T_r is r/λ and its variance is r/λ^2.

For many statistical purposes, where we are trying to find a *pdf* that matches a set of observed data, the *gamma probability law* proves useful. From Euler's formula we have

$$\Gamma(n) = \int_0^\infty x^{n-1} e^{-x}\, dx$$

$$= \int_0^\infty (\lambda u)^{n-1} e^{-\lambda u} (\lambda\, du),$$

using the substitution $x = \lambda u$; it follows that

$$\int_0^\infty \frac{\lambda^n u^{n-1}}{\Gamma(n)} e^{-\lambda u}\, du = 1.$$

Thus the function

$$f_U(u) = \frac{\lambda^n u^{n-1}}{\Gamma(n)} e^{-\lambda u}, \qquad \text{for } u > 0,$$

where both parameters n, and λ are positive, could serve as the *pdf* for a continuous random variable. The random variable U that has this *pdf*, with $n > 0$ not necessarily an integer, is called a gamma random variable, as given in the following definition.

DEFINITION 4.1

A random variable U whose *pdf* is

$$f_U(u) = \frac{\lambda^n u^{n-1}}{\Gamma(n)} e^{-\lambda u}, \qquad \text{for } u > 0,$$

where $n > 0$ and $\lambda > 0$, is called a *gamma random variable with parameters n, λ*.

This gamma probability law has been found useful in modeling such things as the length of time needed to diagnose and repair an automobile engine, the lengths of time between orders for the same item from a catalog, and the yearly growth observed for Monterey pine trees, among others. The Erlang random variable is a particular case of a gamma random variable (or the Erlang family of densities is a subfamily of the gamma family), where the parameter n is restricted to the integer values $r = 1, 2, 3, \ldots$; for a general gamma random variable the parameter n can vary continuously. In particular, the gamma *pdf* can be defined for $0 < n < 1$, in which case the *pdf* becomes unbounded at 0; for $n \geq 1$ the gamma *pdf*s have shapes like the Erlang pictured earlier in Fig. 4.3. For either probability law, the parameter λ can be any positive number. The same integration carried out for the Erlang probability law shows that the mean of a gamma random variable U with parameters n, λ is $\mu_U = n/\lambda$ and the variance is $\sigma_U^2 = n/\lambda^2$.

Example 4.5

As we saw earlier, the Erlang *cdf* can be expressed in terms of the Poisson *cdf*,

$$F_{T_r}(t) = 1 - F_{X_{\lambda t}}(r - 1),$$

which enables one to evaluate Erlang probabilities from Poisson tables or from values for the Poisson probability function. Unfortunately, if Y is a gamma random variable with parameters n, λ, where n is not an integer, there is no corresponding simple way to exactly evaluate the gamma *cdf*, so numerical integration is called for; however, it is not hard to see that

$$F_{T_{\lfloor n+1}}(t) < F_Y(t) < F_{T_{\lfloor n}}(t), \qquad \text{for any } t > 0, \tag{4}$$

where $\lfloor n$ is the largest integer in n. That is, we can easily bound the exact gamma probability by two Erlang probabilities having the integer values for r that bound the gamma parameter n.

For example, suppose Y is the time (in hours) required for you to drive to work on a given day, and assume that Y is gamma with parameters $n = 2.5$ and $\lambda = 5$. Then your expected time to drive to work is $\mu_Y = 2.5/5 = \frac{1}{2}$ hour, and the variance of your driving time is $\sigma_Y^2 = 2.5/5^2 = .1$. To evaluate the probability that your driving time is less than, say, $\frac{1}{2}$ hour, the largest integer in $n = 2.5$ is $\lfloor n = 2$,

$$
\begin{aligned}
F_{T_2}(1/2) &= 1 - F_{X_{2.5}}(1) = 1 - 3.5e^{-2.5} = .7127, \\
F_{T_3}(1/2) &= 1 - F_{X_{2.5}}(2) = 1 - 6.625e^{-2.5} = .4562.
\end{aligned}
\tag{5}
$$

so the above bound says $.4562 < P(Y \leq 1/2) < .7127$. Numeric evaluation gives the actual value $P(Y \leq \frac{1}{2}) = .5841$, which is about halfway between these two bounds.

EXERCISES 4.1

1. Assume that, in the county in which you live, fatal automobile accidents occur like events in a Poisson process at a rate of 1 per (7-day) week.
   a. Suppose we let $t = 0$ represent the end of the last day of some month. (It also then marks the beginning of the first day of the following month). If T_1 is the time of occurrence of the first such accident after $t = 0$ (measured in weeks, so $t = 1$ labels the end of the first week), what is the probability law for T_1?
   b. Suppose, instead, that the time scale is measured in days; what is the probability law for T_1?
   c. Using either method of measuring time (days or weeks), verify that you get the same value for the probability of the first such accident occurring during the eighth day, and evaluate this probability.

2. Assume that earthquakes of magnitude 3.5 or more (on the Richter scale) occur along a 100-mile stretch of the San Andreas fault at a rate of 1 every 40 hours; you begin monitoring this activity at $t = 0$, measuring time in hours, and T_1 is the time of occurrence of the first quake of magnitude 3.5 or more.
   a. What is the expected time of the first occurrence?
   b. What is the median time of the first occurrence?
   c. What is the probability that at least 3 days will pass before the first occurrence?

3. Assume the time of the first million-dollar slot machine payoff in a Las Vegas casino, measured from 12:01 AM next New Year's day, is an Erlang random variable Y with parameters $r = 3$ and $\lambda = 5$ (measuring time in years). What is the probability this payoff occurs before July 1? that it occurs in the month of August? What is the expected date of this payoff?

4. Recall that the Poisson probability law was derived as the limit of the binomial, with parameters n and $p = \mu/n = \lambda t/n$, as $n \to \infty$. Let X be geometric with parameter $p = \mu/n = \lambda t/n$, where λ and t are fixed, while T_1 is the time of the first event (i.e., first success). Then $T_1 > t$ and $X > n$ are equivalent events, if the continuous interval $(0, t]$ is split into n equal-length "Bernoulli trials"; the time of the first event exceeds t only if there are no successes in the n trials. Show that
$$\lim_{n \to \infty} P(X > n) = e^{-\lambda t} = P(T_1 > t).$$

5. Let T_1 be the time (in minutes) that you must wait for your friend to arrive for lunch; assume T_1 is exponential with parameter λ. Granted that $P(T_1 > 1) = \frac{1}{2}$, evaluate $P(T_1 > 2)$ and $E[T_1]$.

6. A quality control specialist at an industrial plant was given some data on the occurrence of injury-causing accidents at the plant. Using these data, she found the interquartile range to be 5 weeks; she also thought an exponential probability law would ably describe this phenomenon.
   a. Let W be an exponential random variable with $r_{iq} = 5$ weeks, and evaluate the mean and median for W.
   b. Find the length $t_{.975} - t_{.025}$ of the interval that brackets the "middle" 95% of the probability distribution for W.
   c. Evaluate the probability that an injury-causing accident occurs in this plant next week.

7. What is the probability law for a "discretized" exponential random variable? That is, if X is exponential with parameter λ, what is the probability law for $Y = \lceil X = 1 + \lfloor X$, the next integer larger than X?

8. Assume, as in Exercise 2, that earthquakes occur at a rate of 1 every 40 hours like events in a Poisson process.
   a. Let T_2 be the time of occurrence of the second earthquake. Evaluate the mean for T_2 and the probability that T_2 exceeds 60 hours.
   b. Let T_3 be the time of occurrence of the third earthquake. Evaluate the mean for T_3 and the probability that T_3 exceeds 60 hours.

9. Let Y be the length of time a hot-air balloon ride lasts (which is restricted by how long it takes to get to a certain area for landing) and assume Y is Erlang with $r = 2$ and $\lambda = 4$ (measured in hours).

a. Evaluate the probability that this ride lasts more than 30 minutes.

b. Evaluate the probability that this ride lasts more than 1 hour, given that it lasts more than 30 minutes. (This illustrates that Y is not memoryless.)

10. Generalize part b of Exercise 9; that is, if Y is Erlang with parameters $2, \lambda$, show that Y does not have the memoryless property.

11. *For the mathematically inclined* Show that $\Gamma(\frac{1}{2}) = \sqrt{\pi}$. (*Hint:* In the integral

$$\Gamma(\tfrac{1}{2}) = \int_0^\infty \frac{e^{-t}}{\sqrt{t}} \, dt,$$

make the change of variable $\sqrt{t} = u$ to get

$$\Gamma(\tfrac{1}{2}) = \int_0^\infty 2e^{-u^2} \, du.$$

Then $(\Gamma(\frac{1}{2}))^2$ is the product of two such integrals and can be looked at as a double integral. Use polar coordinates $u = r\cos\theta$, and $v = r\sin\theta$ to find $(\Gamma(\frac{1}{2}))^2 = \pi$.)

12. The time for a repair facility to fix a given item it has received is a gamma random variable with mean 2.5 and variance 1 (measured in days). What are the parameters n and λ for this gamma probability law?

13. As pointed out in the text, the quantiles for the exponential probability law are simple functions of k and λ; this is not true for the Erlang probability law with $r > 1$ nor for the gamma probability law with $n \neq 1$. Let T_2 be Erlang with $r = 2$ and an arbitrary value of $\lambda > 0$, and write down the equation defining the $100k$th quantile for T_2. This equation can be solved numerically for t_k, for any given value of k, but there is no simple closed-form solution possible with elementary functions, as there is with $r = 1$.

14. The time for a customer to be served by a bank teller is assumed to be a gamma random variable with $n = 2.1$ and $\lambda = 1.4$ (measured in minutes). What is the average time taken to serve such a customer, and what is the standard deviation of this time?

15. A battery-operated transmitter in a remote polar location is serviced by human beings only one time per year; if the battery being used is exhausted, the transmitter is immediately (and automatically) provided with power from the next battery in line, as long as there is at least one charged battery remaining. Assume that the length of time any particular battery will last is an exponential random variable with mean 45 (days), that their lifetimes are independent, and that nine fresh batteries (counting the one being used) are available for the transmitter after it has been serviced. What is the probability that the transmitter will be able to remain active for a full year (365 days)?

16. How many batteries are required in Exercise 15 to make the probability (at least) .9 that the transmitter will operate for a full year?

4.2 | The Normal Probability Distribution

The normal probability law is one of the most frequently employed continuous distributions. As will be seen in Chapter 7, a major reason for its frequent use is given by the Central Limit Theorem, a result showing that many different discrete and continuous probability laws all converge to the normal through certain limiting operations. This in turn allows us to approximate many distributions by the normal. We begin with the definition of the normal distribution.

DEFINITION 4.2 ────────────────────────────

A random variable X with *pdf*

$$f_X(t) = \frac{1}{\sigma\sqrt{2\pi}}e^{-(t-\mu)^2/2\sigma^2}, \qquad \text{for } -\infty < t < \infty,$$

is called a *normal random variable with parameters μ and $\sigma > 0$.*

This normal *pdf* is maximized at $t = \mu$, the value that makes the exponent $(t - \mu)^2/2\sigma^2 = 0$; as t differs from μ, either positively or negatively, the *pdf* declines symmetrically generating a bell-shaped curve centered at μ. Figure 4.4 pictures three normal *pdf* s, each with $\mu = 20$, but with different values of σ^2. The value of μ controls the location of the center of the bell, no matter what the value of σ, and σ controls the relative peakedness of the bell, no matter what the value for μ. Thus if three normal *pdf* s were pictured, all with $\mu = 100$, say, and the same σ-values as those given in Fig. 4.4, the *pdf* s look identical, except they would all be centered at 100.

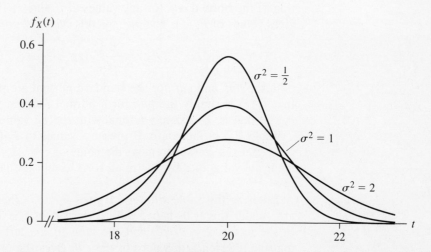

Figure 4.4   Normal *pdf* s

Granted $f_X(t)$ is a legitimate *pdf*, the total area under $f_X(t)$ must be 1. To see that this is the case

$$\int_{-\infty}^{\infty} \frac{1}{\sigma\sqrt{2\pi}} e^{-(t-\mu)^2/2\sigma^2} \, dt = \int_{-\infty}^{\infty} \frac{1}{\sqrt{2\pi}} e^{-z^2/2} \, dz \qquad \left(\text{where } z = \frac{t-\mu}{\sigma}\right)$$

$$= \frac{2}{\sqrt{2\pi}} \int_0^{\infty} e^{-z^2/2} \, dz \qquad \begin{array}{l}\text{(recognizing that the}\\ \text{integrand is symmet-}\\ \text{ric about } z = 0)\end{array}$$

$$= \sqrt{\frac{2}{\pi}} \int_0^{\infty} e^{-u} \frac{du}{\sqrt{2u}} = \frac{1}{\sqrt{\pi}} \Gamma(\tfrac{1}{2}) = 1.$$

(The last line of this equation results from letting $u = z^2/2$ and recognizing that $\Gamma(\tfrac{1}{2}) = \sqrt{\pi}$.) We can see that the normal *pdf* must balance at μ, since it is symmetric about that value, as long as $E[X]$ exists. You will be asked in Exercises 4.2 to show that this is indeed the case so the parameter μ is in fact $E[X]$, the mean for X; you will also be asked to show that the standard deviation for X is σ, the second parameter in the *pdf* (which is the reason for using these symbols for the parameters).

The *cdf* for a normal random variable with mean μ and standard deviation σ is

$$F_X(t) = P(X \le t) = \int_{-\infty}^{t} \frac{1}{\sigma\sqrt{2\pi}} e^{-(x-\mu)^2/2\sigma^2} \, dx; \qquad (1)$$

unfortunately, there is no elementary function whose derivative is the *pdf* $f_X(x)$, so the value for this integral must be determined numerically. If we make the change of variable $z = (x - \mu)/\sigma$ in Eq. (1), we have

$$F_X(t) = \int_{-\infty}^{\frac{(t-\mu)}{\sigma}} \frac{1}{\sqrt{2\pi}} e^{-z^2/2} \, dz = F_Z\left(\frac{t-\mu}{\sigma}\right),$$

no matter what the values for μ and σ. Thus a table of $F_Z(z)$ allows evaluation of the normal *cdf* for any values of μ and σ; Table A.3 in Appendix A gives values of $F_Z(z)$. The *pdf* for this *standard normal random variable* is

$$f_Z(z) = \frac{d}{dz} F_Z(z) = \frac{1}{\sqrt{2\pi}} e^{-z^2/2};$$

both the *pdf* and *cdf* for the standard normal are pictured in Fig. 4.5. Note that $f_Z(z)$ is simply the *pdf* for a normal random variable with $\mu = 0$ and $\sigma = 1$; that is, Z is simply normal with special values for the parameters.

Table B.3 in Appendix B presents values of $F_Z(z)$ for $z = 0$ to $z = 3.9$. Note that the rows increment the value of z by .1, while the columns increment the value of z by .01 (in all rows except the last) and by .1 (in the last row).

Entering this table with $z = 1$ gives $F_Z(1) = P(Z \le 1) = .8413$; the total area under this bell-shaped *pdf* to the left of 1 is .8413. Similarly, the area to the left of 2 is .9772, and the area to the left of 3 is .9987. Recalling that the difference of values of the *cdf* gives the probability of the random variable lying in an interval, it follows that $P(1 < Z < 2) = .9772 - .8413 =$

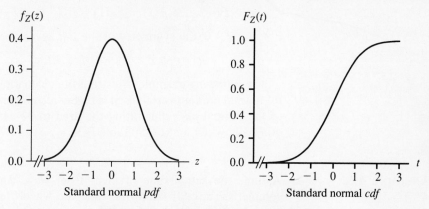

Figure 4.5 Standard normal probability law

.1359 and $P(1 < Z < 1.9) = .9713 - .8413 = .1300$, while $P(3 < Z < 3.9) = 1 - .9987 = .0013$. Z is much more likely to lie in the interval of length .9, starting at $z = 1$, than in the interval of the same length starting at $z = 3$. Since $F_Z(3.9) = 1.0000$ and any *cdf* is nondecreasing, it follows that $F_Z(z) = 1.0000$ for *all* $z > 3.9$ as well.

Table B.3 provides values for the *cdf* of Z only for nonnegative values of z; however, we can take advantage of the symmetry of the *pdf*, to evaluate $F_Z(z)$ for $z < 0$ using values in the same table. First, it is easy to see that $F_Z(0) = .5$, from the symmetry of the *pdf* or from the table; by using symmetry (refer to Fig. 4.6) we can also see that

$$F_Z(-z) = 1 - F_Z(z). \tag{2}$$

Combining Eq. (2) with Table B.3 we find, for example, that

$$P(Z \le -1) = F_Z(-1) = 1 - F_Z(1) = .1587,$$
$$P(-1.6 < Z < -.1) = F_Z(-.1) - F_Z(-1.6) = F_Z(1.6) - F_Z(.1)$$
$$= .9452 - .5398 = .4054,$$

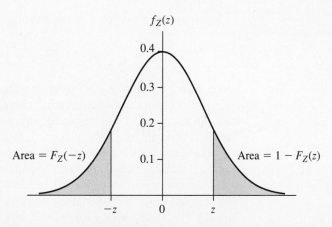

Figure 4.6 $F_Z(-z) = 1 - F_Z(z)$

$$P(-2 < Z < 1) = F_Z(1) - F_Z(-2) = F_Z(1) + F_Z(2) - 1 = .8185.$$

For any $z > 0$, the symmetry of the *pdf* gives

$$P(-z < Z < z) = F_Z(z) - F_Z(-z) = 2F_Z(z) - 1.$$

The following examples illustrate the evaluation of probability statements for normal random variables; it is highly recommended that you draw pictures of the normal *pdf*, illustrating the area to be evaluated similar to Fig. 4.6, especially when first using Table B.3.

Example 4.6

Assume the actual time of flight from New York to San Francisco for a commercial aircraft (from the time the wheels leave the ground until they touch back down) is a normal random variable X with mean $\mu = 270$ and standard deviation $\sigma = 10$, both measured in minutes. The *pdf* for X is then the normal bell-shaped curve centered at 270, and half the flight times are less than 270 minutes. The probability that a flight takes less than 285 minutes is $P(X \leq 285) = F_Z((285 - 270)/10) = F_Z(1.5) = .9332$, while the probability that a flight takes at least 260 minutes is $P(X \geq 260) = 1 - F_X(260) = 1 - F_Z((260 - 270)/10) = F_Z(1) = .8413$; the proportion of flight times between 265 and 275 minutes is $P(265 < X < 275) = F_X(275) - F_X(265) = F_Z((275 - 270)/10) - F_Z((265 - 270)/10) = .3830$.

Example 4.7

Five-pound bags of flour are filled by machine, by weight; the actual weights produced, however, vary slightly. If X is the "true" net weight of one of these bags, and X is normal with mean $\mu = 80$, and $\sigma = \frac{1}{2}$ (in ounces), the proportion of net weights that are less than 79 ounces is $P(X < 79) = F_Z(-2) = .0228$, and the proportion of net weights that exceed 79.5 ounces is $P(X > 79.5) = 1 - F_X(79.5) = F_Z(1) = .8413$.

The *standard form* for a normal random variable X is

$$Z = \frac{X - \mu}{\sigma}, \tag{3}$$

which then has mean $\mu_Z = 0$ and standard deviation $\sigma_Z = 1$. This gives the formal description of the preceding manipulation allowing the standard normal *cdf* to be used in evaluating any normal *cdf*. Equation (3) also says that if Z is standard normal, then $X = \mu + \sigma Z$ (the inverse of the preceding transformation) is normal with mean μ and variance σ^2. An implication is that *any* linear function (that is not a constant) of a normal random variable again is a normal random variable. That is, if $Y = a + bX$, where a and $b \neq 0$ are constants and X is normal with parameters μ, σ, then $Y = a + b(\mu + \sigma Z) = (a + b\mu) + b\sigma Z$ is a linear function of a standard normal random variable and must itself be normal with mean $a + b\mu$ and variance $b^2\sigma^2$. This frequently used result is stated in the following theorem.

> **THEOREM 4.4** If X is a normal random variable with mean μ and variance σ^2, and if a and $b \neq 0$ are any constants, then $Y = a + bX$ is also normal with mean $a + b\mu$ and standard deviation $|b|\sigma$.

Example 4.8

It is assumed that the high temperature to be observed at a given location, during any specific 24-hour period, is a normal random variable X with $\mu_X = 65$ and $\sigma_X = 5$, both given in degrees Fahrenheit. The high temperature on the Celsius scale is $Y = \frac{5}{9}(X - 32)$; Y then is also a normal random variable with $\mu_Y = 18\frac{1}{3}$ and $\sigma_Y = 2\frac{7}{9}$ degrees Celsius.

If x_k is the $100k$th quantile for a normal random variable X with mean μ and standard deviation σ, and z_k is the $100k$th quantile of the standard normal random variable, then $x_k = \mu + \sigma z_k$, as shown in Section 2.4. Table B.3, with inverse table lookup, gives approximate values for the standard normal quantiles; these can then also be used to get approximate values for the quantiles of any normal random variable. For example, since $F_Z(0) = .5$, then $z_{.5} = 0$ is the median for Z and $x_{.5} = \mu + \sigma \cdot 0 = \mu$ is the median for X. Using linear interpolation, $z_{.75} = .675$, $z_{.25} = -.675$ so the interquartile range for a standard normal is $r_{iq} = 1.35$, while the interquartile range for X is $r_{iq} = 1.35\sigma_X$.

Example 4.9

A commercial airline wants the published flight time in its schedule to be the amount of time that its flights are able to meet (or beat) 90% of the time. If, as in an earlier example, the flight time X from New York to San Francisco is normal with mean 270 and standard deviation 10, the 90th quantile for X is $x_{.9} = 270 + 10z_{.9} = 270 + 10(1.283) = 282.83$. The published flight time should then be about 283 minutes; (a little more than) 90% of the flights should be equal to or less than this time.

The Central Limit Theorem will be discussed in Chapter 7; among other things it states that the binomial *cdf* converges to the normal as $n \to \infty$. Thus, if the value of n is "large" the exact value for the binomial *cdf* at any point should be well approximated by the appropriate normal.

So, what is the appropriate normal? Let X be binomial with parameters n, p, where $n > 9 \max(q/p, p/q)$, and let t be any element of $\{0, 1, \ldots, n\}$. Then

$$F_X(t) \doteq F_Z\left(\frac{t + .5 - np}{\sqrt{npq}}\right).$$

Since X is discrete,

$$P(a < X \leq b) \equiv P(a + 1 \leq X \leq b)$$
$$\doteq F_Z\left(\frac{b + .5 - np}{\sqrt{npq}}\right) - F_Z\left(\frac{a + .5 - np}{\sqrt{npq}}\right).$$

In particular, then, given $n > 9 \max\left(q/p, p/q\right)$,

$$P(X = b) \doteq F_Z\left(\frac{b + .5 - np}{\sqrt{npq}}\right) - F_Z\left(\frac{b - .5 - np}{\sqrt{npq}}\right).$$

This approximation is best for values of t in the vicinity of the mean np, for any given n; for any given p it improves the larger n becomes.

To illustrate, suppose X is binomial with $n = 10$ and $p = .5$, so $p/q = q/p = 1$, giving $9 \max\left(q/p, p/q\right) = 9$ and $10 > 9$. From Table B.1, $F_X(6) = .8281, F_X(5) = .6230$, and $F_X(4) = .3770$. The normal approximations for these values use $F_Z(z)$ with $z = (t + .5 - 5)/1.5811$, giving (rounded) arguments of $.32, -.32$, and $.95$. The corresponding values from Table B.3 are $.8289, .6255$, and $.3745$, which are quite close to the exact binomial values, even for values of n this small. Note as well that $F_X(0) = .0010$ and $F_X(1) = .0107$, which are approximated by $.0022$ and $.0136$, respectively; the tail values are not as well approximated on a percentage basis. This approximation is especially useful for large values of n, where exact binomial tables may not be available; many modern computer packages that evaluate binomial probabilities, rather than computing the exact values, appear to use this approximation as well for large n.

Example 4.10

A small firm with 50 unskilled employees is located in an area in which 50% of the population is minority (non-Caucasian); if this firm hired $n = 50$ people totally at random (and the proportions of applicants mirrored the population), the number X of minority employees would be binomial with $n = 50$ and $p = .5$. The expected number of minority employees thus would be $np = 25$. The probability that this firm would have no more than 20 minority employees is then $F_X(20) \doteq F_Z(-1.27) = .1020$, and the probability that it would have exactly 25 minority employees is $P(X = 25) \doteq F_Z(.14) - F_Z(-.14) = .1113$. The exact binomial values for these probabilities are $.1013$ and $.1123$, respectively.

The Central Limit Theorem also implies that the normal distribution gives good approximations to the Poisson *cdf* for relatively large values of μ. Specifically, if X is Poisson with $\mu > 10$, then

$$F_X(t) \doteq F_Z\left(\frac{t + .5 - \mu}{\sqrt{\mu}}\right) \quad \text{for } t \text{ a nonnegative integer.}$$

If X is Poisson with $\mu = 11$, Table B.2 gives $F_X(11) = .5793$, and $F_X(12) = .6887$, so $P(X = 12) = .1094$. This (rounded) normal approximation uses $z = .15$ and $.45$, giving $F_X(11) \doteq F_Z(.15) = 0.5596, F_X(12) \doteq F_Z(.45) = 0.6736$, and $P(X = 12) \doteq .1140$. The normal approximation is better the larger the value for the Poisson parameter μ.

Example 4.11

Customers enter a retail store during busy periods like events in a Poisson process with rate $\lambda = 1$ per minute. If X is the number to enter in a busy hour, then X is Poisson, with $\mu = 60$. The approximate probability that no more than 50 people enter during this hour is $F_X(50) \doteq F_Z(-1.23) = .1093$, and the probability of at least 55 customers entering is $P(X \geq 55) \doteq 1 - F_Z(-.71) = .7611$. The exact values for these Poisson probabilities are .1077 and .7579.

EXERCISES 4.2

1. A rifle round is fired at a bull's-eye target. Let Z represent the *horizontal* miss distance measured from the center of the bull's-eye. Thus $Z < 0$ if the round hits to the left of the center, and $Z > 0$ if the round hits to the right of the center. Assume that Z is standard normal, and evaluate the probabilities of the following events:
   a. $Z > 1.25$ b. $Z < -.67$ c. $|Z| < .5$ d. $|Z| > 1.55$
   e. $|Z| > 1$ or $|Z| < .4$ f. $|Z| > 1$ and $|Z| < 2$.
2. Z is standard normal; find the value for c such that $P(|Z| < c) = \gamma$, where $0 < \gamma < 1$ is given.
3. Because of the timing cycle on a personal computer, the exact time required for a program to complete a complex computation is a normal random variable with $\mu = 165$ and $\sigma = 15$ (in microseconds). What is the probability that this computation requires more than 200 microseconds? less than 150 microseconds?
4. The monthly January rainfall, recorded at a west coast site, is a normal random variable with $\mu = 5.32$, and $\sigma = .82$ (in inches).
   a. What is the probability the recorded January rainfall at this site will exceed 7 inches next year?
   b. What is the 10th quantile for the number of inches recorded at this site in January?
   c. Evaluate the interquartile range for this distribution.
5. A commercial firm that sells wheat flour packages its product in 5-pound bags, filled by a machine. Assume the actual weight of the flour in one of these bags is a normal random variable X with mean 5.01 and standard deviation .05 (pounds).
   a. Evaluate the probability that one of these bags contains more than 5 pounds of flour.
   b. What proportion of these bags contain less than 4.95 pounds?
   c. If you buy three of these bags, what is the probability that all of them contain more than 5 pounds of flour?
6. Assume the height that a college high-jumper will clear (or the height he will jump) on any given attempt is a normal random variable H with $\mu_H = 78$, and $\sigma_H = 2$ (in inches).
   a. What is the probability he will clear at least 6 feet, 3 inches?
   b. What height will he clear with probability .95?
7. Compare the exact values for $P(|Z| < k)$ with the values given by the Chebyshev inequality, for $k = 2, 3$.

8. The standard normal *pdf* is positive for all real z so there is an unlimited number of intervals (a, b) such that $P(a < Z < b) = .9$.
   a. Define three such intervals.
   b. What is the *shortest* interval (a, b) such that $P(a < Z < b) = .9$?

9. X is a normal random variable with $P(X < 10) = .75$, and $P(X > 8) = .95$. What are the values for μ_X and σ_X?

10. Assume that the square footage covered by 1 gallon of a nationally advertised brand of paint is a normal random variable X with mean $\mu = 380$ and standard deviation $\sigma = 20$ (in square feet).
    a. What is the probability that 1 gallon of this paint will prove sufficient to cover (at least) 400 square feet?
    b. What is the probability that it will cover no more than 350 square feet?
    c. How many gallons of paint (and fractions of a gallon) are required to have probability .9 of covering 600 square feet?

11. Scores made by students on "standardized" tests are frequently transformed to give a specified mean and standard deviation. Suppose X is normal with mean μ and standard deviation σ. If $Y = a + bX$, what are the values for a and b such that $\mu_Y = 500$ and $\sigma_Y = 100$?

12. If X is a normal random variable with mean μ and variance σ^2, show that the points of inflection for the *pdf* for X occur at $\mu \pm \sigma$.

13. What happens to the probability law for Y if $b = 0$ and we define $Y = a + bX$, where X is normal with parameters μ_X, σ_X?

14. Show that the mean and variance of the standard normal distribution are 0 and 1. Hence, if μ and $\sigma > 0$ are constants then $X = \mu + \sigma Z$ has mean μ and variance σ^2.

15. Many commercial airlines overbook their flights in the hope of filling their planes. For example, if an aircraft has 130 seats, the airline may issue tickets for 140 reserved seats. Assume this is the case for a particular flight, and that these 140 persons will or will not arrive to take the flight independently, each with probability .9.
    a. How many ticket holders would you expect to arrive for the flight?
    b. What is the (approximate) probability that the airline can provide a seat for each ticket holder who arrives for the flight?
    c. How many reserved-seat tickets should the airline issue if it assumes the common probability is .8 for each person arriving for the flight, and they want to have probability .95 or higher of providing a seat for all ticket holders who arrive for the flight?

16. Telephone calls to an emergency 911 number occur like events in a Poisson process with parameter $\lambda = 5$ per hour between 12 midnight and 5 AM. If X is the number of such calls to occur during this period on a given day, approximate the probability of at least 30 calls arriving. Also approximate the probability of fewer than 20 calls arriving.

17. After a forest fire the seeds of a specific tree are scattered independently and at random over the area at a rate of 2 per square foot. How many square feet (t) must be examined to have $P(X \geq 100) = .9$, where X is the count of seed in the area of t square feet?

4.3 | The Beta Probability Distribution

The gamma probability law is quite versatile and provides a number of different shapes for *pdf*s defined on the positive half of the real line. These prove useful in modeling the behavior of random variables that are positive but unbounded in possible value. Some random variables are by their nature positive but bounded, limited to a finite range. Proportions and percentages fall in this category, being restricted to the interval $(0, 1)$, as do physical quantities like a voltage passed through a limiter. As an example, suppose two chemicals A and B are mixed with a catalyst; the resulting reaction produces C and some by-products. The amount of C produced is frequently described as a proportion X, which in theory can then vary from 0 to 1. Or consider modeling the lethality of an insecticide prepared to kill flies (or some other pest). If a large group of flies is sprayed with the insecticide, a natural variable to observe is the proportion X of flies that are killed, again a variable that must take on a value in the interval $(0, 1)$.

The continuous uniform distribution with $a = 0, b = 1$ provides a *pdf* for a random variable whose range is $R = \{0 < x < 1\}$; its *pdf* is constant on this interval. In this section we shall introduce the beta distribution, whose *pdf* is capable of many shapes over the interval $(0, 1)$. A simple transform of the beta random variable gives a variety of possible shapes for the *pdf*s of random variables restricted to any finite interval.

The gamma function $\Gamma(n)$ occurs in the evaluation of many different definite integrals; in particular it can be shown (although we shall not do so) that

$$\int_0^1 x^{\alpha-1}(1-x)^{\beta-1}dx = \frac{\Gamma(\alpha)\Gamma(\beta)}{\Gamma(\alpha+\beta)},$$

for any $\alpha > 0, \beta > 0$. Thus, since $x^{\alpha-1}(1-x)^{\beta-1} > 0$ for all $0 < x < 1$, the function

$$f_X(x) = \begin{cases} \dfrac{\Gamma(\alpha+\beta)}{\Gamma(\alpha)\Gamma(\beta)} x^{\alpha-1}(1-x)^{\beta-1}, & \text{for } 0 < x < 1, \\ 0, & \text{otherwise,} \end{cases}$$

satisfies the rules for a *pdf*; this *pdf* defines the *beta probability distribution with parameters* α, β. This *pdf* is quite versatile and can take on many different shapes for $0 < x < 1$, four of which are illustrated in Fig. 4.7. Note that using $\alpha = \beta = 1$ makes $f_X(x) = 1$ for $0 < x < 1$, so the uniform probability law on the interval $(0, 1)$ is a special case of a beta random variable. If the parameter α is fixed, increasing the value of β increases the *pdf* for values close to 0; with β fixed, increasing the value of α increases the *pdf* for values close to 1. With $\alpha = \beta$ the *pdf* is symmetric about the line $t = \frac{1}{2}$. With $\alpha < 1$ the *pdf* is unbounded at 0, and with $\beta < 1$ the *pdf* is unbounded at 1. With $\alpha > 1$ and $\beta > 1$ it is easy to see that $f_X(x)$ is largest for $x = (\alpha-1)/(\alpha+\beta-2)$ (you will be asked to verify this in Exercises 4.3); this gives the most likely observed value. The formal definition of the beta probability law follows.

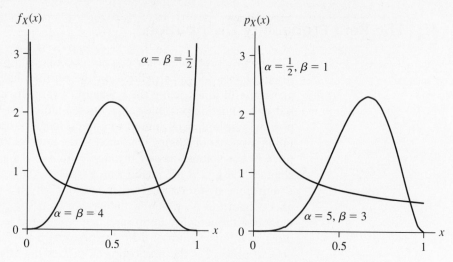

Figure 4.7 Beta probability density functions

DEFINITION 4.3

If X is a continuous random variable whose *pdf* is

$$f_X(x) = \begin{cases} \dfrac{\Gamma(\alpha + \beta)}{\Gamma(\alpha)\Gamma(\beta)} x^{\alpha-1}(1-x)^{\beta-1}, & \text{for } 0 < x < 1, \\ 0, & \text{otherwise,} \end{cases}$$

then X is called a *beta random variable with parameters* $\alpha > 0, \beta > 0$.

Example 4.12

A retail paint store sells paint thinner in any quantities desired, filling the container(s) brought by the customer; it then is important for the retailer to have some idea of how much he might expect to sell in given periods, so he is sure to have stock available when demanded. Assume this retailer keeps a 50-gallon barrel of thinner to fill containers as required. The demand on a given day could be modeled by X, with $R_X = \{0 \le x \le 1\}$, expressed in proportion of 50 gallons (this assumes he will never sell more than a full barrel on a given day, since $X \le 1$). Let the *pdf* for X be $f_X(x) = 30x^4(1-x)$ for $0 \le x \le 1$, a beta distribution with $\alpha = 5$ and $\beta = 2$. This *pdf* is 0 at both $x = 0$ and $x = 1$, with its maximum occurring at $x = \frac{4}{5}$. Because these parameters are integers, it is also easy to find the *cdf* for X: $F_X(t) = 6x^5 - 5x^6$ for $0 \le t \le 1$. Thus the probability he will sell less than 50% of his barrel's capacity on a given day is $P(X \le .5) = F_X(.5) = .109$, while $P(X > .95) = .033$ and $P(.6 < X < .9) = .653$.

Example 4.13

For many theories of our economy, an important economic variable is X, the proportion of disposable income devoted to savings. Over the total population such a proportion varies from one individual to another while always lying on the interval $[0, 1]$. (Some people might argue that the lower endpoint should be a negative number, not 0, to reflect spending more than one's income; we shall ignore this practical point.) If one person is selected at random from the population, with savings X expressed as proportion of disposable income, a natural model for X is the beta distribution. We arbitrarily take the parameters to be $\alpha = 2$ and $\beta = 19$, giving $f_X(x) = 380x(1-x)^{18}$ for $0 \leq x \leq 1$. This *pdf* is maximized for $x = \frac{1}{19} = .053$; our assumptions then imply that .053 is the most likely proportion of savings for a person selected at random. The *cdf* for X is

$$F_X(t) = \int_0^t 380x(1-x)^{18}\,dx = 380 \int_{1-t}^1 (1-u)u^{18}\,du$$

$$= 1 - (1-t)^{19}(1 + 19t), \qquad \text{for } 0 < t < 1.$$

With this assumed *pdf*, the probability that a selected person saves less than 5% of her disposable income is $F_X(.05) = .264$ while the probability she saves at least 10% is $1 - F_X(.1) = .517$.

As with the gamma probability law, the expected value for the kth power of a beta random variable X is straightforward to evaluate for any $k > -\alpha$:

$$E[X^k] = \int_0^1 x^k \frac{\Gamma(\alpha + \beta)}{\Gamma(\alpha)\Gamma(\beta)} x^{\alpha-1}(1-x)^{\beta-1}\,dx,$$

$$= \frac{\Gamma(\alpha + \beta)}{\Gamma(\alpha)\Gamma(\beta)} \int_0^1 x^{k+\alpha-1}(1-x)^{\beta-1}\,dx$$

$$= \frac{\Gamma(\alpha + \beta)}{\Gamma(\alpha)\Gamma(\beta)} \frac{\Gamma(\alpha + k)\Gamma(\beta)}{\Gamma(\alpha + \beta + k)}$$

$$= \frac{\Gamma(\alpha + \beta)\Gamma(\alpha + k)}{\Gamma(\alpha)\Gamma(\alpha + \beta + k)}.$$

Thus the mean for a beta random variable is

$$\mu_X = \frac{\Gamma(\alpha + \beta)\Gamma(\alpha + 1)}{\Gamma(\alpha)\Gamma(\alpha + \beta + 1)} = \frac{\alpha}{\alpha + \beta},$$

and the expected value for X^2 is

$$E[X^2] = \frac{\Gamma(\alpha + \beta)\Gamma(\alpha + 2)}{\Gamma(\alpha)\Gamma(\alpha + \beta + 2)} = \frac{\alpha(\alpha + 1)}{(\alpha + \beta)(\alpha + \beta + 1)}.$$

The resulting variance for the beta probability law is

$$\sigma_X^2 = \frac{\alpha\beta}{(\alpha + \beta)^2(\alpha + \beta + 1)}.$$

In Example 4.12, X was a beta random variable with $\alpha = 5$ and $\beta = 2$; the expected amount sold is then $\mu_X = \frac{5}{7}$ with standard deviation $\sigma_X = \frac{1}{7}\sqrt{\frac{10}{8}} = .160$. The amount saved as a proportion of disposable income was a beta random variable with $\alpha = 2$ and $\beta = 19$. Its mean is $\frac{2}{21}$, and its standard deviation is .063.

Example 4.14

The proportion of insects killed after being sprayed by an insecticide is a beta random variable X with mean .6 and standard deviation .2. Since both the mean and variance for a beta random variable are functions of its two parameters, these values for μ_X and σ_X give the two equations:

$$\frac{\alpha}{\alpha + \beta} = .6, \tag{1}$$

$$\frac{\alpha\beta}{(\alpha + \beta)^2(\alpha + \beta + 1)} = .2^2. \tag{2}$$

Equation (1) gives $\alpha = 1.5\beta$; and substituting this value into Eq. (2) yields $\beta = 2$, and thus $\alpha = 3$, so the *pdf* for X is $f_X(t) = 12t^2(1 - t) = 12t^2 - 12t^3$. The probability that at least half the insects sprayed are killed is then $P(X > .5) = \int_{.5}^{1} 12t^2 - 12t^3 \, dt = \frac{15}{16}$. The probability that less than 80% are killed is $P(X < .8) = \int_0^{.8} f_X(t) \, dt = .8192$. The probability that X is less than its mean is $P(X < .6) = \int_0^{.6} f_X(t) \, dt = .4752$, and the median value of X is $t_{.5} = .6143$ (which can be determined by solving a quartic equation or by numerical methods).

Many personal computer software packages contain a "random number generator." Such a generator will return a number between 0 and 1 (a different one each time). The "random" in the name is meant to imply that the number U produced is uniform on $(0, 1)$ with *pdf* $f_U(u) = 1$ for $0 < u < 1$ and *cdf* $F_U(t) = t$ for $0 \leq t \leq 1$. If this generator produces n successive numbers $U_1, U_2, \ldots, U_n$, these values can be converted into n Bernoulli trials by defining success as the outcome $U \leq t$, where t is some fixed number between 0 and 1. If the generator really does produce numbers that follow the uniform $(0, 1)$ distribution, the probability of a success on each trial is then $P(U \leq t) = F_U(t) = t$ for any $0 \leq t \leq 1$. Granted the resulting Bernoulli trials are independent, the number of successes observed, $Y_{n,t}$, is binomial with parameters n, t.

Now let $U_{(1)}, U_{(2)}, \ldots, U_{(n)}$ denote the ranked numbers generated, from smallest $(U_{(1)})$ to largest $(U_{(n)})$. These ranked values are called the *order statistics* of the n numbers generated. With $n = 5$, say, $U_{(5)}$ then is the largest of the 5 numbers produced by the random number generator. The probability distribution for $U_{(5)}$ describes the behavior of the biggest number generated in groups of 5.

What is the probability distribution for $U_{(j)}$, where $j = 1, 2, \ldots, n$? For any fixed t, the event $\{U_{(j)} \leq t\}$ occurs if and only if we observe j or more successes; this event is equivalent to $\{Y_{n,t} \geq j\}$. Thus the *cdf* for $U_{(j)}$ is

$$F_{U_{(j)}}(t) = P(U_{(j)} \leq t) \equiv P(Y_{n,t} \geq j)$$

$$= \sum_{i=j}^{n} \binom{n}{i} t^i (1 - t)^{n-i}, \qquad \text{for } 0 < t < 1.$$

The *pdf* for $U_{(j)}$ is given by the derivative of this *cdf*; note that

$$\frac{d}{dt}P(Y_{n,t} = i) = \binom{n}{i}\frac{d}{dt}t^i(1-t)^{n-i}$$

$$= n\left[\binom{n-1}{i-1}(1-t) - \binom{n-1}{i}t\right]t^{i-1}(1-t)^{n-i-1},$$

for $1 \le i \le n-1$. Summing these terms involves the same type of cancellation from term to term as was noted in deriving the *pdf* for the Erlang random variable; the only term to survive is the first part of the first term giving the *pdf* for $U_{(j)}$:

$$f_{U_{(j)}}(t) = n\binom{n-1}{j-1}t^{j-1}(1-t)^{n-j}$$

$$= \frac{\Gamma(n+1)}{\Gamma(j)\Gamma(n-j+1)}t^{j-1}(1-t)^{n-j}. \tag{3}$$

We have replaced the factorials in Eq. (3) by their gamma function equivalents to facilitate recognition that this is a beta *pdf* with $\alpha = j$, and $\beta = n - j + 1$. The expected value for $U_{(j)}$ is then $j/(n+1)$, and its variance is $j(n-j+1)/(n+1)^2(n+2)$.

Example 4.15

A random number generator will produce five numbers on the interval $(0, 1)$; $U_{(1)}$ is the smallest, and $U_{(5)}$ is the largest of these numbers. The *pdf* for $U_{(1)}$ is $f_{U_{(1)}}(t) = 5(1-t)^4$ for $0 < t < 1$; this *pdf* is largest at $t = 0$, declining to value 0 at $t = 1$, indicating that values close to 0 are more likely than values close to 1 (not surprising since this describes the behavior of the smallest of a group of five uniform variables). The *pdf* for $U_{(5)}$ is $f_{U_{(5)}}(t) = 5t^4$, the mirror image of $f_{U_{(1)}}(t)$ about the line $t = \frac{1}{2}$. Both of these *pdf*s are pictured in Fig. 4.8, along with the *pdf* for $U_{(3)}$, the middle value of the five numbers. This middle value is more likely to be close to $\frac{1}{2}$ than to either 0 or 1.

The *cdf* for the beta probability law is

$$F_X(t) = \int_0^t \frac{\Gamma(\alpha+\beta)}{\Gamma(\alpha)\Gamma(\beta)}x^{\alpha-1}(1-x)^{\beta-1}\,dx, \qquad \text{for any } 0 < t < 1.$$

As with the normal or gamma random variables, there is no elementary function whose derivative gives the beta *pdf* for all possible $\alpha > 0, \beta > 0$, so numerical integration may be required to evaluate $F_X(t)$ for general parameter values. Recall that the *cdf*s for the uniform order statistics can be expressed in terms of a binomial *cdf*; the *j*th-order statistic $U_{(j)}$ has the beta probability law with parameters $\alpha = j$ and $\beta = n - j + 1$, which we saw can be evaluated from the probability law of the binomial $Y_{n,t} = Y_{\alpha+\beta-1,t}$. Thus if X is a beta random variable in which *both* α and β are integers, then the

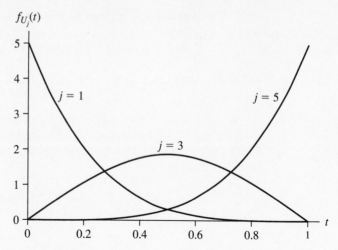

Figure 4.8   *pdf*s for $U_{(j)}$ and $n = 5$, with $j = 1, 3, 5$

cdf for X can be evaluated from the *cdf* for the binomial random variable $Y_{\alpha+\beta-1, t}$:

$$F_X(t) = 1 - F_{Y_{\alpha+\beta-1, t}}(\alpha - 1)$$

$$= \sum_{i=\alpha}^{\alpha+\beta-1} \binom{\alpha + \beta - 1}{i} t^i (1 - t)^{\alpha+\beta-1-i}.$$

Example 4.16

If the proportion of minority wage earners, in a firm selected at random from those in a given industry, is a beta random variable X with $\alpha = 4$ and $\beta = 7$, then some of the values of the *cdf* for X can be evaluated from Table B.1. For example, $F_X(.4) \equiv P(Y_{10,.4} \geq 4) = 1 - F_{Y_{10,.4}}(3) = .6177$, while $P(X \geq .3) = 1 - F_X(.3) = F_{Y_{10,.3}}(3) = .6496$.

 **EXERCISES 4.3**

1. Suppose $n = 2$ independent uniform random variables are generated from the interval $(0, 1)$, and let $X_{(1)}$ be the smaller of these two. Show that the *pdf* for $X_{(1)}$ is triangular, and evaluate its mean and variance.

2. Let $X_{(2)}$ be the larger of two independent uniform $(0, 1)$ random variables. Show that its *pdf* is triangular, and evaluate its mean and variance.

3. The yield X of the desired product in a chemical reaction is a beta random variable with $\alpha = 3$ and $\beta = 5$. Evaluate the mean and variance for X.

4. If X is a beta random variable with $\mu = .6$ and $\sigma^2 = .1$, find the values for the parameters α, β.

5. Let $X_{(1)}, X_{(2)}, X_{(3)}, X_{(4)}$ be the order statistics for four independent uniform $(0, 1)$ random variables. Sketch their *pdf*s and find their means and variances.

6. If X has the beta probability law with parameters α, β, what value of x maximizes its *pdf* $f_X(x)$?

7. A computer is used to generate n uniform $(0, 1)$ random variables. Assume that n is an odd integer so the sample median of these values is $U_{((n+1)/2)}$, the middle ranked value. Show that the *pdf* for $U_{((n+1)/2)}$ is symmetric about $t = \frac{1}{2}$; thus both its mean and median equal .5 and its *pdf* is maximized at this value.

8. Suppose a gambling casino generates nine (independent) uniform $(0, 1)$ values and pays a prize to the person guessing the number closest to the middle ranked value. (See Exercise 7.) What number would you select? If the prize were awarded for the guess closest to $X_{(8)}$, what number would you select?

9. If X is beta with $\alpha = \beta = 2$, evaluate the mean and variance for X. Compare $P(|X - \mu_X| < 2\sigma_X)$ with the Chebyshev bound for this probability.

10. A consumer research group investigated the proportion of household income spent on housing, for households in the United States. The proportions found were well represented by a beta random variable with $\alpha = 10$ and $\beta = 20$. What is the mean household proportion of income spent for housing? What is the standard deviation?

11. The relative humidity of the air is measured on a percentage scale and thus can be treated like a proportion. Assume that a meteorological model uses the *pdf* $f_H(t) = ct^{4.5}$, where c is a constant and $0 < t < 1$, to set the value for the humidity H at a given time and place. What must be the value for c? What is the median humidity at this location?

4.4 | Transformations of Random Variables

Applications frequently require the use of functions of random variables when switching scales of measurement, such as translating temperature readings from the Fahrenheit to the Celsius scale. Many of these translations of scale are linear, going from pounds to ounces, feet or yards to kilometers, and so on. We have previously seen some important results for such linear functions; if $Y = a + bX$, where a and $b \neq 0$ are constants, then the mean for Y is $\mu_Y = a + b\mu_X$ and its standard deviation is $\sigma_Y = |b|\sigma_X$. If Y is a continuous random variable, then its $100k$th quantile is $y_k = a + bx_k$ or $y_k = a + bx_{1-k}$, depending on the sign of b. Indeed, if X is a normal random variable, then so is $Y = a + bX$.

Many applications employ nonlinear functions of random variables. The decibel is a commonly used unit of relative power; if the "standard" power output is p_1 and a second power output is p_2, then their difference in level is $d = 10 \log_{10}(p_2/p_1) = 10(\log_{10} p_2 - \log_{10} p_1)$ measured in decibels. Thus if an original power output p_1 is changed to a (random) level P_2, the difference in level is also random and, measured in decibels, is $D = 10(\log_{10} P_2 - \log_{10} p_1)$. If we know or can specify the probability law for P_2, what is the resulting probability law for D? As another example, a life scientist, in studying the efficacy of using a synthetic female insect pheromone to attract males of the same species, assumes that the pheromone will radiate out a random distance R (depending on the temperature and humidity) in all directions from the source. The area of attraction for the pheromone is then the area A of the circle, centered at the location of the pheromone, with random radius R; hence $A = \pi R^2$ is also random. If we know the probability law for R, what then is the probability law for A?

If X is a random variable with a specified probability law, and we define $W = g(X)$ to be a function of X, then W in turn is a random variable with its own probability law. If, for example, $W = g(X) = 5 + \ln X$ and the range for X is, say, $R_X = \{x > 1\}$, then an observed value of $x = 2$ gives the observed value for W as $w = 5 + \ln 2 = 5.693$; similarly, the observed value of $x = 10$ yields $w = 5 + \ln 10 = 7.303$. Since the observed value for W is determined by the observed value for X, the probability law for W is determined by (or can be derived from) the probability law for X. The discussion of finding the probability law for $W = g(X)$ will consist of two parts, depending on whether X is discrete or continuous.

First let us address the case in which X is a discrete random variable with specified probability function $p_X(x) > 0$ for $x \in R_X$, the range of X. $g(\cdot)$ is a real function of a real variable whose domain of definition includes R_X, and we wish to find the probability law for $W = g(X)$. This case is quite straightforward and follows directly from the theorem of total probability (see Chapter 1). The set R_X gives the totality of possible observed values for X; the range R_W, for $W = g(X)$, necessarily consists of all possible values for $w = g(x)$, where $x \in R_X$, which itself then is a discrete set granted R_X is discrete. For each possible fixed $w \in R_W$, we can define the set of values for x, call it $B_w \subset R_X$, that result in $g(x) = w$. Immediately, then, the theorem of total probability gives

$$P(W = w) = p_W(w) = \sum_{x \in B_w} p_X(x),$$

the probability function for W. This is illustrated in the following two examples.

Example 4.17

A casino employs three biased coins, each of which has probability .6 of producing a head when flipped (so the probability of a tail occurring is .4 for each). These three coins are flipped one time; you as the bettor win \$1 if an even number of heads occurs (0 or 2) and will lose \$1 if the number of heads is odd. If X is the number of heads to occur on one flip of these three coins, then X is binomial with parameters $n = 3$, and $p = .6$, so

$$p_X(x) = \binom{3}{x}(.6)^x(.4)^{3-x}, \qquad \text{for } x \in R_X = \{0, 1, 2, 3\}.$$

Let W be the amount you win (i.e., the net change in your assets) for one play of this game; the range for W is then $R_W = \{-1, 1\}$, since these are the only two values that could occur. Thus there are only two different elements in R_W, meaning there are only two different sets $B_w \subset R_X$, using the above notation. These two sets are $B_{-1} = \{1, 3\}$ (meaning the observed value for W is -1 if the observed X is either 1 or 3) and $B_1 = \{0, 2\}$ (the observed value for W is 1 if the observed X is either 0 or 2). The probability function for W is

$$p_W(-1) = \sum_{x \in B_{-1}} p_X(x) = p_X(1) + p_X(3) = .504,$$

$$p_W(1) = \sum_{x \in B_1} p_X(x) = p_X(0) + p_X(2) = .496,$$

$$p_W(w) = 0, \qquad \text{otherwise.}$$

Then the mean for W is $\mu_W = (-1)(.504) + (1)(.496) = -.008$ (i.e., you could expect to lose \$0.008 for each play), and the variance for W is $\sigma_W^2 = .999936$.

Example 4.18

The owner of a small bakery bakes three loaves of white sandwich bread each day for walk-in customers; the number of these he sells depends on the number demanded. Let D represent the number of such loaves demanded, on any given day, and assume the probability function for D is given in Table 4.1.

Table 4.1
Probability Function for D

$d =$	0	1	2	3	4
$p_D(d) =$	.05	.10	.30	.30	.25

Let S be the number of these loaves that he will *sell* on this same day; once the three loaves are made, the range of S then certainly is $R_S = \{0, 1, 2, 3\}$, and S is a function of D, the number of loaves demanded. Specifically, $S = D$ for $D \leq 3$, and $S = 3$ if $D > 3$; thus $p_S(s) = p_D(s)$ for $s = 0, 1, 2$, and $p_S(3) = p_D(3) + p_D(4)$, giving the probability function in Table 4.2.

Table 4.2
Probability Function for S

$s =$	0	1	2	3
$p_S(s) =$	.05	.10	.30	.55

The probability that the baker will have one or more of these loaves not sold is then $P(S \leq 2) = .45$.

If X is a discrete random variable, it is quite straightforward to find the probability function for $W = g(X)$ using $p_W(w) = P(B_w)$ as illustrated in these two examples. If X is a continuous random variable, a much richer collection of possibilities exists, in large part because R_X is then continuous; as we have seen, if R_X is continuous, then our probability model allows us to compute the probability that the observed value for X lies in specific *intervals*. Granted that we define $W = g(X)$ in such a case, we then need to transfer the probability content from intervals, not points, in order to find the probability law for W.

The *cdf* proves invaluable in deriving the probability law for $W = g(X)$ from the probability law for X. The basic idea of this transfer is very simple; if t is any fixed number and $W = g(X)$, then the event $\{W \leq t\}$ is equivalent to the event $\{g(X) \leq t\}$. Thus the *cdf* for W is given by $F_W(t) = P(g(X) \leq t)$ and, if we can easily evaluate this latter probability, this gives the *cdf* for W. Evaluation of $P(g(X) \leq t)$ is especially straightforward for functions $g(x)$ that are *monotonic*, a type of function we shall now discuss.

Many functions of random variables employed in practical problems are *monotonic*. A *monotonic function* $g(x)$ has the property that its first derivative exists and always has the same sign for the intervals of x-values considered. Thus, if for some interval of x-values we have $dg(x)/dx > 0$, the function is *monotonic increasing over the interval* (since the slope is positive for all values in the interval). If $g(x)$ is defined for all values of x in some interval, and $dg(x)/dx < 0$ for all such x, then $g(x)$ is *monotonic decreasing over the interval* (since the slope is negative). If the random variable W is defined by $W = g(X)$, where $g(x)$ is either monotonic increasing or monotonic decreasing for $x \in R_X$, it is generally quite simple to express the *cdf* for W in terms of the *cdf* for X. The *pdf* for W is then given by the derivative of its *cdf*. Such functions as

$$g(x) = 20 + 15x, \qquad \text{for all real } x,$$
$$g(x) = \ln x, \qquad \text{for } x > 0,$$
$$g(x) = x^2, \qquad \text{for } x > 0,$$

are all monotonic increasing (for the ranges given for x), while such functions as

$$g(x) = 20 - 15x, \qquad \text{for all real } x,$$
$$g(x) = e^{-x}, \qquad \text{for all real } x,$$
$$g(x) = x^2, \qquad \text{for } x < 0,$$

are all monotonic decreasing. Note that the same function (such as x^2) may be increasing for one range of values and decreasing for another. As will become evident, in finding the probability law for $W = g(X)$ the only thing of importance is the behavior of $g(\cdot)$ for $x \in R_X$.

Example 4.19

Monotonic functions $w = g(x)$ have the property that there is one and only one value of x corresponding to each possible w and vice versa; this allows us to then also define the inverse function, denoted by $x = g^{-1}(w)$. For example, suppose we define $w = g(x) = \ln x$ for $x > 0$; this function is monotonic increasing since $dg(x)/dx = 1/x > 0$ for $x > 0$. This equation can in turn be solved for x in terms of w giving $x = e^w = g^{-1}(w)$. If we define $w = g(x) = x^2$ for $x < 0$, then $dg(x)/dx = 2x < 0$ for all such x, so this function is monotonic decreasing over this range. Again we can solve for the inverse function, giving $x = g^{-1}(w) = -\sqrt{w}$. (The sign is negative since the function $g(x)$ was defined for negative x.)

If $w = g(x)$ is monotonic (either increasing or decreasing), the inverse function $x = g^{-1}(w)$ defines the value of x that results in a given value for w; that is, $g(g^{-1}(w)) = w$. Recall that $dw/dx \equiv (dx/dw)^{-1}$, the reciprocal of the derivative of x with respect to w. Thus, if $dw/dx > 0$, then $dx/dw > 0$; so g^{-1} is increasing if $g(\cdot)$ is increasing and decreasing if $g(\cdot)$ is decreasing. Figure 4.9 pictures monotonic increasing and decreasing functions; it also points out that $w = g(x) \le t$ is equivalent to $x \le g^{-1}(t)$, if $g(\cdot)$ is increasing and equivalent to $x \ge g^{-1}(t)$, if $g(\cdot)$ is decreasing.

It is quite straightforward to derive the *cdf* for $W = g(X)$, from the *cdf* for X, when $g(\cdot)$ is monotonic for $x \in R_X$. If $g(\cdot)$ is monotonic increasing, for any fixed t the event $\{W \le t\}$ is equivalent to $\{g(X) \le t\}$, which in turn is equivalent to the event $\{X \le g^{-1}(t)\}$. Since probabilities of occurrence of equivalent events must be equal, we have

$$F_W(t) = P(W \le t) = P(g(X) \le t)$$
$$= P(X \le g^{-1}(t)) = F_X(g^{-1}(t)).$$

If $g(\cdot)$ is decreasing, for any fixed t the event $\{W \le t\}$ is equivalent to $\{g(X) \le t\}$, which in turn is equivalent to the event $\{X \ge g^{-1}(t)\}$. Thus for decreasing functions this gives

$$F_W(t) = P(W \le t) = P(g(X) \le t) = P\left(X \ge g^{-1}(t)\right)$$
$$= 1 - P\left(X < g^{-1}(t)\right)$$
$$= 1 - \left(F_X(g^{-1}(t)) - P(X = g^{-1}(t))\right). \tag{1}$$

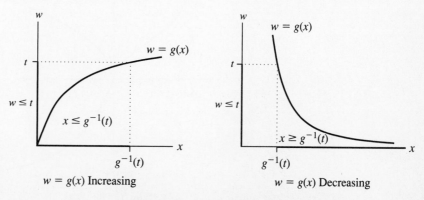

Figure 4.9   Monotonic functions

The term $P(X = g^{-1}(t))$ is included to cover the cases in which X is discrete (since this probability may be positive); if X is continuous, this term is always equal to 0. This establishes the following theorem.

THEOREM 4.5 If X is a random variable with *cdf* $F_X(t)$ and if $g(\cdot)$ is a monotonic function for $x \in R_X$, the *cdf* for $W = g(X)$ is given by

$$F_W(t) = \begin{cases} F_X\left(g^{-1}(t)\right), & \text{if } g(\cdot) \text{ is increasing,} \\ 1 - \left(F_X(g^{-1}(t)) - P(X = g^{-1}(t))\right), & \text{if } g(\cdot) \text{ is decreasing.} \end{cases}$$

This result holds whether X is discrete or continuous; the previously discussed procedure of simply transferring the point probabilities, however, is generally more straightforward when X is discrete.

If X is a continuous random variable, then $P(X = g^{-1}(t)) = 0$ for all values of t, and $F_W(t) = 1 - (F_X(g^{-1}(t)))$, if $g(\cdot)$ is decreasing. Only the behavior of $g(x)$ for $x \in R_X$ is cogent in deriving the probability law for $W = g(X)$. For example, if $g(x)$ is monotonic increasing for, say, $x \in R_X$ but not for other real x, that fact has no effect on the *cdf* for $W = g(X)$.

Let us illustrate this reasoning by deriving the *Rayleigh probability law* from the exponential probability law. If X is exponential with parameter λ, then $R = \sqrt{X}$ has the *Rayleigh probability law*. The square root function is monotonic increasing, so the *cdf* for R is given by

$$F_R(t) = P(R \le t) = P(\sqrt{X} \le t) = P(X \le t^2)$$
$$= F_X(t^2) = 1 - e^{-\lambda t^2}, \qquad \text{for } t > 0.$$

This transformation, or transfer of probability from X to R is illustrated in Fig. 4.10, in a way suggested by Clark Pritchard. Three graphs are combined

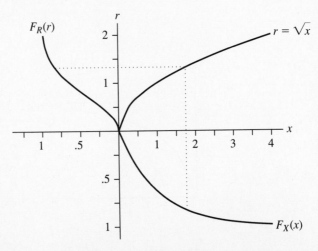

Figure 4.10 $R = \sqrt{X}$, with X exponential

into one in this figure, illustrating the transformation $r = \sqrt{x}$, and also giving the original *cdf* $F_X(x)$ and the derived *cdf* for R, $F_R(r)$. As pictured, the height of the exponential *cdf* at a given value (such as $x = 1.75$) transfers directly to be the height of the Rayleigh *cdf*, but at a transformed location ($r = 1.323$ in this case). Note that this manipulation results in the shape of $F_R(r)$ being different from the shape of $F_X(x)$ (because $r = \sqrt{x}$ is nonlinear).

The *pdf* for a continuous random variable is given by the derivative of its *cdf*. Thus if $g(\cdot)$ is an increasing function and $W = g(X)$ is continuous, its *pdf* is related to the *pdf* for X by

$$f_W(t) = \frac{d}{dt} F_W(t) = \frac{d}{dt} F_X(g^{-1}(t)) = f_X(g^{-1}(t)) \frac{dg^{-1}(t)}{dt},$$

using the chain rule for derivatives. If $g(\cdot)$ is a decreasing function, the *pdf* for $W = g(X)$ is

$$f_W(t) = \frac{d}{dt} F_W(t) = \frac{d}{dt} \left(1 - F_X(g^{-1}(t)) \right) = -f_X(g^{-1}(t)) \frac{dg^{-1}(t)}{dt},$$

again using the chain rule. Since the derivative $dg^{-1}(t)/dt$ is negative for a decreasing function $g(\cdot)$, we can summarize both these results in the following corollary.

COROLLARY 4.1 If X is a continuous random variable and we define $W = g(X)$, where $g(\cdot)$ is monotonic (either increasing or decreasing) for $x \in R_X$, then the *pdf* for W is given by

$$f_W(t) = f_X(g^{-1}(t)) \left| \frac{dg^{-1}(t)}{dt} \right|.$$

The following examples illustrate the use, and power, of Theorem 4.5 and its Corollary 4.1.

Example 4.20

Theorem 4.5 was in fact used in deriving the probability law for $X = \mu + \sigma Z$, where Z was a standard normal random variable. *Any* linear function is monotonic, depending on whether the slope is positive (increasing) or negative (decreasing). Suppose X is an exponential random variable with $\lambda = 3$, so $F_X(t) = 1 - e^{-3t}$ for $t \geq 0$, and the *pdf* for X is $f_X(t) = 3e^{-3t}$. Let us find the probability law for $W = 5 - 3X$, a decreasing linear function of X. The range for X is $R_X = \{x \geq 0\}$; for $x \in R_X$, $w = 5 - 3x$ varies from 5 to $-\infty$, so the range for W is $R_W = \{w \leq 5\}$. The inverse function for $w = 5 - 3x$ is gotten by simply solving for x in terms of w, giving $x = (5 - w)/3 = g^{-1}(w)$; thus the *cdf* for W is

$$F_W(t) = \begin{cases} 1 - F_X\left(g^{-1}(t)\right) = 1 - (1 - e^{-3(5-t)/3}) = e^{t-5}, & \text{for } t \in R_W, \\ 1, & \text{for } t \geq 5. \end{cases}$$

The *pdf* for W is

$$f_W(t) = \frac{d}{dt} F_W(t) = e^{t-5},$$

by simply differentiating $F_X(t)$. The same result can be gotten using Corollary 4.1:

$$g^{-1}(t) = \frac{5-t}{3},$$

$$\frac{d}{dt} g^{-1}(t) = -\frac{1}{3},$$

$$f_X(g^{-1}(t)) \left| \frac{d}{dt} g^{-1}(t) \right| = 3e^{-3(5-t)/3} \left| -\frac{1}{3} \right|$$

$$= e^{t-5}, \qquad t \leq 5.$$

We could use this *pdf* to evaluate the mean and variance for W or, since W is a linear function of X, it is perhaps easier to use our earlier results, $\mu_W = 5 - 3\mu_X = 4$ and $\sigma_W = |-3|\sigma_X = 1$.

Suppose X and $W = g(X)$ are both continuous; various properties of their probability laws can then be described by their quantiles, as we know, and certainly the quantiles for the two must be related. Suppose $g(x)$ is monotonic increasing and we let x_k represent the 100kth quantile for X, while w_k represents the 100kth quantile for $W = g(X)$. Then, from the definition of the quantiles of a probability law, $F_W(w_k) = k$ and $F_X(x_k) = k$. From Theorem 4.5, $F_W(w_k) = F_X(g^{-1}(w_k)) = k$ implies that $g^{-1}(w_k) = x_k$, so $w_k = g(x_k)$, giving a straightforward and easily used relationship between quantiles for monotonic increasing functions. You will be asked in Exercises 4.4 to find the relationship between the quantiles of X and $W = g(X)$ when $g(\cdot)$ is monotonic decreasing.

Example 4.21

We earlier mentioned that $R = \sqrt{X}$, where X is exponential, follows the Rayleigh probability law; we found the *cdf* for R to be $F_R(t) = 1 - e^{-\lambda t^2}$ for $t > 0$. The 100kth quantile for R is the solution to $1 - e^{-\lambda r_k^2} = k$, which is easily solved to find

$$r_k = \sqrt{-\frac{1}{\lambda} \ln(1-k)} = \sqrt{x_k},$$

where x_k is the 100kth quantile for X. Thus the median for R is $.833/\sqrt{\lambda}$, and its interquartile range is $.641/\sqrt{\lambda}$.

The *pdf* for R is

$$f_R(t) = \frac{d}{dt}\left(1 - e^{-\lambda t^2}\right) = 2\lambda t e^{-\lambda t^2}, \quad \text{for } t > 0.$$

Thus the mean for R is

$$\mu_R = \int_0^\infty t\, 2\lambda t e^{-\lambda t^2}\, dt = \int_0^\infty 2u e^{-u}\, \frac{du}{2\sqrt{\lambda u}}$$

$$= \frac{\Gamma(\frac{3}{2})}{\sqrt{\lambda}} = \frac{\sqrt{\pi}}{2\sqrt{\lambda}} = \frac{.886}{\sqrt{\lambda}},$$

where we have used the substitution $u = \lambda t^2$; since the square root function is not linear, $\mu_R \neq g(\mu_X) = 1/\sqrt{\mu_X}$. By evaluating $\mathrm{E}[R^2] = \mathrm{E}[X] = 1/\lambda$ we find $\sigma_R^2 = (1 - \pi/4)/\lambda$.

Example 4.22

The exponential probability law has had wide use in the theory of life testing, modeling the length of time X until an item fails. Since this probability law has only one parameter, the general shapes of the exponential *pdf*s are the same, differing only by the value for λ, which controls the height of the *pdf* at 0 and the subsequent spread of the *pdf*. To provide a richer set of possibilities for the shape of the *pdf*, the *Weibull probability law* is also frequently employed in life testing applications. It can be derived as the probability law for a simple polynomial function of an exponential random variable; to find this probability law, assume that X is exponential with parameter $\lambda = 1$, and let $W = \alpha X^{\frac{1}{\beta}}$, where $\alpha > 0$ and $\beta > 0$ are constants. The inverse transformation is then given by $X = (W/\alpha)^\beta$ and the *cdf* for W is

$$F_W(t) = F_X\left(\left(\frac{t}{\alpha}\right)^\beta\right) = 1 - e^{-(t/\alpha)^\beta},$$

for $t \in R_W = R_X = \{t \geq 0\}$; the *pdf* for W is then

$$f_W(t) = \frac{\beta t^{\beta-1}}{\alpha^\beta} e^{-(t/\alpha)^\beta}, \qquad \text{for } t > 0. \tag{2}$$

For any $k > \beta$, it is easy to see that $\mathrm{E}[W^k] = \alpha^k \Gamma(1 + k/\beta)$, from which we can evaluate the mean and variance of W. Note that $\beta = 1$ makes W exponential with parameter $\alpha = 1/\lambda$; with $\beta < 1$ this *pdf* is unbounded at 0; and with $\beta > 1$ the *pdf* is 0 at $w = 0$, going through a maximum at $t = \alpha (1-1/\beta)^{1/\beta}$, then asymptotically going to 0 as $w \to \infty$. The $100k$th quantile for the Weibull probability law is

$$w_k = \alpha x_k^{1/\beta} = \alpha\left(-\frac{1}{\lambda}\ln(1-k)\right)^{1/\beta}.$$

It is common in some areas of applied statistics to take the natural logarithms of observed data values and then to assume that the resulting values are normally distributed (have a normal distribution). This corresponds, in the language of random variables, to saying we have a random variable X such that $Y = \ln X$ is normal with mean μ_Y and variance σ_Y^2; the original random variable X is said to follow the lognormal probability law. Granted that

$Y = \ln X$, which is $Y = g^{-1}(X)$, the inverse transformation in our preceding discussion, we see that $X = e^Y = g(Y)$ has the lognormal distribution. The *cdf* for X, with $t > 0$, is then $F_X(t) = F_Y(\ln t) = F_Z((\ln t - \mu_Y)/\sigma_Y)$, where $F_Z(\cdot)$ is the standard normal *cdf*.

Since $X = e^Y$ is not a linear function of Y, the mean and standard deviation of X are not linear functions of μ_Y and σ_Y. We can find μ_X and σ_X^2 from the *pdf* for X or, equivalently, we could evaluate them from the *pdf* for Y since $X = g(Y)$ is a function of Y. Since $Y = \mu_Y + \sigma_Y Z$ is a linear function of a standard normal random variable Z, we also have $X = e^{\mu_Y + \sigma_Y Z} = h(Z)$, where X is a function of Z and Z is standard normal. We shall exploit this fact in finding μ_X. If Z is a standard normal random variable and t is any real number, then

$$
\begin{aligned}
E[e^{tZ}] &= \int_{-\infty}^{\infty} e^{tz} \frac{1}{\sqrt{2\pi}} e^{-z^2/2} \, dz \\
&= \int_{-\infty}^{\infty} \frac{1}{\sqrt{2\pi}} e^{-(z^2 - 2tz + t^2 - t^2)/2} \, dz \\
&= e^{t^2/2} \int_{-\infty}^{\infty} \frac{1}{\sqrt{2\pi}} e^{-(z-t)^2/2} \, dz = e^{t^2/2},
\end{aligned}
\tag{3}
$$

by completing the square in the exponent, and recognizing that the last integral simply gives the area under a normal *pdf* with mean t and variance 1 (so the definite integral equals 1). Since

$$
\mu_X = E[X] = E[e^{\mu_Y + \sigma_Y Z}] = e^{\mu_Y} E[e^{\sigma_Y Z}],
$$

we immediately have $\mu_X = e^{\mu_Y} e^{\sigma_Y^2/2} = e^{\mu_Y + \sigma_Y^2/2}$, using $t = \sigma_Y$ in Eq. (3). You can verify that the variance of the lognormal distribution is $e^{2\mu_Y + \sigma_Y^2}(e^{\sigma_Y^2} - 1)$ (by evaluating $E[X^2]$). The $100k$th quantile for X is $x_k = e^{y_k} = e^{\mu_Y + \sigma_Y z_k}$, where z_k is the corresponding standard normal quantile.

Recall that the range of a continuous random variable X is defined to be the collection of x-values such that $f_X(x) > 0$; this in turn means that R_X is the set of x-values for which the slope of the *cdf* $F_X(x)$ is increasing, so $W = F_X(X)$ is then a monotonic increasing function of X. What is the probability law for W? The inverse to the function $w = F_X(x)$ is $x = F_X^{-1}(w)$; the range of $W = F_X(X)$ is $R_W = \{0 \le w \le 1\}$ since $0 \le F_X(x) \le 1$ for all values of x. Thus for any $0 \le t \le 1$ we have

$$
F_W(t) = F_X\left(F_X^{-1}(t)\right) = t,
$$

which is the *cdf* for a uniform $(0, 1)$ random variable. That is, $W = F_X(X)$ is uniform on the interval $(0, 1)$, *no matter which continuous $F_X(\cdot)$ we employ* in the transformation.

The inverse of an increasing function is also an increasing function; that is, if $w = F_X(x)$ is monotonic increasing, so is $x = F_X^{-1}(w)$. Now suppose W is uniform on the interval $(0, 1)$, so its *cdf* is $F_W(t) = t$, for $0 \le t \le 1$, and define $X = F_X^{-1}(W)$. (Note that the inverse function is then $W = F_X(X)$, since the

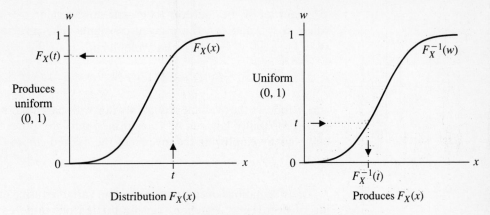

Figure 4.11 Probability integral transform, continuous *cdf*

inverse of the inverse returns the original function.) What is the probability law for X? Since $F_X^{-1}(\cdot)$ is increasing, we have

$$P(X \le t) = F_W(F_X(t)) = F_X(t), \qquad \text{for } 0 \le F_X(t) \le 1,$$

that is, for any $t \in R_X$; thus, if we start with a uniform $(0, 1)$ random variable W and transform it into $X = F_X^{-1}(W)$, the resulting random variable has *cdf* $F_X(t)$, the *cdf* whose inverse was employed in the transformation. This result is called the *probability integral transform* and shows how any continuous probability law can be generated by transforming a uniform random variable. The following theorem summarizes this discussion.

THEOREM 4.6 Let $F_X(\cdot)$ be a continuous *cdf*, defining the probability law for a random variable X. The random variable $W = F_X(X)$ then is uniform on the interval $(0, 1)$. Conversely, if W is uniform on $(0, 1)$ the random variable $X = F_X^{-1}(W)$ has the probability law defined by $F_X(\cdot)$.

Figure 4.11 illustrates the results stated in Theorem 4.6. This transformation is discussed further in the following example.

Example 4.23

Consider the *cdf* $F_X(t) = 1 - 1/(t + 1)$ for $t \ge 0$, and $F_X(t) = 0$ for $t < 0$. Thus the range of X is $R_X = \{x > 0\}$ and, if we define $w = F_X(x) = 1 - 1/(1 + x)$, the inverse function is $x = F_X^{-1}(w) = w/(1 - w)$. To illustrate the probability integral transform, suppose W is uniform on the interval $(0, 1)$ and we define $X = W/(1 - W)$, which is the inverse of the *cdf* $F_X(t)$. Then we also have $W = 1 - 1/(X + 1)$ and

$$F_X(t) = F_W\left(1 - \frac{1}{t+1}\right) = 1 - \frac{1}{t+1}, \qquad \text{for } 0 \le 1 - \frac{1}{t+1} \le 1,$$

that is, for $t \ge 0$; the probability law for X has *cdf* $F_X(t)$ as claimed.

Again let W be uniform $(0,1)$ and now define $X = \alpha \left(-\ln(1 - W)\right)^{1/\beta}$, where $\alpha > 0$ and $\beta > 0$ are given constants. The inverse transformation is $W = 1 - e^{-(X/\alpha)^\beta}$, which is the Weibull *cdf*. Thus the *cdf* for X is

$$F_X(t) = F_W(1 - e^{-(t/\alpha)^\beta}) = 1 - e^{-(t/\alpha)^\beta},$$

and X follows the Weibull probability law with parameters α, β. The uniform $(0,1)$ probability law can be transformed into *any* other continuous probability law by employing the inverse of the desired *cdf* in the transformation.

Transformations of continuous random variables using monotonic (increasing or decreasing) functions can be carried out quite easily. There are, of course, other commonly employed transformations that may not have this monotonic property for all $x \in R_X$, the range of the random variable being transformed. Perhaps the most frequently employed of these are X^2 and $|X|$, where R_X includes intervals of both positive and negative numbers; in certain applied areas, trigonometric transformations ($\sin X$, $\tan^{-1} X$, and so forth) are used and for some R_X may be nonmonotonic. Still, if $W = g(X)$, it must be true that $P(W \le t) = P(g(X) \le t)$; with $g(\cdot)$ nonmonotonic, this latter probability may involve X lying in some union of intervals. The theorem of total probability assures us that $P(W \le t)$ is then given by the sum of the probability contents of those intervals for which $g(x) \le t$.

Let us illustrate this discussion by letting Z be standard normal, so R_Z is the whole real line, and define $W = Z^2$. The range for W is then $R_W = \{w \ge 0\}$, and $F_W(t) = 0$ for $t < 0$. For $t \ge 0$,

$$F_W(t) = P(W \le t) = P(Z^2 \le t) = P\left(-\sqrt{t} \le Z \le \sqrt{t}\right)$$
$$= F_Z\left(\sqrt{t}\right) - F_Z\left(-\sqrt{t}\right) = 2F_Z\left(\sqrt{t}\right) - 1,$$

from the symmetry of the standard normal; the *cdf* for W can be evaluated directly from the table of the standard normal *cdf* (Table B.3 in Appendix B). Thus the *pdf* for W is, for $t \ge 0$,

$$f_W(t) = \frac{d}{dt} F_W(t) = \frac{d}{dt}\left(2F_Z\left(\sqrt{t}\right) - 1\right)$$
$$= \frac{1}{2\sqrt{t}} 2f_Z\left(\sqrt{t}\right) = \frac{1}{\sqrt{2t\pi}} e^{-t/2};$$

W is called a χ^2 (chi-square) random variable with one degree of freedom. As will be seen, the standard normal is frequently used in many applications, as is its square. This *pdf* for W is actually a gamma *pdf* with $\lambda = \frac{1}{2} = n$; that is, the χ^2 probability law with one degree of freedom is a particular case of the gamma probability law (as is the χ^2 with more degrees of freedom).

The parameter of the χ^2 distribution is called *degrees of freedom*. With $W = Z^2$, the square of a single standard normal random variable, the χ^2 law has *one* degree of freedom; if $W = Z_1^2 + Z_2^2 + \cdots + Z_k^2$, where $Z_1, Z_2, \ldots, Z_k$

are independent standard normals (a concept to be discussed in Chapter 7), then again the probability law for W is χ^2, now with k degrees of freedom. This parameter called "degrees of freedom" counts the number of independent standard normals which have been squared and added to get the value for W.

Example 4.24

To illustrate this nonmonotonic case further, assume that X is uniform on the interval $(-1, 2)$ and let $W = |X|$; the range for X is then $R_X = \{-1 \le x \le 2\}$ and its *cdf* is $F_X(t) = (t+1)/3$ for $t \in R_X$. The range for W is $R_W = \{0 \le w \le 2\}$, so $F_W(t) = 0$ for $t < 0$, and $F_W(t) = 1$ for $t > 2$. The *cdf* for $t \in R_W$ is

$$F_W(t) = P(|X| \le t) = P(-t \le X \le t) = F_X(t) - F_X(-t);$$

for $0 \le t \le 1$, then $F_W(t) = (t+1)/3 - (-t+1)/3 = 2t/3$, while for $1 < t \le 2$, $F_W(t) = (t+1)/3 - 0 = (t+1)/3$ since $F_X(-t) = 0$ for $t > 1$. The *pdf* for W then equals $\frac{2}{3}$, for $0 < w < 1$ and equals $\frac{1}{3}$, for $1 < w < 2$. Thus W is not uniform; the probability that it lies between 0 and 1 is twice the probability that it lies between 1 and 2. This occurs, of course, because $W = |X|$ and all the probability of X lying between -1 and 1 gets mapped into W lying between 0 and 1.

Transformations of continuous random variables can also produce discrete distributions. One of the simplest and most useful transformations of this type is provided by the "probability integral transform" for a discrete probability law, a procedure generating any desired *discrete* distribution from the continuous uniform $(0, 1)$ probability law. Suppose $F_X(t)$ is a discrete *cdf*, describing the probability law for a discrete random variable X with range R_X. Recall then that $F_X(t)$ is a step function with jumps occurring at each $x \in R_X$, with the size of the jump being given by $p_X(x)$, the amount of probability located at x. For example, take $R_X = \{2, 4, 6, 10\}$ with the values for $p_X(x)$ being .4, .3, .2, and .1, respectively, at these four observed x values. The *cdf* for X then is

$$F_X(t) = \begin{cases} .4, & \text{for } 2 \le t < 4, \\ .7, & \text{for } 4 \le t < 6, \\ .9, & \text{for } 6 \le t < 10, \\ 1, & \text{for } 10 \le t. \end{cases}$$

Let W be uniform on the interval $(0, 1)$, and define

$$X = \begin{cases} 2, & \text{if } 0 \le W \le .4, \\ 4, & \text{if } .4 < W \le .7, \\ 6, & \text{if } .7 < W \le .9, \\ 10, & \text{if } .9 < W. \end{cases}$$

We then would have $P(X = 2) = P(0 \le W \le .4) = .4$, $P(X = 6) = P(.7 < W \le .9) = .2$, and so forth. Defined this way the random variable X has the

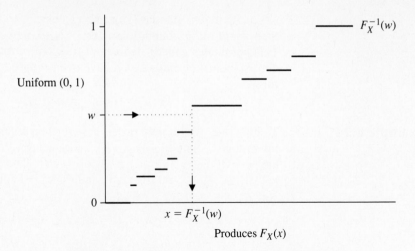

Figure 4.12   Probability integral transform, discrete *cdf*

discrete range R_X defined earlier in the paragraph and the same probability function.

If we have any discrete *cdf*, generate a uniform random variable W on the interval $(0, 1)$, and then define the observed value for X to be the element of the discrete range that matches the "gap" in $F_X(t)$ into which W falls, then the random variable generated from W will have the discrete *cdf* $F_X(t)$. A little more exactly, if W is uniform on the interval $(0, 1)$, then all observed values for W will lie between 0 and 1. Define X to be the *smallest* value in R_X such that $F_X(X) \geq W$ and (rather sloppily) let this value be denoted as $X = F_X^{-1}(W)$. The possible observed values for this X are then precisely the elements of R_X, since $F_X(t)$ is discontinuous (and jumps in value) at the elements of R_X. If we let $x_1 < x_2$ denote two *successive* elements in R_X, this rule will give $X = x_2$ if and only if $F_X(x_1) < W \leq F_X(x_2)$; thus, since $F_W(t) = t$ for $0 \leq t \leq 1$, we have $P(X = x_2) = F_W(F_X(x_2)) - F_W(F_X(x_1)) = F_X(x_2) - F_X(x_1)$. That is, $P(X = x_2)$ is then given by the size of the jump in $F_X(t)$ at $t = x_2$; in short, $X = F_X^{-1}(W)$ then has *cdf* $F_X(t)$, just as in the continuous case discussed earlier. This procedure is illustrated in Fig. 4.12 and is applied in the following example.

Example 4.25

Suppose a computer simulation of the social effects of gambling includes the play of a state lottery game; in this game scratch-off tickets are purchased and the winners are determined by the message under the material scratched off by the purchaser. Let us also assume that individual tickets are well modeled as independent Bernoulli trials with probability of success (winning some prize) equal to $\frac{1}{9}$. The number of tickets an individual must purchase to *first* win some prize is a geometric random variable X with parameter $p = \frac{1}{9}$. The *cdf* for X thus is

$$F_X(t) = 1 - \left(\frac{8}{9}\right)^{\lfloor t \rfloor}, \qquad \text{for any } t > 0,$$

where $\lfloor t$ represents the largest integer in t (so $\lfloor t$ is the largest integer such that $\lfloor t \leq t$). As before, assume that W is a continuous uniform random variable on the interval $(0, 1)$, and let $X = F_X^{-1}(W)$ be the *smallest* integer such that

$$F_X(X) = 1 - \left(\frac{8}{9}\right)^{\lfloor X} = 1 - \left(\frac{8}{9}\right)^{X} \geq W,$$

or $X \geq (\ln(1 - W))/\ln \frac{8}{9}$ since $\ln \frac{8}{9} < 0$. Thus the value for X is $1 + \lfloor((\ln(1 - W)]/\ln \frac{8}{9})$, the largest integer in the ratio $(\ln(1 - W))/\ln \frac{8}{9}$ plus 1. One of the problems in Exercises 4.4 asks you to verify the fact that $1 - W$ is uniform on $(0, 1)$ if W is uniform on this interval. Thus we can just as well replace $1 - W$ by W so $X = \lfloor 1 + (\ln W)/\ln \frac{8}{9}$. If the observed uniform value is $w = .7678$, say, this transforms into the observed value of X as $x = \lfloor 1 + (\ln .7678/\ln \frac{8}{9}) = 3$, while if $w = .0214$ is observed, then $x = \lfloor 1 + (\ln .0214)/\ln \frac{8}{9} = 33$. The discrete integer values for X generated by this procedure follow the geometric probability law with $p = \frac{8}{9}$.

 **EXERCISES 4.4**

1. A marksman fires five rounds at a target, each of which either does or does not hit the bull's-eye. Assume that the probability of each round hitting the bull's-eye is .8 and that the individual shots are independent.
   a. What is the probability law for X, the number of rounds to hit the bull's-eye?
   b. What is the probability law for the number of rounds M to miss the bull's-eye, that is, $M = 5 - X$?

2. Assume, as in Example 4.17, that a casino will flip three coins and that you win \$1 if the number of heads is even and lose \$1 if the number of heads is odd. This time assume that the probability of a head occurring is $\frac{3}{4}$, and find the probability function for W, the amount you will win on one play. Evaluate $E[W]$.

3. A pair of fair dice is rolled one time; let $W = -1$ if the sum of the two numbers is odd and $W = 1$ if the sum of the two numbers is even. Find the probability function for W and evaluate μ_W, and σ_W^2.

4. Let X be exponential with parameter λ and define $W = X + a$; find the *pdf* for W. W is called a *shifted exponential random variable*, since its values are shifted a units from X.

5. Find the *pdf* for $Y = 5 + 3X$, granted that X is exponential with $\mu_X = 4$.

6. Assume that W is uniform on the interval $(0, 1)$, and show that $U = 1 - W$ has the same probability law.

7. Generalize the result from Exercise 6. Let W be uniform on the interval (a, b), and find the probability law for $U = b - W$.

8. In a bartering situation, the price paid for an item could be modeled as a random variable Y (idealized as being continuous). The seller's first offering price (the largest possible value) is to be b, and the smallest possible selling price is a. You want the average selling price to be μ_Y and the standard deviation to be σ_Y.

    a. Let X be a beta random variable with parameters α, β. If $Y = g(X)$, find the function $g(\cdot)$ that gives the desired range for Y.

    b. What are the required values for α and β to give the desired values for μ_Y and σ_Y?

9. *For the modeler* A searchlight can swing through a $180°$ angle; label these extremes $-\pi/2$ and $\pi/2$ (using radians). One yard directly in front of the searchlight, perpendicular to the line from the searchlight at 0 radians, is a straight (infinitely long) wall extending from $-\infty$ to ∞. Suppose the searchlight is set at an angle Θ selected "randomly" from $(-\pi/2, \pi/2)$ and then turned on. The spot illuminated on the wall is then $W = \tan \Theta$. Find the *cdf* and *pdf* for W. This is called the *Cauchy probability law.*

10. Let X be a continuous random variable, and define $W = g(X)$, where $g(\cdot)$ is monotonic decreasing. Show that $w_k = g(x_{1-k})$ where w_k is the $100k$th quantile for W and x_{1-k} is the $100(1-k)$th quantile for X.

11. The random variable U is uniform on $(0, 1)$; define $V = \ln 1/U^2$, and find the probability law for V.

12. Let X be exponential, with parameter λ, and define $W = (e^X - 1)/\beta$, where $\beta > 0$ is a constant. Find the *cdf* for W. This is one form of the *Pareto probability law.*

13. Evaluate the $100k$th quantile for the Pareto probability law (see Exercise 12) with parameters λ, β.

14. Suppose X is a normal random variable with mean μ_X and variance σ_X^2; let $W = (X - \mu_X)^2/\sigma_X^2$, the square of the standard form for X. What is the probability law for W?

15. Let W be a χ^2 random variable with one degree of freedom. Evaluate
    a. $P(W < 1)$ b. $P(W > \frac{1}{4})$ c. μ_W and σ_W.

16. Let W be a Weibull random variable with parameters α, β. Evaluate the $100k$th quantile for W.

17. If Z is standard normal, $Y = |Z|$ is called a *folded standard normal random variable*. Find the *cdf* for Y.

18. A round is fired at a target. The horizontal miss distance is Z, a standard normal random variable, where the target lies at the origin $z = 0$. Let X be the magnitude of the miss distance (ignoring whether the round fell to the left or right of the target). What is the probability law for X? Evaluate $P(X < .9)$.

19. Express the quantiles of the folded standard normal distribution (see Exercise 17) in terms of the quantiles for the standard normal distribution.

20. Express the quantiles of the χ^2 one degree of freedom distribution in terms of the quantiles for the standard normal distribution.

21. A round is fired at a target. The horizontal miss distance is X, a normal random variable with $\mu_X = .5$, and $\sigma_X = 1$, and the target lies at the origin $x = 0$. Let Y be the magnitude of the miss distance (ignoring whether the round fell to the left or right of the target). Evaluate $P(Y < .9)$.

22. The random variable X is exponential with parameter λ. Define $Y = \lceil X$ to be the value gotten by always rounding X up to the next integer (e.g., $\lceil \pi = 4, \lceil e = 3$, and $\lceil 5 = 5$). What is the probability law for Y?

23. Assume X is binomial with parameters $n = 3$ and $p = \frac{1}{2}$ and define $U = F_X(X)$. What is the probability law for U?
24. Let X be Weibull with parameters $\alpha = 1$ and $\beta = 3$, and define $Y = \lfloor X$, where $\lfloor X$ is the integer part of X. Find the probability law for Y.

4.5 | Summary

The continuous uniform probability law describes a continuous random variable with range $R_X = \{a < x < b\}$ and a constant *pdf*.

CONTINUOUS UNIFORM RANDOM VARIABLE

Density function: $f_X(x) = \dfrac{1}{b-a}$

Range: $R_X = \{a < x < b\}$

Parameters: $-\infty < a < b < \infty$

Mean: $\dfrac{b+a}{2}$

Variance: $\dfrac{(b-a)^2}{12}$

The Poisson process assumptions imply that the time to the first occurrence, starting from an arbitrary origin $t = 0$, is exponential, with range $R_X = \{x > 0\}$.

EXPONENTIAL RANDOM VARIABLE

Density function: $f_X(x) = \lambda e^{-\lambda x}$

Range: $R_X = \{x > 0\}$

Parameter: $\lambda > 0$

Mean: $\dfrac{1}{\lambda}$

Variance: $\dfrac{1}{\lambda^2}$

The time of occurrence of the rth event in a Poisson process with parameter λ has the Erlang distribution with range $R_X = \{x > 0\}$.

ERLANG RANDOM VARIABLE

Density function: $f_X(x) = \dfrac{\lambda^r x^{r-1}}{(r-1)!} e^{-\lambda x}$

Range: $R_X = \{x > 0\}$

Parameters: $r = 1, 2, 3, \ldots, \qquad \lambda > 0$

Mean: $\dfrac{r}{\lambda}$

Variance: $\dfrac{r}{\lambda^2}$

The gamma function is $\Gamma(n) = \int_0^\infty u^{n-1} e^{-u}\, du$ for $n > 0$. In normalized form it is used to define the *pdf* for the gamma probability law.

GAMMA RANDOM VARIABLE

Density function: $f_X(x) = \dfrac{\lambda^n x^{n-1}}{\Gamma(n)} e^{-\lambda x}$

Range: $R_X = \{x > 0\}$

Parameters: $n > 0, \lambda > 0$

Mean: $\dfrac{n}{\lambda}$

Variance: $\dfrac{n}{\lambda^2}$

The normal probability law is used extensively in many statistical applications and has a symmetric bell-shaped *pdf* with range $R_X = \{-\infty < x < \infty\}$.

NORMAL RANDOM VARIABLE

Density function: $f_X(x) = \dfrac{1}{\sigma\sqrt{2\pi}} e^{-(x-\mu)^2/2\sigma^2}$

Range: $R_X = \{-\infty < x < \infty\}$

Parameters: $-\infty < \mu < \infty,\ 0 < \sigma$

Mean: μ

Variance: σ^2

The beta random variable has the restricted range $R_X = \{0 < x < 1\}$, with a *pdf* that can take on a wide variety of shapes over this interval.

BETA RANDOM VARIABLE

Density function: $f_X(x) = \dfrac{\Gamma(\alpha + \beta)}{\Gamma(\alpha)\Gamma(\beta)} x^{\alpha-1}(1-x)^{\beta-1}$

Range: $R_X = \{0 < x < 1\}$

Parameters: $\alpha > 0,\ \beta > 0$

Mean: $\dfrac{\alpha}{\alpha + \beta}$

Variance: $\dfrac{\alpha\beta}{(\alpha + \beta)^2(\alpha + \beta + 1)}$

Granted the probability law is specified for a random variable X, the probability law for $W = g(X)$ can be derived from that for X. If X is discrete the probability content for observed values $x \in R_X$ is transferred directly to $w \in R_W$, using the theorem of total probability as needed. If X is continuous, the probability law for $W = g(X)$ is generally most easily found from the basic equivalence of events $\{W \le t\} \equiv \{g(X) \le t\}$. For monotonic functions this leads to

$$F_W(t) = \begin{cases} F_X(g^{-1}(t)), & \text{if } g(\cdot) \text{ is increasing,} \\ 1 - F_X(g^{-1}(t)), & \text{if } g(\cdot) \text{ is decreasing.} \end{cases}$$

If $F_X(t)$ is the *cdf* for a continuous random variable, then $W = F_X(X)$ is uniform on the interval $(0, 1)$ and the *cdf* of $X = F_X^{-1}(W)$ is $F_X(t)$. This probability integral transform can be used to generate observed values from any discrete or continuous probability law, given observed values from a uniform $(0, 1)$ distribution.

 **EXERCISES 4.5**

1. Let U be a uniform random variable on $(0, 2)$ and find the probability law for $V = |U - 1|$.

2. A plaque beside an elevator reads "Capacity: 950 pounds, 6 persons." Assume that the total weight of six persons, randomly selected from the group that usually employs this elevator, is a normal random variable W with mean 900 and standard deviation 45. If a random group of six of these persons is on the elevator at one time, what is the probability their combined weight exceeds the stated limit?

3. The time to failure for a piece of equipment is assumed to be exponential with median $t_{.5} = 18$ months. What is the probability this piece of equipment will fail in its first year of operation?

4. Each of five persons independently select an integer at random between 1 and 9 inclusive (so each selection can be taken as discrete uniform, parameter 9). Let $Y_{(1)}$ be the smallest of the integers selected. Evaluate $P(Y_{(1)} \le 2)$.

5. Let $Y_{(5)}$ be the largest of the integers selected in Exercise 4, and evaluate $P(Y_{(5)} \le 2)$.

6. Telephone calls arrive at an 800 number like events in a Poisson process with parameter $\lambda = 30$ per hour between the hours of 8 AM and 5 PM, weekdays. What is the probability that the first call on a weekday arrives after 8:05 AM? that it arrives between 8:03 AM and 8:06 AM? What is the expected time of this first call?

7. Let T_2 be the time at which the second call arrives (after 8 AM on a weekday), for the situation described in Exercise 6. Evaluate the probability this second call arrives after 8:05 AM and the probability it arrives between 8:03 AM and 8:06 AM. What is the expected time of this second call?

8. Let T be exponential with parameter λ, and define $U = X^2$. Find the probability law for U.

9. Assume that the weight of the cargo loaded onto a moving van, when it is ready to go on an intercity trip, is a normal random variable with mean 15,000 and standard deviation 1200 (both in pounds). What proportion of intercity trips are made with total loaded weight in excess of 14,000 pounds? What proportion of these trips are made with total loaded weight less than 12,000 pounds?

10. A computer algorithm generates n independent uniform $(0, 1)$ random variables. What is the probability that the smallest of these is less than $1/n$? What is the limiting value of this probability as $n \to \infty$?

11. A Geiger counter is set to produce an audible beep for every 10th particle which it registers. If it is exposed to a particle source emitting at the rate of 15 particles per hour, what is the expected time until its first beep?

12. The *pdf* for X is $f_X(x) = \frac{1}{8}(x + 3)$ for $-3 < x < 1$. Let $W = |X|$, and find the *cdf* for W. Also find the *pdf* for W.

13. Assume that U is continuous uniform on the interval $(0, 1)$, and let $X = a + (b - a)U$, where a and b are constants such that $b > a$. Show that X is uniform on the interval (a, b).

CHAPTER 5

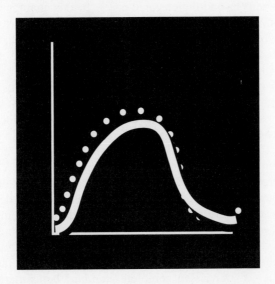

JOINTLY DISTRIBUTED
RANDOM VARIABLES

We have studied many results concerning random variables, both discrete and continuous, and have seen several alternative methods for describing their probability laws; the idea of describing certain aspects of a probability law, such as the location of its "middle" or the extent of its "variability," using single numbers, has also been explored. In all of the cases discussed to this point, random variables were treated singly or separately. The value for a single random variable was to be observed, and the set of possible observed values, the range R_X for the random variable X, was then one-dimensional. In many cases we may want to model the simultaneous behavior of two or more random variables. We could call the cases treated in the previous two chapters one-dimensional, since the range of the random variable could always be taken as a subset of the real line, which is a one-dimensional space. If we want to model the behavior of two, three, or, more generally, n random variables, we can picture the range (the set of *possible* observed values) as being a subset of a higher-dimensional set of real numbers (two-dimensional, three-dimensional, and so on, depending on the number of random variables involved). This increase in dimensionality brings with it a certain richness of possibilities, in several senses, that is not present in one-dimensional cases. In this chapter we shall treat some of the simpler aspects of these higher-dimensional cases.

To make this discussion a little more concrete, suppose four fair dice are tossed onto a table. The probability law for the number X_1 of 1's to occur is binomial with $n = 4$ and $p = \frac{1}{6}$, as is the probability law for the number X_2 of 2's to occur. What is the *joint* probability law for the vector (or 2-tuple) (X_1, X_2) (typically written simply X_1, X_2 without the parentheses)? Or suppose a household in a given state is selected at random; we might like to model the behavior of random variables like $X_1 = $ the annual income of the household, $X_2 = $ the number of persons in the household, $X_3 = $ the number of regular job holders in the household, $X_4, X_5, \ldots = $ the ages of the persons in the household, along with a myriad of other similar variables. The household selected, then, would simultaneously determine the values for all these variables. The joint behavior of these random variables would generally be of interest. If a newly designed microcomputer is to be constructed, a model of the cost per unit X_1 for production models, the number X_2 of units demanded by customers per year, and in addition variables like the times $X_3, X_4, \ldots$ to complete standard benchmark computations would be very useful.

In Section 5.1 we shall explore methods for describing the joint probability law for several random variables and introduce the marginal probability laws. The important concepts of independence of random variables and conditional distributions for random variables are introduced in Section 5.2. In Section 5.3 we discuss the multinomial and bivariate normal random variables.

5.1 | Joint and Marginal Probability Laws

We have seen in previous chapters that the probability model for a single random variable can be specified by its probability function (for discrete random variables) or its *pdf* (for continuous random variables). These same functions are used in modeling the behavior of $X_1, X_2, \ldots, X_n$, situations in which n random variables are jointly observed. The most commonly occurring cases are those in which all n of the random variables are discrete or all n are continuous; these are the only cases we shall explicitly discuss, but it is quite simple to extend these concepts to cases in which some of the variables are discrete and others are continuous.

DEFINITION 5.1

If the observed values for two or more random variables are simultaneously determined by the same random mechanism, they are said to be *jointly distributed*. Their (joint) range is the set of possible observed values (*n*-tuples).

Most of our definitions and examples will use $n = 2$ random variables. The extension to higher-dimensional cases is generally quite straightforward; comments will be made about this situation in necessary places.

If X_1, X_2 are two random variables that are observed on the same experiment, the distinction between the discrete and continuous case again hinges totally on the (two-dimensional) range for the random variables, the set of possible observed values. This range R_{X_1,X_2} is a collection of 2-tuples, with the first element giving the observed value for X_1 and the second giving the observed value for X_2. If R_{X_1,X_2} consists of a discrete set of points in the (x_1, x_2)-plane, then X_1, X_2 is discrete; if R_{X_1,X_2} contains all the points in some region (or union of regions) with nonzero area in the (x_1, x_2)-plane, then X_1, X_2 is continuous.

DEFINITION 5.2

If R_{X_1,X_2} is discrete, the *probability function* for X_1, X_2 is $p_{X_1,X_2}(x_1, x_2)$, giving the probability of x_1, x_2 being the observed values for X_1, X_2. If X_1, X_2 is continuous, the *pdf* for X_1, X_2 is $f_{X_1,X_2}(x_1, x_2)$, where this *pdf* is positive for $(x_1, x_2) \in R_{X_1,X_2}$; the probability that X_1, X_2 lies in some region $\mathcal{A}$ is given by integrating $f_{X_1,X_2}(x_1, x_2)$ over $\mathcal{A}$.

The probability function and *pdf* both directly parallel their counterparts for single random variables. Any function of two variables, $p_{X_1,X_2}(x_1, x_2)$, can serve as a probability function as long as

1. $p_{X_1,X_2}(x_1, x_2) \geq 0$ for all x_1, x_2;

2. $\displaystyle\sum_{(x_1,x_2) \in R_{X_1,X_2}} p_{X_1,X_2}(x_1, x_2) = 1.$

In like manner any function of two variables, $f_{X_1,X_2}(x_1, x_2)$, can serve as a *pdf* as long as

1. $f_{X_1,X_2}(x_1, x_2) \geq 0$ for all x_1, x_2;

2. $\displaystyle\int_{(x_1,x_2) \in R_{X_1,X_2}} f_{X_1,X_2}(x_1, x_2)\, dx_1\, dx_2 = 1.$

These requirements come directly from the Kolmogorov probability axioms, as they did earlier when we were discussing the single-variable case.

Example 5.1

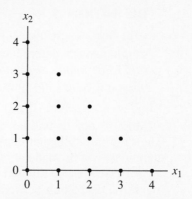

Figure 5.1   Range R_{X_1,X_2} for X_1, X_2

Suppose four fair dice are rolled one time, and we let X_1 be the number of dice to show one spot, while X_2 is the number of dice to show two spots. Then X_1, X_2 is discrete with the (joint) range $R_{X_1,X_2} = \{(x_1, x_2) : x_1 = 0, 1, \ldots, 4, x_2 = 0, 1, \ldots, 4, \text{ such that } x_1 + x_2 \leq 4\}$, as pictured in Fig. 5.1.

Each of these two random variables can individually range through the integers 0, 1, 2, 3, 4, but their sum must be no larger than 4; this is a discrete set of points. The probability function for X_1, X_2 then is positive only for points in this joint range (and is 0 for all others).

We can categorize the outcome for each die into three classes: a 1, a 2, or some other number. The probabilities of occurrence of these classes are then $\frac{1}{6}, \frac{1}{6}, \frac{2}{3}$, respectively, for a fair die. Since the outcomes shown by the four dice are presumably independent, we have a situation similar to independent Bernoulli trials. The only difference is that now we will observe any one of three possibilities, rather than only two as with Bernoulli trials, for each of the dice.

The probability that we will observe x_1 1's and x_2 2's (and necessarily then we also observe $4 - x_1 - x_2$ outcomes that are neither 1 nor 2) in any order is $(\frac{1}{6})^{x_1}(\frac{1}{6})^{x_2}(\frac{2}{3})^{4-x_1-x_2}$ because of the independence of the outcomes on the dice (i.e., independence of the trials). Recall from Chapter 1 that n symbols, of which x_1 are of type 1 (a 1 is rolled), x_2 are of type 2 (a 2 is rolled), and $4 - x_1 - x_2$ are of a third type (neither 1 nor 2 is rolled) can be permuted in $\binom{4}{x_1, x_2, 4-x_1-x_2}$ ways. It follows then that the probability function for X_1, X_2 is

$$p_{X_1,X_2}(x_1, x_2) = \binom{4}{x_1, \, x_2, \, 4 - x_1 - x_2} \left(\frac{1}{6}\right)^{x_1} \left(\frac{1}{6}\right)^{x_2} \left(\frac{2}{3}\right)^{4-x_1-x_2},$$

whose values are given in Table 5.1.

Thus the probability of finding that $X_1 = 1$ and $X_2 = 2$ is $12(\frac{1}{6})^3(\frac{2}{3}) = \frac{1}{27} = .0370$, as given in Table 5.1. The probability that $X_1 = X_2$ is given by the sum of the probabilities that they both equal 0, 1, or 2; that is, $.1975 + .1481 + .0046 = .3502$, the sum of the diagonal elements. This probability law for X_1, X_2 is a special case of the multinomial, to be discussed in more detail later in this chapter.

Table 5.1

$$p_{X_1,X_2}(x_1,x_2)$$

		$x_1 =$				
	0	**1**	**2**	**3**	**4**	**Total**
0	.1975	.1975	.0741	.0123	.0008	.4823
1	.1975	.1481	.0370	.0031	0	.3858
$x_2 =$ **2**	.0741	.0370	.0046	0	0	.1157
3	.0123	.0031	0	0	0	.0154
4	.0008	0	0	0	0	.0008
Total	.4823	.3858	.1157	.0154	.0008	1.000

Suppose X_1, X_2 is discrete with range R_{X_1,X_2} and probability function $p_{X_1,X_2}(x_1,x_2)$; we then can easily find the one-dimensional probability function for X_1 (as well as the probability function for X_2). The range for X_1 consists of all the first values of the 2-tuples in R_{X_1,X_2}, which give all the possible observed values for X_1. If x_1 is one of these values, the probability that $X_1 = x_1$ is the sum of $p_{X_1,X_2}(x_1,x_2)$ over all possible values of x_2 for this given value of x_1, from the Theorem of Total Probability. In like manner, summing $p_{X_1,X_2}(x_1,x_2)$ over all x_1 for a given x_2 gives the total probability of finding that $X_2 = x_2$. These one-dimensional probability functions are called the *marginal probability functions* since they occur as the marginal totals for tables of the two-dimensional probability function. This gives the following theorem.

THEOREM 5.1   If X_1, X_2 are discrete with probability function $p_{X_1,X_2}(x_1,x_2)$ and range R_{X_1,X_2}, then the marginal range for X_1 is R_{X_1}, the set of first elements of $(x_1,x_2) \in R_{X_1,X_2}$. The marginal probability function for X_1 is

$$p_{X_1}(x_1) = \sum_{x_2} p_{X_1,X_2}(x_1,x_2), \qquad \text{for } x_1 \in R_{X_1}.$$

Interchanging X_1 and X_2 gives the marginal probability function for X_2.

In Example 5.1 these two marginal probability functions are given by the marginal totals in Table 5.1. Note that these two marginal probability functions are the same; that is, $P(X_1 = x) = P(X_2 = x)$ for $x = 0,1,2,3,4$. It is necessary for this to be the case, since the probability law for the number of 1's to occur on four rolls of a fair die is identical to the probability law for the number of 2's to occur. In fact, each of these random variables is binomial, with $n = 4$ and $p = \frac{1}{6}$. (Notice in Table 5.1 that the sums for rows 1 and 2 are actually .4822 and .3857, respectively; the values .4823 and .3858 are the rounded probabilities of the binomial, $n = 4$, $p = \frac{1}{6}$, equaling 0 or 1 respectively.)

In many senses the simplest possible two-dimensional continuous case is the uniform one, which has a constant *pdf* over its range (just like the one-dimensional case). X_1, X_2 is uniform with range R_{X_1,X_2} if $f_{X_1,X_2}(x_1, x_2) = c$, a constant for $(x_1, x_2) \in R_{X_1,X_2}$. Let $\mathcal{R}$ be the area of R_{X_1,X_2} in the (x_1, x_2)-plane. Then

$$\iint_{R_{X_1,X_2}} f_{X_1,X_2}(x_1, x_2)\, dx_1\, dx_2 = c\mathcal{R},$$

and it follows that the constant c must be the reciprocal of $\mathcal{R}$, so that the total volume under the *pdf* is 1.

Example 5.2

Imagine that a Cartesian coordinate system has been superimposed on a circular dart board with the origin at the center of the bull's-eye and with one unit (in either direction) equal to the radius of the board. One dart is thrown at the board; the point that it hits will have coordinates (x, y), which we shall model as being the observed value for a two-dimensional random variable X, Y whose *pdf* is $f_{X,Y}(x, y)$. Quite arbitrarily, let us assume X, Y to be uniform over the dart board; this assumption brings with it two important points. First, this says that the dart is certain to hit the board somewhere; the probability of missing the board altogether is 0, which may not be appropriate for all dart throwers (including the author). Second, the uniformity of the *pdf* implies that *any* small fixed area on the board has the same probability of including the impact point (x, y), regardless of where it is located; this, of course, is very possibly not appropriate for any particular thrower, but gives a simple illustrative structure.

The area of a circle with radius r is πr^2; since the unit employed for our example sets $r = 1$ as the radius of the dart board, the *pdf* for X, Y is simply $f_{X,Y}(x, y) = 1/\pi$ for $x^2 + y^2 \leq 1$, defining a cylinder with height $1/\pi$. Suppose the bull's-eye itself is defined by a circle with radius $\frac{1}{8}$ centered at the origin $(0, 0)$; the probability that the dart hits the bull's-eye, with our assumptions, then is the volume under $f_{X,Y}(x, y)$ over this circle. Since $f_{X,Y}(x, y)$ is constant, this probability is given by the ratio of the area of the bull's-eye to the area of the whole board, $(\pi/64)/\pi = 1/64$. Because of the assumed uniformity, this is also the probability that the dart will land in *any* circle with radius $\frac{1}{8}$ contained on the dart board. We also have

$$P(X \leq .5, Y \leq 0) = \frac{1}{4} + \int_0^{.5} \int_{-\sqrt{1-x^2}}^0 \frac{1}{\pi}\, dy\, dx$$

$$= \frac{1}{4} + \frac{1}{2\pi}\left(\frac{\sqrt{3}}{4} + \frac{\pi}{6}\right) = .4022,$$

and

$$P(X > Y + \tfrac{1}{2}) = P(Y < X - \tfrac{1}{2})$$

$$= \iint_A \frac{1}{\pi}\, dy\, dx$$

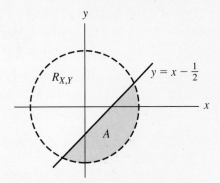

Figure 5.2 Range for X, Y and region $\mathcal{A}$

$$= \left[\frac{\pi}{2} - \left(\frac{\sqrt{7}}{8} + \arcsin \frac{1}{\sqrt{8}} \right) \right] \frac{1}{\pi} = .2797. \qquad (1)$$

Figure 5.2 shows the circular dart board which defines $R_{X,Y}$ and the region of integration A for Eq. (1).

We have seen that summing the joint probability function, over the range for one of the two discrete random variables, gives the marginal or one-dimensional probability function for the other. The same approach—integrating the joint *pdf* over one variable while the other variable is held constant—gives the marginal *pdf* for the variable held constant. To see that this must be the case, suppose X_1, X_2 are jointly continuous with *pdf* $f_{X_1,X_2}(x_1,x_2)$. Let a and b be two constants such that $a < b$, and consider the event $\{a < X_1 < b\}$ pictured in Fig. 5.3.

Since this event leaves X_2 unrestricted in the (x_1, x_2)-plane, we have

$$P(a < X_1 < b) = \int_a^b \left\{ \int_{-\infty}^\infty f_{X_1,X_2}(x_1,x_2)\,dx_2 \right\} dx_1;$$

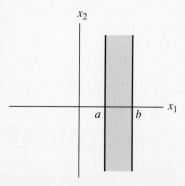

Figure 5.3 The event $a < X_1 < b$

that is, to evaluate $P(a < X_1 < b)$ we must integrate

$$\int_{-\infty}^{\infty} f_{X_1,X_2}(x_1, x_2)\, dx_2$$

over the interval (a, b). But $P(a < X_1 < b)$ is evaluated by integrating $f_{X_1}(x_1)$, the *pdf* for X_1, over (a, b); it follows that the value of this integral must be $f_{X_1}(x_1)$ for any given x_1:

$$f_{X_1}(x_1) = \int_{-\infty}^{\infty} f_{X_1,X_2}(x_1, x_2)\, dx_2,$$

so integrating $f_{X_1,X_2}(x_1, x_2)$ over x_2 gives the marginal *pdf* for X_1. Interchanging the roles of X_1 and X_2 in this discussion shows that the marginal *pdf* for X_2 must be

$$f_{X_2}(x_2) = \int_{-\infty}^{\infty} f_{X_1,X_2}(x_1, x_2)\, dx_1.$$

The joint *pdf* for X_1, X_2 also gives the marginal *pdf*'s. These results establish the following theorem.

THEOREM 5.2 If X_1, X_2 are continuous with *pdf* $f_{X_1,X_2}(x_1, x_2)$ and range R_{X_1,X_2}, then the (marginal) range for X_1 is R_{X_1}, the set of first elements of $(x_1, x_2) \in R_{X_1,X_2}$. The marginal *pdf* for X_1 is

$$f_{X_1}(x_1) = \int_{x_2 \in R_{X_1,X_2}} p_{X_1,X_2}(x_1, x_2)\, dx_2, \qquad \text{for } x_1 \in R_{X_1}.$$

Interchanging X_1 and X_2 gives the marginal *pdf* for X_2.

Note this is the same manipulation done with the joint probability function to get the marginals for the individual discrete variables; instead of summing, its continuous counterpart (integration) is applied to the joint *pdf*.

For the circular dart board discussed in Example 5.2, the marginal *pdf* for X is

$$f_X(x) = \int_{-\infty}^{\infty} f_{X,Y}(x, y)\, dy$$
$$= \int_{-\sqrt{1-x^2}}^{\sqrt{1-x^2}} \frac{1}{\pi}\, dy = \frac{2}{\pi}\sqrt{1-x^2}, \tag{2}$$

for $-1 < x < 1$, since $f_{X,Y}(x, y) > 0$ only for $x^2 + y^2 \leq 1$. This *pdf* is *not* uniform on $(-1, 1)$. The possible observed y-values (distance above or below the middle) are affected by the observed x-value (the closer $|x|$ is to 1, the more restricted the possibilities for y). By symmetry, Eq. (2) is also the *pdf* for Y and is pictured in Fig. 5.4.

In any two-dimensional case, the marginal probability laws can be derived from the joint probability law, by summing or integrating as appropriate. It is important to realize that this procedure goes only one way; we can find

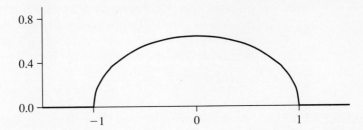

Figure 5.4 Marginal *pdf* for circular uniform

the marginal probability laws from the joint probability law, but it is not in general possible to find the joint probability law given only a specification of the marginal distributions. Said another way, there are many different joint probability laws that have the same marginal distributions.

To illustrate this for the discrete case, recall Example 5.1 in which we defined X_1 to be the number of 1's to occur when four fair dice are thrown, and X_2 to be the number of 2's to occur (on the same four rolls). Then both X_1 and X_2 are binomial, with $n = 4$, and $p = \frac{1}{6}$, and their joint probability function is given in Table 5.1. Table 5.2 has the same marginals, but the entries in the body are different, so the joint probability functions are different. One of the problems in Exercises 5.1 suggests a simple continuous two-dimensional case that illustrates the same point; knowledge of the marginal *pdf*s is not in general sufficient to specify the joint *pdf*.

We have discussed the probability laws for discrete and continuous two-dimensional random variables; the same basic ideas carry over to higher-dimensional cases. The joint probability law for $X_1, X_2, \ldots, X_n$ is specified by either a probability function or a *pdf*, defined over an *n*-dimensional space, depending on whether the individual random variables are discrete or continuous. In each case, the joint probability law brings with it the specification of the marginals of all orders; that is, granted the probability law for X_1, X_2, X_3, X_4, say, we can then also find the one-dimensional marginals for each random variable separately, the joint (marginal) for X_1, X_2 and for X_2, X_3, X_4, and so forth. Our concern with cases where $n \geq 3$ will primarily

Table 5.2

$$p_{X_1,X_2}(x_1, x_2)$$

		$x_1 =$					
		0	*1*	*2*	*3*	*4*	*Total*
	0	.2327	.1860	.0558	.0074	.0004	.4823
	1	.1860	.1488	.0447	.0060	.0003	.3858
$x_2 =$	*2*	.0558	.0447	.0134	.0018	.0001	.1157
	3	.0074	.0060	.0018	.0002	0	.0154
	4	.0004	.0003	.0001	0	0	.0008
Total		.4823	.3858	.1157	.0154	.0008	1.000

be in the special situation of independent random variables, to be introduced in the next section. Independence of random variables, like independence of events, brings with it a number of simplifications in the structure of joint distributions. This important concept is used extensively in statistical methods.

EXERCISES 5.1

1. A fair coin is flipped three times; let X_1 be the number of heads to occur on the first two flips, and let X_2 be the number of heads to occur on the last two flips. Find the probability function and the marginal probability functions for X_1, X_2. Evaluate $P(X_1 = X_2)$.

2. Experience has shown that 10% of the people to take a standardized test will score greater than 90, 30% will score between 75 and 90, while the remainder will score less than 75. Assume that 15 persons from the same school take this test and evaluate the following probabilities:
   a. All score less than 90.
   b. Five score greater than 90, five score between 75 and 90, and five score less than 75.
   What assumptions have you made?

3. The time to failure for an electronic item is assumed to be an exponential random variable with parameter $\lambda = .001$, with time measured in hours. Eight of these items are tested until they fail. Evaluate the probability that four fail in the first 500 hours, two fail between 500 and 1000 hours, while the other two failure times exceed 1000 hours. What assumptions are behind your answer?

4. Suppose 10 fair dice are tossed onto a table.
   a. Evaluate $P(X_1 = X_2 = 1)$.
   b. Evaluate $P(X_1 = X_2)$.
   c. If the 10 dice are tossed onto the table and it is found that there are *equal* numbers of 1's and 2's ($X_1 = X_2 = x$), what is the most probable value for x?

5. There are many popular five-dice games, some of which involve poker-like constructions. Suppose five fair dice are rolled one time.
   a. What is the probability of getting five of a kind (i.e., all show the same face)?
   b. What is the probability of getting four of a kind (i.e., four show the same face, the fifth does not)?
   c. What is the probability of getting a full house (i.e., three show one face, two show a second face)?

6. A sequence of independent Bernoulli trials is performed, each with parameter p. Let X_1 be the trial number of the first success, while X_2 is the trial number of the second success. Find the joint probability function for X_1, X_2, as well as their marginal probability functions.

7. A discrete random vector has the probability function specified by $P(X_1 = 1, X_2 = 2, X_3 = 3) = \frac{1}{2}, P(X_1 = 2, X_2 = 3, X_3 = 1) = \frac{1}{3}, P(X_1 = 3, X_2 = 1, X_3 = 2) = \frac{1}{6}$, with all other 3-tuples (x_1, x_2, x_3) having probability 0. Evaluate all six marginal probability functions.

8. Two transistors are soldered onto the same circuit board. Their times to failure are X, Y whose joint *pdf* is $f_{X,Y}(x,y) = 6e^{-2x-3y}$ for $x > 0$ and $y > 0$. Find the marginal *pdf*s for X and for Y. Evaluate $P(X > Y)$.

9. Suppose you throw a dart at a square dart board. On this dart board is an (x,y) coordinate system having its origin at the center of the board, with each side of the board 2 units in length parallel to the two coordinate axes. Also assume that the impact point X, Y is uniform over this square. What is the joint *pdf* for X, Y? Evaluate the marginal *pdf*s. As was done in Example 5.2, evaluate $P(X \le .5, Y \le 0)$ and $P(X > Y + \frac{1}{2})$.

10. Now assume the square dart board from Example 9 is rotated through a 45° angle relative to the (x,y) coordinate system, so that the corners of the board are at $(x,y) = (-\sqrt{2},0), (0,-\sqrt{2}), (0,\sqrt{2}), (\sqrt{2},0)$. Find the marginal *pdf*s for X and Y, granted the joint *pdf* is again uniform.

11. Assume the joint *pdf* for X, Y is given by

$$f_{X,Y}(x,y) = \frac{x+y+1}{2}, \qquad \text{for } 0 < x < 1, \quad 0 < y < 1,$$

and find the marginal *pdf*s for X and Y.

12. Assume the joint *pdf* for X, Y, Z is

$$f_{X,Y,Z}(x,y,z) = \frac{x+y+1}{2}, \qquad \text{for } 0 < x < 1, \quad 0 < y < 1, \quad 0 < z < 1.$$

a. Find the marginal *pdf* for X, Y.
b. Find the marginal *pdf* for X, Z.
c. Find the marginal *pdf* for Z.

5.2 | Conditional Distributions and Independence

Conditional probability is a useful tool, as we have seen earlier, and is equally useful for random variables. Recall that if A and B are two events with $P(B) > 0$, then the conditional probability of A occurring, given B has occurred, is defined by the ratio

$$P(A \mid B) = \frac{P(A \cap B)}{P(B)}. \tag{1}$$

Table 5.3 reproduces the joint probability law for X_1, X_2, the number of dice showing one and two spots, respectively, when four fair dice are rolled. The first row of this table gives the probabilities of observing zero through four dice with one spot, together with $x_2 = 0$ dice showing two spots. The sum of the values in this first row is $P(X_2 = 0) = .4823$, the total probability of observing no dice with two spots. If we divide each entry in this first row by this total, and recall the definition of conditional probability given by Eq. (1), the resulting entries give the *conditional* probabilities of X_1 equaling the

Table 5.3

$$p_{X_1, X_2}(x_1, x_2)$$
$$x_1 =$$

		0	1	2	3	4	Total
	0	.1975	.1975	.0741	.0123	.0008	.4823
	1	.1975	.1481	.0370	.0031	0	.3858
$x_2 =$	**2**	.0741	.0370	.0046	0	0	.1157
	3	.0123	.0031	0	0	0	.0154
	4	.0008	0	0	0	0	.0008
Total		.4823	.3858	.1157	.0154	.0008	1.000

integers 0 through 4, respectively, given that $X_2 = 0$. These values specify the conditional probability function for X_1 given that $X_2 = 0$.

Table 5.4 has "normalized" the entries from Table 5.3, dividing each entry by the corresponding row total (and rounding), thus giving rows that must sum to 1. (Actually, the entries in this table are the ratios of the exact probability for the cell, divided by the exact row probability.) These values are the conditional probability functions for X_1, the number of dice with one spot given that x_2 dice show two spots. The conditional probability distribution for X_1 is different for each different given value x_2. If we are given that 0 dice show two spots, then we are equally likely to find 0 or 1 dice with one spot (and these two values are the most probable), while if we are given that the number of dice showing two spots is any number bigger than 0, the most likely number showing one spot is 0. Indeed, if we are given $x_2 = 4$, the only possible value for x_1 is 0 and its conditional probability then is 1.

If we divide the *column* entries in Table 5.3 by the corresponding column totals, the ratios produced specify the conditional probability distributions for X_2 given the different possible values for x_1. Since the entries in Table 5.3 are symmetric, these conditional probability laws for X_2 are the same as their counterparts given in Table 5.4. The definition of the conditional probability function for X_1, given $X_2 = x_2$, follows. Interchanging the roles of X_1 and X_2 gives the conditional probability function for X_2 given a value for X_1.

Table 5.4

$$p_{X_1 | X_2}(x_1 | x_2)$$

$$x_1 =$$

		0	1	2	3	4
	0	.4096	.4096	.1536	.0256	.0016
	1	.5120	.3840	.0960	.0080	0
$x_2 =$	**2**	.6400	.3200	.0400	0	0
	3	.8000	.2000	0	0	0
	4	1.0000	0	0	0	0

DEFINITION 5.3

Let X_1, X_2 be jointly distributed with probability function $p_{X_1,X_2}(x_1,x_2)$. The *conditional probability function for X_1, given $X_2 = x_2$*, is

$$p_{X_1|X_2}(x_1 \,|\, x_2) = \frac{p_{X_1,X_2}(x_1,x_2)}{p_{X_2}(x_2)}, \qquad \text{where } x_1 \in R_{X_1}, \ x_2 \in R_{X_2}.$$

As we have seen above, these conditional probability functions may be different, depending on the value given for the conditioning variable. Definition 5.3 also implies that

$$p_{X_1,X_2}(x_1,x_2) = p_{X_1|X_2}(x_1 \,|\, x_2)p_{X_2}(x_2). \tag{2}$$

That is, we can find the joint probability function for X_1, X_2 if we know the marginal for one random variable and the conditional distributions for the other. Equation (2) can be exploited in finding joint probability laws, as illustrated in the following example.

Example 5.3

Assume that injury-causing automobile accidents occur in a certain area at a rate of 2 per month, like events in a Poisson process. Let us also assume the probability is .05 that each such accident will involve one or more fatalities, independently from one accident to another. What is the probability law for the number of fatal accidents (one or more fatalities involved) per month?

Let X_1 be the number of injury-causing accidents to occur in the given month; then the marginal probability law for X_1 is Poisson with $\mu = 2$ from the information given. The conditional probability function for the number X_2 of fatal accidents, given $X_1 = x_1$, is binomial with $n = x_1$, and $p = .05$ with the above assumptions; its conditional probability function is

$$p_{X_2|X_1}(x_2 \,|\, x_1) = \binom{x_1}{x_2}(.05)^{x_2}(.95)^{x_1-x_2}$$

for $x_2 = 0, 1, \ldots, x_1$. It follows then that the joint probability function for X_1, X_2 is

$$\begin{aligned} p_{X_1,X_2}(x_1,x_2) &= p_{X_2|X_1}(x_2 \,|\, x_1)p_{X_1}(x_1) \\ &= \binom{x_1}{x_2}(.05)^{x_2}(.95)^{x_1-x_2}\frac{(2)^{x_1}}{x_1!}e^{-2}, \end{aligned} \tag{3}$$

for $x_1 = 0, 1, 2, \ldots$, and $x_2 = 0, 1, \ldots, x_1$. By canceling the $x_1!$ term, combining the two terms that contain x_2 in their exponents, and introducing the term $2^{x_2} \times 2^{-x_2}$, we can rewrite this joint *pdf* as

$$p_{X_1,X_2}(x_1,x_2) = \frac{1}{x_2!}(.1)^{x_2}e^{-2}\frac{(1.9)^{x_1-x_2}}{(x_1-x_2)!}. \tag{4}$$

We find the marginal probability function for X_2 by summing Eq. (4) over the range for x_1 (which consists of the integers $x_2, x_2+1, \ldots$) for any possible value of x_2, giving

$$p_{X_2}(x_2) = \frac{1}{x_2!}(.1)^{x_2}e^{-2} \times e^{1.9} = \frac{1}{x_2!}(.1)^{x_2}e^{-.1}, \tag{5}$$

for $x_2 = 0, 1, 2, \ldots$. Since Eq. (5) is a Poisson probability function, X_2 itself is Poisson, with parameter .1.

Example 5.4

To measure the "worth" of a recreational area, each person utilizing the area over a period of time is encouraged to pick up a questionnaire, fill it out, and send it back to the research team conducting the investigation. Let us assume that n people utilize this area over the time period in question; let us further assume that these people behave like independent Bernoulli trials with parameter p_1, where p_1 is the probability a person will take a questionnaire. If X_1 is the number of persons taking a questionnaire over this period, then X_1 is binomial with parameters n, p_1.

It is well known that people taking questionnaires do not always fill them out and return them. To model this phenomenon, let us assume that each person who takes a questionnaire has probability p_2 of filling it out and returning it to the research team, and that these persons act independently. Thus, if x_1 persons choose to pick up questionnaires, we will let X_2 be the number of these who actually fill them out and return them. With the assumptions just stated, then, X_2 is binomial with parameters x_1, p_2, granted x_1 persons initially picked up the questionnaire; that is, the conditional probability function for X_2, given $X_1 = x_1$, is binomial with parameters x_1, p_2. What then is the marginal probability function law for X_2, the number of returned questionnaires?

The joint probability function for X_1, X_2 is the product

$$p_{X_1,X_2}(x_1, x_2) = \binom{n}{x_1}p_1^{x_1}(1-p_1)^{n-x_1}\binom{x_1}{x_2}p_2^{x_2}(1-p_2)^{x_1-x_2}, \tag{6}$$

for $x_1 = 0, 1, 2, \ldots, n$ and $x_2 = 0, 1, 2, \ldots, x_1$. The marginal probability function for X_2 is given by summing Eq. (6) over the range of possible x_1-values for the given x_2 (i.e., over $R_{X_1|x_2} = \{x_1 = x_2, x_2 + 1, \ldots, n\}$). Rearranging terms, canceling $x_1!$, and introducing $(n-x_2)!/(n-x_2)!$ into Eq. (6) then gives

$$p_{X_2}(x_2) = \binom{n}{x_2}\left(\frac{p_2}{1-p_2}\right)^{x_2}\sum_{x_1=x_2}^{n}\binom{n-x_2}{x_1-x_2}(p_1(1-p_2))^{x_1}(1-p_1)^{n-x_1}$$

$$= \binom{n}{x_2}\left(\frac{p_2}{1-p_2}\right)^{x_2}\sum_{j=0}^{n-x_2}\binom{n-x_2}{j}(p_1(1-p_2))^{j+x_2}(1-p_1)^{n-x_2-j}$$

$$= \binom{n}{x_2}\left(\frac{p_2}{1-p_2}\right)^{x_2}(p_1(1-p_2))^{x_2}(p_1(1-p_2)+(1-p_1))^{n-x_2}$$

$$= \binom{n}{x_2}(p_1p_2)^{x_2}(1-p_1p_2)^{n-x_2}.$$

Thus, X_2 is binomial with parameters $n, p_1 p_2$. This fact could well be anticipated (and possibly be more easily recognized) by realizing that the construct used (independent Bernoulli trials with probability p_1, with successes then again independent Bernoulli trials with parameter p_2) is equivalent to independent Bernoulli trials with probability of success given by $p_1 p_2$. That is, calling a returned questionnaire a success, the probability of a success is given by the product of p_1 (the probability that the questionnaire is taken) and p_2, the probability that it is then filled out and returned.

The use of conditional probability with discrete random variables proceeds directly from the original definition of the conditional probability for the occurrence of event A, given that B has occurred. If X_1 and X_2 are discrete and $x_2 \in R_{X_2}$ then $p_{X_1}(x_2) = P(X_2 = x_2) > 0$; the ratio

$$\frac{p_{X_1,X_2} x_1, x_2}{p_{X_2}(x_2)}$$

is then well defined for all $x_2 \in R_{X_2}$. Conditional probability is also very useful with continuous random variables, but in the continuous case the probability of occurrence of individual points is necessarily 0, as we know. For example, if we want to model the arrival times X_1, X_2, of the first two telephone calls (after an arbitrary origin 0) and have observed that $X_1 = 5$ (i.e., the first call arrived at the time labeled 5 minutes), then our previous discussion does not directly cover the conditional probability law for X_2 given $X_1 = 5$, since $P(X_1 = 5) = 0$. However, very similar reasoning leads to the definition of the conditional *pdf* for X_2 given $X_1 = 5$.

Recall that for a continuous random variable, the probability that the observed value for X_1 falls in a small interval of length Δx_1, centered at 5, is approximately given by the product $f_{X_1}(5) \Delta x_1$, the height of the *pdf* times the width of the interval. Let B be the event $\{X_1 = 5\}$, so $P(B) \doteq f_{X_1}(5) \Delta x_1$. In like manner the probability that the observed value for the pair X_1, X_2 lies in a small rectangle of area $\Delta x_1 \Delta x_2$ centered at $(5, x_2)$ is approximately given by the product $f_{X_1,X_2}(5, x_2) \Delta x_1 \Delta x_2$. Let A be the event $\{X_2 = x_2\}$, where $x_2 \in R_{X_2}$ is some fixed value; then $P(A \cap B) \doteq f_{X_1,X_2}(5, x_2) \Delta x_1 \Delta x_2$. Thus the conditional probability that $X_2 = x_2$, given $X_1 = 5$, is approximated by the ratio

$$P(A \mid B) \doteq \frac{f_{X_1,X_2}(5, x_2) \Delta x_1 \Delta x_2}{f_{X_1}(5) \Delta x_1} = \frac{f_{X_1,X_2}(5, x_2)}{f_{X_1}(5)} \Delta x_2.$$

The ratio $f_{X_1,X_2}(5, x_2)/f_{X_1}(5)$, for $(5, x_2) \in R_{X_1,X_2}$, will be called the conditional *pdf* for X_2 given $X_1 = 5$. Equivalently, we may arrive at this same result by realizing that $f_{X_1,X_2}(5, x_2)$ traces the height of the joint *pdf* over the line $x_1 = 5$ in the (x_1, x_2)-plane. If we have observed (exactly) 5 as the value for x_1, areas under $f_{X_1,X_2}(5, x_2)$ should be indicative of the relative probabilities of X_2 lying in various intervals. Dividing $f_{X_1,X_2}(5, x_2)$ by $f_{X_1}(5)$ simply normalizes the resulting curve so that its total area is 1 (it is then a *pdf*).

Since

$$\int_{-\infty}^{\infty} \frac{f_{X_1,X_2}(5,x_2)}{f_{X_1}(5)} \, dx_2 = \frac{f_{X_1}(5)}{f_{X_1}(5)} = 1, \tag{7}$$

and this ratio is positive for $(5, x_2) \in R_{X_1,X_2}$, Eq. (7) is in fact a legitimate *pdf*. Notice that this is the identical manipulation, using *pdf*s rather than probability functions, that is employed with discrete random variables; the result is again a *pdf*, which then can be used to answer any questions of interest, as with any other *pdf*. Our formal definition of the conditional *pdf* for a continuous random variable follows.

DEFINITION 5.4

Let X_1, X_2 be continuous jointly distributed random variables. The *conditional pdf for X_1, given $X_2 = x_2$*, is

$$f_{X_1|X_2}(x_1 \mid x_2) = \frac{f_{X_1,X_2}(x_1,x_2)}{f_{X_2}(x_2)}, \qquad \text{where } x_1 \in R_{X_1}, \ x_2 \in R_{X_2}.$$

As in the discrete case, this conditional *pdf* for X_1 may change with the given value of x_2. The following example employs Definition 5.4 to evaluate some conditional *pdf*s.

Example 5.5

Suppose X, Y is uniform on the unit square with corners

$$(-1, -1), \ (-1, 1), \ (1, -1), \ (1, 1)$$

so the joint *pdf* is $f_{X,Y}(x, y) = \frac{1}{4}$ for (x, y) in this square. The resulting marginal *pdf* for X (as well as for Y) is $f_X(x) = \frac{1}{2}$ for $-1 < x < 1$. The conditional *pdf* for Y, given $X = .1$, is then given by the ratio

$$f_{Y|X}(y \mid x = .1) = \frac{f_{X,Y}(.1, y)}{f_X(.1)} = \frac{1}{2}, \qquad \text{for } -1 < y < 1. \tag{8}$$

It is easy to see that Eq. (8), the conditional *pdf* for Y, is the same as the marginal *pdf* for Y, no matter what the given value $-1 < x < 1$ for X. That is, the conditional *pdf* for Y, given $X = x$ is uniform on $(-1, 1)$ for all x. We can also easily verify that Eq. (8) is the conditional *pdf* for X, given $Y = y$ where $-1 < y < 1$. These conditional probability laws are uniform, of course, because the original joint *pdf* for X, Y was taken to be uniform.

In Example 5.2 we discussed the impact point of a dart thrown at a circular dart board with radius 1. The joint *pdf* for the point hit was taken to be $f_{X,Y}(x, y) = 1/\pi$ for $x^2 + y^2 \leq 1$, and the marginal *pdf* for X was *not* uniform; in fact, it was found (see Section 5.1, Eq. (2)) to be

$$f_X(x) = \frac{2}{\pi} \sqrt{1 - x^2}, \qquad \text{for } -1 < x < 1.$$

The conditional *pdf* for Y, given $X = x$, is given by the ratio

$$f_{Y|X}(y \mid x) = \frac{1/\pi}{2\sqrt{1-x^2}/\pi}$$

$$= \frac{1}{2\sqrt{1-x^2}}, \qquad \text{for } -\sqrt{1-x^2} < y < \sqrt{1-x^2}. \qquad (9)$$

The given value for x can be any point in the interval $(-1, 1)$. This conditional *pdf* is again uniform (remember the given x-value is a constant), since the joint *pdf* was uniform, but now the length of the interval over which Y is uniform is $2\sqrt{1-x^2}$, which varies with the given value of x. Thus in this case, the conditional *pdf* for Y is *not* the same for the different possible x-values. Because of symmetry, Eq. (9) also gives the conditional *pdf* for X, given a value for Y.

As in the discrete case, the joint *pdf* for two continuous random variables can be expressed as the product of a marginal *pdf* and a conditional *pdf*; that is, multiplying the defining equation by $f_{X_2}(x_2)$ gives

$$f_{X_1,X_2}(x_1, x_2) = f_{X_2}(x_2) f_{X_1 \mid X_2}(x_1 \mid x_2). \qquad (10)$$

This then gives an easy way to evaluate the joint *pdf* for X_1, X_2, granted these two functions are known.

Example 5.6

Suppose telephone calls arrive at the switchboard of a large corporation like events in a Poisson process with parameter λ. Starting at an arbitrary time (labeled 0), let T_1 be the time of occurrence of the first call to arrive; T_1 then is exponential with parameter λ. The length of time from the observed time t_1 of the first call until the arrival of the next call should again be exponential with parameter λ. If we denote this *elapsed* time by Y, then Y has the same probability law as T_1, and it is also exponential with parameter λ. If T_2 is the time of occurrence of this second call, measured from the same origin as T_1, then $T_2 = t_1 + Y$, so $T_2 - t_1 = Y$ is exponential; that is, given $T_1 = t_1$, the conditional *pdf* for T_2 is a *shifted exponential*, the *pdf* of an exponential random variable plus the constant t_1, the time of occurrence of the first call. Thus this conditional *pdf* is

$$f_{T_2 \mid T_1}(t_2 \mid t_1) = \lambda e^{-\lambda(t_2 - t_1)}, \qquad \text{for } t_2 - t_1 > 0. \qquad (11)$$

The joint *pdf* for T_1, T_2 is then the product of the marginal *pdf* for T_1 and this conditional *pdf* for T_2 given $T_1 = t_1$:

$$f_{T_1,T_2}(t_1, t_2) = \left\{ \lambda e^{-\lambda t_1} \right\} \left\{ \lambda e^{-\lambda(t_2 - t_1)} \right\}, \qquad \text{for } t_1 > 0, \quad t_2 - t_1 > 0$$

$$= \lambda^2 e^{-\lambda t_2}, \qquad \text{for } 0 < t_1 < t_2.$$

The marginal *pdf* for T_2 is then

$$f_{T_2}(t_2) = \int_0^{t_2} \lambda^2 e^{-\lambda t_2} \, dt_1 = \lambda^2 t_2 e^{-\lambda t_2}, \qquad \text{for } t_2 > 0.$$

Thus, by recognizing the form of the *pdf*, we can see that T_2 is Erlang with parameters $r = 2$ and λ. Given $T_2 = t_2$, the conditional *pdf* for T_1 is given by the ratio

$$f_{T_1 \mid T_2}(t_1 \mid t_2) = \frac{f_{T_1, T_2}(t_1, t_2)}{f_{T_2}(t_2)}$$

$$= \frac{\lambda^2 e^{-\lambda t_2}}{\lambda^2 t_2 e^{-\lambda t_2}} = \frac{1}{t_2}, \qquad \text{for } 0 < t_1 < t_2,$$

for any $t_2 > 0$; thus the conditional probability law for T_1, given $T_2 = t_2 > 0$, is *uniform* on the interval $(0, t_2)$, since this is the *pdf* for a uniform random variable. Granted the time t_2 of the second call is known, the time of the first occurrence is uniformly distributed on the interval $(0, t_2)$, regardless of the value of the parameter λ, which may or may not strike you as being intuitive. In a sense this result is driven by the memoryless property of the exponential; knowing the second call occurred at time t_2 gives no information about the time of the first call, beyond the fact that it must lie between 0 and t_2 (uniformly, with the assumptions made).

Recall the important idea of independent events, discussed in Chapter 1; A and B are called independent if the conditional probability of A occurring, granted B has occurred, is the same as the unconditional probability of A occurring, or $P(A \mid B) = P(A)$. It follows that the occurrence (or nonoccurrence) of the event B gives no information about whether A may also have occurred. This concept of independence is easily extended to random variables and is very frequently employed in many applied contexts. Suppose we have two jointly distributed discrete random variables X, Y such that $p_{X \mid Y}(x \mid y) = p_X(x)$, for all $x \in R_X$ and for each possible given value y; that is, the conditional probability function for X is identical to the marginal probability function for X, no matter what the given value is for Y. Then it must certainly be true that $P(a < X < b \mid Y = y) = P(a < X < b)$, for any interval (a, b); the fact that Y has been observed to equal y gives no information about the observed value for X. In such a case we shall call the random variables X, Y independent. Note then that

$$p_{X \mid Y}(x \mid y) = \frac{p_{X, Y}(x, y)}{p_Y(y)} = p_X(x), \qquad \text{for all } (x, y),$$

implies that the joint probability function must be given by the product of the marginal probability functions.

In the same way, we shall call two continuous random variables independent if and only if, for all possible pairs (x, y), their joint *pdf* factors into the product of the two marginal *pdf*s. This discussion is summarized in the following definition.

DEFINITION 5.5

The two random variables X, Y are *independent* if and only if

$$p_{X,Y}(x,y) = p_X(x)p_Y(y), \qquad \text{if } X, Y \text{ are discrete,}$$
$$f_{X,Y}(x,y) = f_X(x)f_Y(y), \qquad \text{if } X, Y \text{ are continuous,}$$

for all $(x,y) \in R_{X,Y}$.

Granted we have the joint *pdf* or probability function for X, Y, it is then straightforward to evaluate the marginal *pdf*s or marginal probability functions; checking to see whether the two random variables are independent then reduces to investigating whether the relationship given in Definition 5.5 is satisfied. This is illustrated in the following example.

Example 5.7

Assume that the joint *pdf* for X, Y is

$$f_{X,Y}(x,y) = \begin{cases} e^{-x-y}, & \text{for } x > 0, y > 0, \\ 0, & \text{otherwise.} \end{cases}$$

Then the marginal *pdf* for X is

$$f_X(x) = \begin{cases} \int_0^{\infty} e^{-x-y}\, dy = e^{-x} \int_0^{\infty} e^{-y}\, dy = e^{-x}, & \text{for } x > 0, \\ 0, & \text{for } x \le 0. \end{cases}$$

Integrating $f_{X,Y}(x,y)$ with respect to x (from 0 to ∞) gives $f_Y(y) = e^{-y}$ for any $y > 0$, and $f_Y(y) = 0$ for $y \le 0$. Thus we have $f_{X,Y}(x,y) = f_X(x)f_Y(y)$ for all (x,y), so X and Y are independent random variables. Notice that if $a, b, c,$ and d are any constants such that $a < b, c < d$, then

$$P(a < X < b, c < Y < d) = \int_a^b \int_c^d e^{-x-y}\, dx\, dy$$
$$= \int_a^b e^{-x}\, dx \int_c^d e^{-y}\, dy$$
$$= P(a < X < b)P(c < Y < d).$$

This joint probability of X and Y lying in the rectangle defined by $a < x < b, c < y < d$ is given by the product of the two marginal probabilities, no matter where the rectangle is located, and is the same type of multiplication of probabilities that occurs for independent events.

Suppose the joint *pdf* for X, Y is

$$f_{X,Y}(x,y) = \begin{cases} e^{-y}, & \text{for } 0 < x < y < \infty, \\ 0, & \text{otherwise.} \end{cases}$$

This joint *pdf* then is positive only in the wedge defined by $0 < x < y$. The marginal *pdf* for X is

$$f_X(x) = \int_x^\infty e^{-y}\,dy = e^{-x}, \qquad \text{for } x > 0,$$

and the marginal *pdf* for Y is

$$f_Y(y) = \int_0^y e^{-y}\,dx = ye^{-y}, \qquad \text{for } y > 0.$$

Since $f_{X,Y}(2,2) = e^{-2}$, which is not the same as $f_X(2)f_Y(2) = (e^{-2})(2e^{-2})$, X and Y are not independent. The product of the two marginals *pdf*s is equal to the joint *pdf* only on the line $y = e^x$ for these two random variables, not for all (x, y).

It was pointed out earlier that knowing the marginal probability functions (or marginal *pdf*s) is not in general equivalent to knowing the joint probability function (or joint *pdf*). In the case of *independent* random variables, the marginals do specify the joint probability law, since the joint probability law is the simple product of the two marginals. Thus if we are building a model for X and Y and we assume that they are independent, specification of the separate marginal probability laws also gives the joint probability law (as the product of the marginals). This type of construct is used very frequently in statistical modeling of observed data values.

The concept of independence generalizes very easily to any number of random variables; our definition for the independence of n random variables follows.

DEFINITION 5.6

Let $X_1, X_2, \ldots, X_n$ be jointly distributed random variables; these random variables are *independent* if and only if their joint probability function (or *pdf*) is given by the product of the marginal probability functions (or marginal *pdf*s).

Thus if $X_1, X_2, \ldots, X_n$ are discrete, their joint probability function is given by

$$p_{X_1,X_2,\ldots,X_n}(x_1, x_2, \ldots, x_n) = p_{X_1}(x_1)p_{X_2}(x_2) \cdots p_{X_n}(x_n)$$

for all $(x_1, x_2, \ldots, x_n)$. Similarly if $X_1, X_2, \ldots, X_n$ are continuous, their joint *pdf* is given by

$$f_{X_1,X_2,\ldots,X_n}(x_1, x_2, \ldots, x_n) = f_{X_1}(x_1)f_{X_2}(x_2) \cdots f_{X_n}(x_n)$$

for all $(x_1, x_2, \ldots, x_n)$. A consequence of Definition 5.6 is that the independence of $X_1, X_2, \ldots, X_n$ implies the independence of any subset of these n random variables (just as the independence of n events implies the independence of any subset of the n events). As with events, however, this independence implication goes only one way: If $X_1, X_2, \ldots, X_k$ are independent and are each individually independent of X_{k+1}, it does not follow that $X_1, X_2, \ldots, X_{k+1}$ must be independent.

Example 5.8

Many computer packages have built-in routines that claim to generate "random" numbers on the interval $(0, 1)$. If we let X_1 be the number generated by such a routine, we can then assume X_1 is uniform on the interval $(0, 1)$, with *pdf* $f_{X_1}(x) = 1$ for $0 < x < 1$, granted the claim is true. Indeed, if the same routine is called on to generate two successive such numbers, say X_1 and X_2, it is tacitly assumed (if not explicitly stated) that X_1, X_2 are independent random variables, whose joint *pdf* is given by the product of the two marginal *pdf*s:

$$f_{X_1, X_2}(x_1, x_2) = f_{X_1}(x_1)f_{X_2}(x_2) = 1, \qquad \text{for } 0 < x_1 < 1, \quad 0 < x_2 < 1.$$

Indeed, the veracity or worth of such a random-number generator can be checked by comparing the proportions of generated pairs (x_1, x_2) that fall in given regions, with the volumes under $f_{X_1, X_2}(x_1, x_2)$ over these same regions. We need not limit ourselves to pairs, of course; if the generator really does produce independent values selected from this unit interval, then the joint probability function for n such values (which we label $X_1, X_2, \ldots, X_n$, representing their order of occurrence) must be

$$f_{X_1, X_2, \ldots, X_n}(x_1, x_2, \ldots, x_n) = 1, \qquad \text{for } 0 < x_i < 1, \quad i = 1, 2, \ldots, n.$$

You may in fact have a generator that produces numbers from a normal distribution with specified mean μ and variance σ^2, again presumably independently. Thus if X_i is one of these generated values, its *pdf* should be

$$f_{X_i}(x) = \frac{1}{\sigma\sqrt{2\pi}}e^{-(x-\mu)^2/2\sigma^2}.$$

If n values are generated from this same normal distribution, their joint *pdf* again should be given by the product of these normal marginal *pdf*s:

$$f_{X_1, X_2, \ldots, X_n}(x_1, x_2, \ldots, x_n) = \prod_{i=1}^{n} \frac{1}{\sigma\sqrt{2\pi}}e^{-(x_i-\mu)^2/2\sigma^2}$$

$$= \left(\frac{1}{\sigma\sqrt{2\pi}}\right)^n e^{-\sum_{i=1}^{n}(x_i-\mu)^2/2\sigma^2}.$$

A Poisson process generates independent counts in nonoverlapping intervals. Suppose telephone calls arrive at a realtor's office at a rate of 15 per hour during business hours (9 AM – 5 PM), like events in a Poisson process. If we let $X_1, X_2, \ldots, X_8$ be the numbers of these calls to arrive in each of the eight hours of a given business day, then each of these random variables is Poisson with $\mu = 15$, and their joint probability function is given by the product of the marginal probability functions:

$$p_{X_1, X_2, \ldots, X_8}(x_1, x_2, \ldots, x_8) = \prod_{j=1}^{8} \frac{15^{x_i}}{x_i!}e^{-15}$$

$$= \frac{15^{\sum x_i}}{\prod x_i!}e^{-120}, \qquad \text{for } x_i = 0, 1, 2, \ldots, \quad i = 1, 2, \ldots, 8.$$

EXERCISES 5.2

1. Assume that X_1, X_2 is uniform on the square with corners

$$(-1, 0), (0, -1), (0, 1), (1, 1).$$

   a. Find the marginal *pdf* for X_1.
   b. Find the conditional *pdf* for X_2, given $X_1 = x_1$.

2. You select one integer at random from $R_Y = \{1, 2, 3\}$; call the selected integer Y. Then you flip Y fair coins. What is the probability you observe no heads?

3. Assume the joint *pdf* for X_1, X_2 is

$$f_{X_1, X_2}(x_1, x_2) = x_1 e^{-x_1(1 + x_2)}, \qquad \text{for } x_1 > 0, x_2 > 0.$$

   a. Are X_1 and X_2 independent?
   b. Evaluate the conditional *pdf* for X_1, given $X_2 = x_2$, and the conditional *pdf* for X_2, given $X_1 = x_1$.

4. The game of Lotto 6/53 involves drawing 6 numbers from the integers 1, 2,..., 53. Assume that the draw is completely fair and let X_1, X_2 be the *first two* numbers drawn in one play (without replacement).

   a. What is the joint probability law for X_1, X_2?
   b. Evaluate the conditional probability function for X_2, given $X_1 = x_1$.
   c. Evaluate the conditional probability function for X_1, given $X_2 = x_2$.

5. A fair die is rolled three times; let X_1 be the number of 1's to occur on the first two rolls, and let X_2 be the number of 2's to occur on the last two rolls. What is the joint probability function for X_1, X_2?

6. To generate a "continuous" uniform random number on the interval $(0, 1)$, with three-decimal-place accuracy (three digits to the right of the decimal), a computer program actually generates a sequence of three independent digits, each equally likely to take on the values $0, 1, ..., 9$. The number generated then consists of the decimal point followed by these three digits.

   a. Letting D_1, D_2, D_3 represent the three digits generated, what is their joint probability function?
   b. Let $X = .D_1 D_2 D_3$ be the "continuous" uniform random number generated. What would you take as the exact probability law for X?
   c. Evaluate $P(X \leq \frac{1}{3})$ with this procedure.

7. Suppose a person buys a ticket for her state lottery game on a sporadic basis. Assume she flips a fair coin each Saturday; if a head occurs, she buys a ticket, and if a tail occurs she does not. Further assume that the probability a purchased ticket will win some prize is $\frac{1}{9}$. Over a 10-week period, what is the probability distribution for the number of prizes she will win in the lottery?

8. Assume the number of people to enter a store, on a given business day, is a random variable N that is binomial with parameters $\mathcal{N}, \pi$. Also assume the number of cash sales made by the store on any given business day, given that $N = n$ people entered the store that day, is a binomial random variable X with parameters n, p. What is the marginal probability law for X, the number of sales made that day?

9. *Sampling inspection*   A small manufactured item is sold to a retailer in lots of 1000. Assume that each item is a Bernoulli trial (and that the trials are independent) with p being the probability of the item being defective in some manner. The retailer selects a random sample of 10 items from the lot of 1000 and inspects each one. What is the marginal probability law for Y, the number of defectives found in the sample of 10? (*Hint*: Use conditional probability, letting X be the number of defectives in the lot of 1000.)

10. A Geiger counter is used to record the number of radioactive particles emitted by a source; unfortunately, the counter is defective and occasionally particles do not register. More specifically, assume each particle to enter the counter is a Bernoulli trial, with p being the probability the particle will register, and assume these trials are independent. What is the probability law for the number Y of particles counted in an interval of time of length t, granted the particles are emitted like events in a Poisson process with parameter λ? (*Hint*: Use conditional probability, and let X be the number of particles emitted in the given time period.)

11. Suppose a single point in the (x, y)-plane is chosen in the following way: First the x-coordinate is determined as the observed value for a random variable X, which is uniform on $(0, 1)$. Then the y-coordinate is determined by the observed value for a random variable Y, which is uniform on the interval $(0, x)$, where x is the observed value for X.
    a. Find the joint *pdf* for X, Y.
    b. Evaluate the marginal *pdf* for Y; then find the conditional *pdf* for X given $Y = y$.

12. Suppose n independent Bernoulli trials are performed, each with probability of success equal to p. If there are x successes in the first n_1 trials, what is the probability law for the number of successes in the remaining $n - n_1$ trials?

13. Suppose n independent Bernoulli trials are performed, each with probability of success equal to p. Assuming x successes are observed, what is the probability law for the number of successes in the first $n_1 < n$ trials?

14. Let T_1, T_2 have the joint *pdf* $f_{T_1, T_2}(t_1, t_2) = e^{-t_2}$ for $0 < t_1 < t_2$, and evaluate $P(T_1 \leq 1 \mid T_2 > 2)$.

15. Granted a sequence of independent Bernoulli trials, each with parameter p, let X_1 be the trial number of the first success and let X_2 be the trial number of the second success. Evaluate the joint probability function for X_1, X_2. (*Hint*: Consider the conditional probability function for X_2 given $X_1 = x_1$.)

16. Independent Bernoulli trials with parameter p are performed until the occurrence of the second success. If the second success occurs on trial number x_2, what is the probability function for the trial number of the first success?

In this section we shall discuss the multinomial probability law, one of the most frequently used discrete probability laws, and see some of its properties. This probability law is multidimensional, discrete, and is based on a generalization of Bernoulli trials, simple experiments with $k \geq 2$ different outcomes. The multi-hypergeometric probability law will also be discussed in an example and developed further in the Exercises 5.3. We shall also discuss the bivariate normal probability law, a generalization of the one-dimensional normal we have already studied; it is one of the most frequently used two-dimensional continuous probability laws.

The multinomial probability law is a generalization of the binomial; recall from Chapter 3 that the binomial describes the number of successes observed in n Bernoulli trials, which have two different outcomes. A multinomial trial is a simple experiment that has $k \geq 2$ different possible outcomes. For example, the roll of a single die is a multinomial trial with $k = 6$ different possible outcomes. If a new medication becomes available for a specific ailment, and the observed result for each person administered the medication is categorized into the classes "completely cured," "considerably improved," "slightly improved," "no effect," and "condition worsened," then each person could be modeled as a multinomial trial with $k = 5$ different outcomes.

Any experiment whose sample space has k elements can be thought of as a single multinomial trial with k different outcomes. Recall that a Bernoulli trial has a single parameter associated with it, the probability p of observing a success. A multinomial trial has a vector-valued parameter $p_1, p_2, \ldots, p_k$ associated with it, specifying the probabilities of occurrence for the k different types of outcomes possible. Since (exactly) one of these outcomes must be observed on each performance, it is necessary that $\sum_{i=1}^{k} p_i = 1$, and there is some redundancy built into this notation. A Bernoulli trial is a multinomial trial with $k = 2$, and for this case the two parameters are $p_1 = p$ and $p_2 = 1 - p$, the probabilities of the two outcomes success and failure, respectively.

Example 5.9

Let X be an exponential random variable with parameter $\lambda = 1$; then the $100k$th quantile for X is $x_k = -\ln(1 - k)$. Thus $x_{.4} = .511, x_{.7} = 1.207$, and $x_{.9} = 2.303$ (to three decimal places), so $P(X < .511) = .4, P(.511 < X < 1.207) = .3, P(1.207 < X < 2.303) = .2$, and $P(X > 2.303) = .1$. We can use these intervals to convert an observed value for X into a single multinomial trial with $k = 4$ classes. The vector of probabilities for this multinomial trial then is $p_1 = .4, p_2 = .3, p_3 = .2, p_4 = .1$.

Suppose that n identical (meaning each trial has the same vector of parameters) and independent multinomial trials are performed; the sample space S is the Cartesian product $S = T_1 \times T_2 \times \cdots \times T_n$, where $T_i = \{1, 2, \ldots, k\}$, for $i = 1, 2, \ldots, n$, so S is a set of n-tuples. The only difference between this experiment and one that consists of n Bernoulli trials is the fact that each element in an n-tuple belonging to S can now take on k different values rather than only 2. As with an experiment consisting of independent Bernoulli trials, the probability that any single-element event (an n-tuple) will occur is then given by the *product* of the probabilities of the elements in the given n-tuple, because of the assumed independence of the trials. That is, each single-element event probability is a product of n terms, and each of these terms must be p_1, p_2, $\cdots$ or p_k, depending on the outcome observed on the corresponding trial. Thus each single-element event probability has the form $p_1^{x_1} p_2^{x_2} \cdots p_k^{x_k}$, where $\sum_{i=1}^{k} x_i = n$, and x_i counts the number of times that outcome i occurs in the given n-tuple.

If we define the k random variables, $X_i =$ number of times outcome i occurs, $i = 1, 2, \ldots, k$, in the n trials, it is clear that

$$\sum_{i=1}^{k} X_i = n$$

since there are only k different possible outcomes on each trial. Thus it is necessary that

$$X_k = n - \sum_{i=1}^{k-1} X_i$$

and the value for X_k (or *any* single one of $X_1, X_2, \ldots, X_k$) is actually determined by the values for the other $k - 1$ random variables. This means that only $k - 1$ of the variables are random, so the value for X_k is redundant. The same redundancy occurs if we let X_1 be the number of successes and X_2 the number of failures in n Bernoulli trials, since $X_1 + X_2 = n$. In terms of notation, it is useful to employ this redundancy for multinomial trials, but you should remember that the sum of all k random variables is fixed at the number of trials, n, and thus there are really only $k - 1$ random variables involved.

To find the probability function for $X_1, X_2, \ldots, X_k$, we need to count the number of n-tuples in S that contain (exactly) x_i outcomes of type i, $i = 1, 2, \ldots, k$ (where the x_i-values are nonnegative integers satisfying $\sum_i x_i = n$) since each such n-tuple has the same probability of occurrence. If we are given n symbols of which x_i are of type i, then the number of permutations possible (i.e., the number of n-tuples in S of this type) is given by $\binom{n}{x_1, x_2, \ldots, x_k}$, as we know. Thus the joint probability function for $X_1, X_2, \ldots, X_k$ is

$$p_{X_1, X_2, \ldots, X_k}(x_1, x_2, \ldots, x_k) = \binom{n}{x_1, x_2, \ldots, x_k} p_1^{x_1} p_2^{x_2} \cdots p_k^{x_k},$$

for $(x_1, x_2, \ldots, x_k) \in R_{X_1, X_2, \ldots, X_k}$, where $R_{X_1, X_2, \ldots, X_k}$ contains all possible k-tuples, $(x_1, x_2, \ldots, x_k)$, such that $x_i = 0, 1, 2, \ldots, n$ for $i = 1, 2, \ldots, k$, with $\sum_{i=1}^{k} x_i = n$. This establishes the following theorem.

THEOREM 5.3   If an experiment consists of n independent multinomial trials, each with parameters $p_1, p_2, \ldots, p_k$, and we define X_i to be the number of times in the n trials that outcome type i occurs (where $i = 1, 2, \ldots, k$), then $X_1, X_2, \ldots, X_k$ is called a multinomial random variable with parameters $n, p_1, p_2, \ldots, p_k$. The probability function for $X_1, X_2, \ldots, X_k$ is

$$p_{X_1, X_2, \ldots, X_k}(x_1, x_2, \ldots, x_k) = \binom{n}{x_1, x_2, \ldots, x_k} p_1^{x_1} p_2^{x_2} \cdots p_k^{x_k},$$

for $(x_1, x_2, \ldots, x_k) \in R_{X_1, X_2, \ldots, X_k}$, as specified above.

This multinomial probability law reproduces itself, in the sense that any subset of the random variables again has the multinomial probability law. This must happen since single multinomial trials can be "collapsed" into fewer classes, but the independence of the trials remains and the counts of the numbers of times these classes occur is again multinomial.

Example 5.10

Assume 12 fair dice are tossed onto a table; each die is then a multinomial trial with vector of parameters $p_1, p_2, \ldots, p_6$, where each $p_i = \frac{1}{6}$. If the trials are independent (i.e., the face shown on any given die has no influence on the faces of the others), then $X_1, X_2, \ldots, X_6$, where X_i is number of times face i occurs), is multinomial with $n = 12$, and probabilities all equal $\frac{1}{6}$ for the different classes occurring. Thus their joint probability function is

$$p_{X_1, X_2, \ldots, X_6}(x_1, x_2, \ldots, x_6) = \binom{12}{x_1, \ x_2, \ \cdots \ x_6}\left(\frac{1}{6}\right)^{12},$$

for $x_i = 0, 1, \ldots, 6$, $i = 1, 2, \ldots, 6$ and $\sum_{i=1}^{6} x_i = 12$. The probability that all 12 dice show the same face is then

$$p_{X_1, X_2, \ldots, X_6}(12, 0, \ldots, 0) + p_{X_1, X_2, \ldots, X_6}(0, 12, 0, \ldots, 0)$$

$$+ \cdots + p_{X_1, X_2, \ldots, X_6}(0, 0, \ldots, 12) = 6\left(\frac{1}{6}\right)^{12} = \left(\frac{1}{6}\right)^{11}.$$

The probability that each of the faces 1 through 5 occurs exactly once and that face 6 occurs seven times is

$$p_{X_1, X_2, \ldots, X_6}(1, 1, \ldots, 7) = \frac{12!}{7!}\left(\frac{1}{6}\right)^{12} = .00005,$$

while the probability that each face occurs two times is

$$p_{X_1, X_2, \ldots, X_6}(2, 2, \ldots, 2) = \frac{12!}{2^6}\left(\frac{1}{6}\right)^{12} = .0034.$$

If we combine or collapse classes of outcomes, the probability law of the counts again is hypergeometric. For example, let class 1 consist of numbers no greater than 3, let class 2 contain 4 and 5, and let class 3 contain the single number 6; if we let Y_1, Y_2, Y_3 be the numbers of the 12 dice that give results in these three classes, then Y_1, Y_2, Y_3 is multinomial with $n = 12, p_1 = \frac{1}{2}, p_2 = \frac{1}{3}$, and $p_3 = \frac{1}{6}$. The probability function is

$$p_{Y_1,Y_2,Y_3}(y_1, y_2, y_3) = \binom{12}{y_1, y_2, y_3}\left(\frac{1}{2}\right)^{y_1}\left(\frac{1}{3}\right)^{y_2}\left(\frac{1}{6}\right)^{y_3},$$

for $y_i = 0, 1, \ldots, 12$, $i = 1, 2, 3$ such that $\sum_i y_i = 12$. The probability of finding six outcomes in the first class, four in the second, and two in the third is

$$p_{Y_1,Y_2,Y_3}(6, 4, 2) = \frac{12!}{6!4!2!}\left(\frac{1}{2}\right)^6\left(\frac{1}{3}\right)^4\left(\frac{1}{6}\right)^2 = .0743.$$

A single observation of a random variable Y can be converted into a multinomial trial by partitioning its range into k pieces. Then if $Y_1, Y_2, \ldots, Y_n$ are independent, and each with the same probability law, this group of n observations can be converted into n multinomial trials with $p_1, p_2, \ldots, p_k$ giving the probabilities of an individual observation falling into the various parts of the partition. This is illustrated in the following example.

Example 5.11

Suppose $n = 25$ independent uniform $(0, 1)$ random variables are generated on a computer. Let X_1 count the number of these that are less than .4, so the observed value falls with in the interval $(0, .4)$. Similarly, let X_2 count the number in $[.4, .7)$, X_3 count the number in $[.7, .9)$, and X_4 the number in $[.9, 1)$. Then we have converted each uniform random variable into a multinomial trial with parameters $.4, .3, .2, .1$, and X_1, X_2, X_3, X_4 is multinomial with $n = 25$ and with these same probabilities of the four classes occurring:

$$p_{X_1,X_2,X_3,X_4}(x_1, x_2, x_3, x_4) = \binom{25}{x_1, \ x_2, \ x_3, \ x_4,}(.4)^{x_1}(.3)^{x_2}(.2)^{x_3}(.1)^{x_4}.$$

The probability of observing the counts $10, 7, 5, 3$, respectively, in these four classes is then $\binom{25}{10, 7, 5, 3}(.4)^{10}(.3)^7(.2)^5(.1)^3 = .0173.$

The marginal probability law for any subset of $X_1, X_2, \ldots, X_k$, augmented by the count of those variables *not* falling in the subset of classes used, is again multinomial, as has been mentioned. Conditional probability functions for some subset of $X_1, X_2, \ldots, X_k$, given values for the others not in the subset, again are multinomial. This can be verified directly from the definition of a conditional probability function, taking the ratio of the joint probability function to the marginal probability function of the variables with given values. This ratio of two multinomials again is multinomial in form.

This result can perhaps be seen more easily by considering a particular example. Suppose, as in Example 5.11, that $n = 25$ uniform $(0, 1)$ random variables are generated and the counts X_1, X_2, X_3, X_4 are multinomial with $n = 25$ and with the vector of probability values $.4, .3, .2, .1$ for the four classes. What is the *conditional* probability function for X_1, X_2, X_3 given that 6 of the generated uniform values fell in $[.9, 1)$ (i.e., given that $x_4 = 6$)?

The marginal probability that $X_4 = 6$ is given by

$$P(X_4 = 6) = \binom{25}{6}(.1)^6(.9)^{19}, \tag{1}$$

since any single one of these variables is binomial with parameters $n = 25$ and the appropriate p_i (.1 in this case). Dividing $p_{X_1,X_2,X_3,X_4}(x_1,x_2,x_3,6)$ by Eq. (1) then gives the conditional probability law for X_1, X_2, X_3, given $X_4 = 6$. You should do this and cancel common terms, recognizing then that the result is again multinomial.

This same result can be derived by considering this situation directly. We assume that 25 uniform $(0, 1)$ random variables have been observed, and we find that six of them are at least .9. Then the conditional probability a uniform $(0, 1)$ random variable falls in $(0,.4)$ is $\frac{4}{9}$, that it falls in $[.4,.7)$ is $\frac{3}{9}$ and that it falls in $[.7,.9)$ is $\frac{2}{9}$, given it is smaller than .9. If we are given that $X_4 = 6$, then we know that the remaining 19 uniform values are less than .9. Thus, the conditional probability function for X_1, X_2, X_3, given $X_4 = 6$, is multinomial with parameters $19, \frac{4}{9}, \frac{3}{9}, \frac{2}{9}$:

$$p_{X_1,X_2,X_3|X_4}(x_1,x_2,x_3 \mid x_4 = 6) = \binom{19}{x_1, \ x_2, \ x_3}\left(\frac{4}{9}\right)^{x_1}\left(\frac{3}{9}\right)^{x_2}\left(\frac{2}{9}\right)^{x_3},$$

for $x_i = 0, 1, \ldots, 19$, $i = 1, 2, 3$, and $\sum x_i = 19$. This is also the ratio of the joint probability law for all X_1, X_2, X_3, X_4, with $x_4 = 6$, divided by the marginal probability that $X_4 = 6$.

You will recall that the binomial probability law occurs in sampling with replacement, where each item sampled can be classified into two groups; the hypergeometric probability law describes sampling without replacement. If each item sampled can be classified into k groups and a random sample of n items is selected with replacement, then the counts of sampled items falling into the k groups is easily seen to be multinomial with parameters $n, p_1, p_2, \ldots, p_k$, where p_i is the probability of selecting an item from group i on each draw. If, as is more commonly done, the sample is selected without replacement, the *multi-hypergeometric probability law* describes the numbers of counts to occur in the sample. This probability law is illustrated in the following example.

Example 5.12

A corporation has 15 equivalent job openings available at the same time and advertises this fact in the local paper. The ad brings in 210 applicants for the positions, all equally qualified, of whom 93 are from racial group $\mathcal{G}_1$, 68 are from racial group $\mathcal{G}_2$, and the remainder are from racial group $\mathcal{G}_3$. Suppose then that the 15 jobs are allotted at random to 15 of the 210 applicants, and let X_i count the number of these jobs that go to applicants from racial group $\mathcal{G}_i$, $i = 1, 2, 3$. What is the probability law for X_1, X_2, X_3?

If the jobs are allotted at random, then each of the $\binom{210}{15}$ possible subsets of size 15 are equally likely to occur. The 210 applicants are partitioned into three groups of sizes 93, 68, and 49 respectively. Then computing $P(X_1 = x_1, X_2 = x_2, X_3 = x_3)$ for any possible $x_1, x_2,$ and x_3 simply requires evaluating the number of subsets of size 15 that can be constructed, where the numbers occurring from the three groups are x_1, x_2, and x_3. But this is the product $\binom{93}{x_1}\binom{68}{x_2}\binom{49}{x_3}$ where $x_i = 0, 1, \ldots, 15$, $i = 1, 2, 3$, and $\sum x_i = 15$. Thus

$$p_{X_1, X_2, X_3}(x_1, x_2, x_3) = \frac{\binom{93}{x_1}\binom{68}{x_2}\binom{49}{x_3}}{\binom{210}{15}};$$

this is an example of the multi-hypergeometric probability law with parameters $m = 210, r_1 = 93, r_2 = 68, r_3 = 49$, and $n = 15$. Note that $\sum X_i = 15$ is fixed, so we really only have two random variables here and are using the same redundant notation employed with the multinomial. The probability that the 15 jobs are allocated to eight persons from $\mathcal{G}_1$, four persons from $\mathcal{G}_2$, and three persons from $\mathcal{G}_3$ is then $\binom{93}{8}\binom{68}{4}\binom{49}{3}/\binom{210}{15} = .0489$.

If 15 persons were chosen from the 210 *with replacement*, then X_1, X_2, X_3 would be multinomial with parameters $n = 15, \frac{93}{210}, \frac{68}{210}, \frac{49}{210}$. The multinomial probability of this event is .0465, fairly close to the exact multi-hypergeometric value since the sample size employed (15) is fairly small compared with the numbers of people in the different groups.

The bivariate normal probability law has found many uses in both the physical and social sciences. As the name is meant to convey, it is a two-dimensional (bivariate) continuous probability law whose marginal probability laws are normal. It is not the only two-dimensional probability law with this property, but is by far the one most frequently used in applications. We shall introduce this probability law through conditional reasoning, in a standardized form, from which it will be easy to get the full bivariate normal.

Assume X is a standard normal random variable and the conditional probability law for Y, given $X = x$, is normal with mean ρx and variance $1 - \rho^2$, where $-1 < \rho < 1$ is a given constant. It is, of course, required that $|\rho| < 1$ to assure that the variance of Y (for this conditional *pdf*) is positive. That is, the conditional *pdf* for Y, given $X = x$, is

$$f_{Y|X}(y \mid x) = \frac{1}{\sqrt{2\pi(1 - \rho^2)}} e^{-(y - \rho x)^2/2(1 - \rho^2)}. \tag{2}$$

Equation (2) defines a symmetric bell-shaped normal curve centered at ρx with variance $1 - \rho^2$; the closer $|\rho|$ gets to 1, the smaller the variance gets (and the more peaked is the *pdf*).

Granted this marginal for X and this conditional *pdf* for Y, the joint *pdf* for X and Y is then

$$f_{X,Y}(x, y) = f_X(x) f_{Y|X}(y \mid x)$$

$$= \left\{ \frac{1}{\sqrt{2\pi}} e^{-x^2/2} \right\} \left\{ \frac{1}{\sqrt{2\pi(1 - \rho^2)}} e^{-(y - \rho x)^2/2(1 - \rho^2)} \right\}$$

$$= \frac{1}{2\pi\sqrt{1 - \rho^2}} e^{-(x^2 - 2\rho xy + y^2)/2(1 - \rho^2)}. \tag{3}$$

This *pdf* $f_{X,Y}(x,y)$ is symmetric in x and y. Thus we would arrive at the same joint *pdf* by assuming Y is standard normal while the conditional *pdf* for X, given $Y = y$, is normal with mean ρy and variance $1 - \rho^2$. This symmetry implies that the marginal *pdf*s for both X and Y are standard normal. (This can also be verified by integrating Eq. (3).)

The joint *pdf* given by Eq. (3) is a special case of the *bivariate normal probability law*, and the parameter ρ is called the *correlation coefficient*. It will be discussed in more detail in Chapter 6. Note that if $\rho = 0$ then Eq. (3) is the product of two standard normal *pdf*s, while for $\rho \neq 0$ this is not so. With $\rho = 0$ then, this special bivariate normal is actually the probability law for two independent standard normal random variables (since the joint *pdf* is the product of the marginal *pdf*s).

The general bivariate normal probability law has five parameters, not just the correlation coefficient ρ. The additional four parameters are the means for the two variables and the variances of the two variables. These can be introduced by interpreting X and Y as the standard forms for two random variables, say X_1 and X_2, so

$$X = \frac{X_1 - \mu_1}{\sigma_1} \qquad \text{and} \qquad Y = \frac{X_2 - \mu_2}{\sigma_2}.$$

Granted X is standard normal, then X_1 is normal with mean μ_1 and standard deviation σ_1; similarly, since the marginal *pdf* for Y is standard normal, the marginal *pdf* for X_2 is normal with mean μ_2 and standard deviation σ_2. The conditional *pdf* for Y is normal with mean ρx and variance $1 - \rho^2$. If $X = (X_1 - \mu_1)/\sigma_1 = x$, then $X_1 = x_1 = \mu_1 + \sigma_1 x$ is the value conditioned on. Thus the conditional *pdf* for $Y = (X_2 - \mu_2)/\sigma_2$, given $X = (X_1 - \mu_1)/\sigma_1 = x$, is normal with mean $\rho x = \rho(x_1 - \mu_1)/\sigma_1$, and variance $1 - \rho^2$.

The conditional *pdf* for X_2, given $X_1 = x_1$, is then normal with mean $\mu_2 + \rho\frac{\sigma_2}{\sigma_1}(x_1 - \mu_1)$ and variance $(1 - \rho^2)\sigma_2^2$. This *bivariate normal pdf* is

$$f_{X_1,X_2}(x_1,x_2) = \frac{1}{2\pi\sigma_1\sigma_2\sqrt{(1-\rho^2)}}e^{-q(x_1,x_2)/2}, \tag{4}$$

where

$$q(x_1,x_2) = \frac{1}{1-\rho^2}\left[\frac{(x_1-\mu_1)^2}{\sigma_1^2} - 2\rho\left(\frac{x_1-\mu_1}{\sigma_1}\right)\left(\frac{x_2-\mu_2}{\sigma_2}\right) + \frac{(x_2-\mu_2)^2}{\sigma_2^2}\right].$$

The function $q(x_1,x_2)$ which occurs in the exponent of Eq. (4) is a quadratic form in the variables $(x_1 - \mu_1)/\sigma_1$ and $(x_2 - \mu_2)/\sigma_2$; it is nonnegative for all (x_1,x_2) and is equal to 0 only for $x_1 = \mu_1, x_2 = \mu_2$. Thus $(x_1,x_2) = (\mu_1,\mu_2)$ gives the maximum value for the *pdf*. The contours in the (x_1,x_2)-plane where this bivariate normal *pdf* is constant require $q(x_1,x_2)$ to equal a positive constant. This gives the equation of an ellipse in the (x_1,x_2)-plane, centered at (μ_1,μ_2). This bivariate normal *pdf* is thus shaped like a hill with the peak located at (μ_1,μ_2); if $\rho = 0$, then Eq. (4) factors into the product of the two marginal *pdf*s. If $\rho = 0$ and $\sigma_1 = \sigma_2$, the constant contours of the *pdf* are circles and this probability law is called the *circular normal*.

Example 5.13

The bivariate normal probability law given by Eq. (4) is frequently employed in describing scores made on national standardized tests, like the Scholastic Aptitude Test, which you may have taken at some point in time. Such a standardized test will typically have at least two parts. For concreteness, let us assume a particular test has two parts, labeled "verbal" and "mathematics." Furthermore, we shall assume that a randomly selected student who takes this test will make scores X_1 and X_2 on these two parts, and that X_1, X_2 are bivariate normal.

The parameters for this bivariate normal distribution will be taken as $\mu_1 = \mu_2 = 500, \sigma_1 = \sigma_2 = 100$, and $\rho = .4$. The marginal probability a student will score better than 550 on the verbal part (with no information about his mathematics score) is $P(X_1 > 550) = P(Z > \frac{1}{2}) = .3085$, where Z is standard normal. The marginal probability he scores less than 400 on the mathematics portion (with no information about his verbal score) is $P(X_2 < 400) = P(Z < -1) = .1587$.

If we are given that a student scored $x_2 = 600$ on the mathematics portion, then the conditional distribution of her verbal score is normal with mean $\mu_1 + \rho \frac{\sigma_2}{\sigma_1}(600 - \mu_2) = 500 + (.4)\frac{100}{100}(600 - 500) = 540$ and standard deviation $100\sqrt{1 - .4^2} = 91.7$. The conditional probability that her verbal score exceeds 550 is $P(X_1 > 550 \,|\, x_2 = 600) = P(Z > (550 - 540)/91.7) = P(Z > .109) = .4598$, greater than the marginal probability for this same event.

Similarly, if it is given that a student scored $x_1 = 450$ on the verbal portion, the conditional *pdf* for his mathematics score is normal with mean $500 + (.4)\left(\frac{100}{100}\right)(450 - 500) = 480$ and standard deviation 91.7. Thus the probability his mathematics score is less than 400 is $P(X_2 < 400 \,|\, x_1 = 450) = P(Z < (400 - 480)/91.7) = P(Z < -.872) = .1906$.

 **EXERCISES 5.3**

1. Let X_1, X_2, X_3 be multinomial with parameters $n = 9, p_1 = \frac{1}{4}, p_2 = \frac{1}{3}$, and $p_3 = \frac{5}{12}$. Evaluate the probability that the three types of outcomes occurred equally frequently (i.e., $X_1 = X_2 = X_3$). What is the conditional probability law for X_1, X_2 given $X_3 = 3$?

2. A five-dimensional random variable $Y_1, Y_2, \ldots, Y_5$ has probability function

$$p_{Y_1, Y_2, \ldots, Y_5}(y_1, y_2, \ldots, y_5) = \binom{15}{y_1, \; y_2, \; y_3, \; y_4, \; y_5}\left(\frac{1}{5}\right)^{15},$$

for $y_i = 0, 1, \ldots, 15, i = 1, 2, \ldots, 5$, such that $\sum_{i=1}^{5} y_i = 15$.
   a. What is the marginal probability law for Y_3?
   b. What is the marginal probability law for Y_2, Y_4?
   c. Evaluate $P(Y_2 = Y_4)$.
   d. Evaluate $P(Y_1 = Y_2 = Y_3 = Y_4 = Y_5)$.

3. A moderate-size university has 8000 full-time registered undergraduate students, of whom 2600 are classified freshmen, 2100 are sophomores, 1900 are juniors, and the remaining 1400 are seniors. The student council

selects 100 of these students at random. If X_1, X_2, X_3, X_4 represent the numbers of freshmen, sophomores, juniors, and seniors, respectively, in the sample, what is their probability law? What is the probability the sample includes 33 freshmen, 26 sophomores, 24 juniors, and 17 seniors? Approximate this probability by the appropriate multinomial.

4. What are the marginal probability laws for X_1, X_2, X_3, X_4 in Exercise 3?

5. If $X_1, X_2, \ldots, X_k$ is multi-hypergeometric with parameters m, $r_1, r_2, \ldots, r_k$, n, what is the marginal probability law for X_i?

6. Granted 33 freshmen occur in the sample of 100 students discussed in Exercise 3, what is the conditional probability function for X_2, X_3, X_4? Use this to evaluate the conditional probability of getting 26 sophomores, 24 juniors, and 17 seniors in the sample of 100.

7. A bookstore receives 100 copies of a best-selling novel; of these, 50 have a dust jacket with a red background, 30 have a green background, and the remainder have a yellow background.
   a. If 20 of these books are sold in the first day, and X_1, X_2, X_3 are the numbers of these with the different colored jackets in the order given, what is the probability law for X_1, X_2, X_3? Assume those sold were selected at random from the 100.
   b. Evaluate $p_{X_1, X_2, X_3}(10, 6, 4)$.
   c. What is the probability that all 20 have a red dust jacket?

8. A town contains 125 fast-food restaurants, of which 30 belong to chain $\mathcal{D}$, 20 belong to chain $\mathcal{B}$, 40 belong to chain $\mathcal{W}$, and the rest belong to chain $\mathcal{M}$. A total of 400 tourists eat lunch at a fast-food restaurant in this town on a given day. If each of these tourists selects his or her restaurant at random, and X_1, X_2, X_3, X_4 count the numbers to select restaurants from chains $\mathcal{D}, \mathcal{B}, \mathcal{W}$, and $\mathcal{M}$, respectively, what is the probability law for X_1, X_2, X_3, X_4?

9. The game of bridge requires four players, each of whom is dealt 13 cards from a regular 52-card deck. Suppose you are playing this game, and assume that the 13 cards you receive are selected at random from the 52. Define X_1 to be the number of hearts you receive, X_2 the number of diamonds you receive, X_3 the number of spades you receive, and X_4 the number of clubs you receive. What is the probability law for X_1, X_2, X_3, X_4? What is the probability law for X_1?

10. Many different computer algorithms exist for generating observed values from any desired probability law; consider the probability integral transform (Theorem 4.6), as one example. Suppose one of these algorithms is used to generate 200 independent standard normal observations (observed values of standard normal random variables). Let C_1 be the number of these generated values that are smaller than -2, let C_2 count those between -2 and -1, C_3 those between -1 and 0, and so forth up to C_6 counting those that exceed 2, so $\sum_{i=1}^{6} C_i = 200$. What is the probability law for $C_1, C_2, \ldots, C_6$?

11. The bivariate normal probability law is frequently assumed to describe the impact points of rounds fired at a target in a two-dimensional plane.

Thus if we imagine an (x,y) coordinate system centered at the target, the vertical miss distance is Y and the horizontal miss distance is X. Assume that X, Y are bivariate normal with $\mu_X = 1, \mu_Y = .5, \sigma_X = .5, \sigma_Y = 1$, and $\rho = .6$.

    a. Evaluate $P(X \le .5)$ and $P(Y \le .5)$.

    b. Evaluate $P(X \le 0 \mid Y = -.2)$, the probability that the round hits to the left of the target, given it was .2 units below the target.

12. Assume that X_1, X_2 are bivariate normal with $\mu_1 = \mu_2 = 0$, $\sigma_1 = \sigma_2$, and $\rho = 0$; this is a special case of the circular normal, mentioned in the text. The circular probable error (called the *CEP*) is the radius of the circle centered at $(0,0)$ that has probability .5 of including the observed values x_1, x_2. Evaluate the *CEP*.

13. Go through the reasoning necessary to show that if $X_1, X_2, \ldots, X_k$ is multi-hypergeometric with parameters m, $r_1, r_2, \ldots, r_k$, n, then the conditional probability law for $X_1, X_2, \ldots, X_{k-1}$, given $X_k = x_k$, again has the same probability law with parameters $m - r_k$, $x_1, x_2, \ldots, x_{k-1}$, $n - x_k$.

14. Assume that a shipment of $m = 500$ radios is received by a large retailer. To keep things simple, suppose there are only two types of defects these radios may suffer: type I (exterior scratches of some sort) and type II (defective speakers). Also assume that $r_1 = 400$ of these radios have no defects of either type, while $r_2 = 40$ of them have defects of type I only, $r_3 = 50$ have defects of type II only, and the remaining $r_4 = 10$ have both types of defects. If a random sample of $n = 20$ of these radios is selected for inspection (without replacement), let X_1, X_2, X_3, X_4 be the numbers of radios in the sample that fall into these four classes. What is the probability law for X_1, X_2, X_3, X_4? What is the probability law for X_1, X_2, X_3 given $X_4 = 1$?

15. At a particular time, assume that 10 of the television sets in a small town are tuned to the local CBS station; 8, 12, and 5 of the sets are tuned to the local NBC, ABC, and PBS stations, respectively, and the remaining 45 are turned off. If a random sample of 20 of these sets is being monitored at this time, let C, N, A, P, O represent the numbers in the sample that fall into these respective categories. Given that $O \ge 10$, what is the probability that $C = 5$?

16. There are a number of different continuous bivariate distributions other than the bivariate normal whose marginals are standard normal. One of the simplest of these is given by

$$f_{X,Y}(x,y) = \begin{cases} 2\dfrac{1}{2\pi}e^{-(x^2+y^2)/2}, & \text{for } x > 0, y > 0 \quad \text{or} \quad x < 0, y < 0 \\ 0 & \text{otherwise.} \end{cases}$$

(Note that this *pdf* is twice the product of two standard normal *pdf*s, for quadrants I and III, and is 0 otherwise.)

    a. Show that the marginal *pdf*s for X and Y are standard normal.

    b. Are X and Y independent?

5.4 | Summary

The joint probability function for discrete $X_1, X_2, \ldots, X_n$ is positive for elements of the range $R_{X_1, X_2, \ldots, X_n}$ and must satisfy

$$\sum_{x_1, x_2, \ldots, x_n \in R_{X_1, X_2, \ldots, X_n}} p_{X_1, X_2, \ldots, X_n}(x_1, x_2, \ldots, x_n) = 1.$$

The probability that any region will contain the observed values for the random variables is given by the sum of $p_{X_1, X_2, \ldots, X_n}(x_1, x_2, \ldots, x_n)$ for all $x_1, x_2, \ldots, x_n$ in the joint range that are contained in the region. The marginal probability law for any subset of $X_1, X_2, \ldots, X_n$ is given by summing the joint probability function over the observed values for the variables not included in the subset of interest.

The joint *pdf* for continuous $X_1, X_2, \ldots, X_n$ is positive for elements in the range $R_{X_1, X_2, \ldots, X_n}$ and must satisfy

$$\int_{-\infty}^{\infty} \cdots \int_{-\infty}^{\infty} f_{X_1, X_2, \ldots, X_n}(x_1, x_2, \ldots, x_n) \, dx_1 \cdots dx_n = 1.$$

The probability that any region will contain the observed values for the random variables is given by the integral of $f_{X_1, X_2, \ldots, X_n}(x_1, x_2, \ldots, x_n)$ over the region. The marginal probability law for any subset of $X_1, X_2, \ldots, X_n$ is given by integrating the joint *pdf* with respect to the variables not included in the subset of interest.

In either the discrete or continuous case, knowledge of the marginal probability laws alone is not sufficient to specify the joint probability law.

A multinomial trial has k different outcomes whose probabilities are denoted by $p_1, p_2, \ldots, p_k$. If n identical independent multinomial trials are performed and $X_1, X_2, \ldots, X_k$ record the numbers of times the different outcomes occur, then $X_1, X_2, \ldots, X_k$ is called a multinomial random variable (or vector).

MULTINOMIAL RANDOM VARIABLE

Probability function: $\displaystyle \binom{n}{x_1 \; x_2 \; \cdots \; x_k} p_1^{x_1} p_2^{x_2} \cdots p_k^{x_k}$

Range: All k-tuples, $(x_1, x_2, \ldots, x_k)$, such that $x_i = 0, 1, 2, \ldots, n$, where $i = 1, 2, \ldots, k$ and $\sum_{i=1}^{k} x_i = n$

Parameters: $n = 1, 2, \ldots, \; p_1, p_2, \ldots, p_k; \; \sum_i p_i = 1$

Comments: Any subset is multinomial, as are conditional laws.

The multi-hypergeometric probability law describes the results of a random sample of size n selected without replacement from a population of m items, which contains r_i items of type i, where $i = 1, 2, \ldots, k$.

MULTI-HYPERGEOMETRIC RANDOM VARIABLE

Probability function: $\dbinom{r_1}{x_1}\dbinom{r_2}{x_2}\cdots\dbinom{r_k}{x_k}\Big/\dbinom{m}{n}$

Range: All k-tuples, $(x_1, x_2, \ldots, x_k)$, such that $x_i = 0, 1, 2, \ldots, n$, where $i = 1, 2, \ldots, k$, and $\sum_{i=1}^{k} x_i = n$

Parameters: $n = 1, 2, \ldots$; positive integers $r_1, r_2, \ldots, r_k$, such that $\sum_i r_i = m$

Comments: Any subset is multi-hypergeometric, as are conditional laws.

Conditional probability laws are defined by ratios of joint probability laws to marginal probability laws.

CONDITIONAL PROBABILITY FUNCTION

$$p_{X_1|X_2}(x_1 \mid x_2) = \frac{p_{X_1,X_2}(x_1, x_2)}{p_{X_2}(x_2)}, \qquad \text{for } x_1, x_2 \in R_{X_1,X_2}$$

Implication: $p_{X_1,X_2}(x_1, x_2) = p_{X_2}(x_2)p_{X_1|X_2}(x_1|x_2)$

CONDITIONAL *pdf*

$$f_{X_1|X_2}(x_1 \mid x_2) = \frac{f_{X_1,X_2}(x_1, x_2)}{f_{X_2}(x_2)}, \qquad \text{for } x_1, x_2 \in R_{X_1,X_2}$$

Implication: $f_{X_1,X_2}(x_1, x_2) = f_{X_2}(x_2)f_{X_1|X_2}(x_1|x_2)$

The bivariate normal probability law has five parameters: the means and variances for the two random variables and their correlation coefficient ρ, where $-1 < \rho < 1$. The joint *pdf* is maximized at (μ_1, μ_2), the two means, and decreases symmetrically in both variables. The constant contours are ellipses. The marginal *pdf*s are normal with the same mean and variance as in the joint *pdf*. If $\rho = 0$, the two random variables are independent. The conditional probability laws are again normal. The conditional distribution for X_2 given $X_1 = x_1$ has mean $\mu_2 + \rho\frac{\sigma_1}{\sigma_2}(x - \mu_1)$ and variance $\sigma_2^2(1 - \rho^2)$.

BIVARIATE NORMAL *pdf*

$$f_{X_1,X_2}(x_1,x_2) = \frac{1}{2\pi\sigma_1\sigma_2\sqrt{1-\rho^2}} e^{-q(x_1,x_2)/2},$$

$$q(x_1,x_2) = \frac{1}{1-\rho^2}\left[\frac{(x_1-\mu_1)^2}{\sigma_1^2} - 2\rho\frac{x_1-\mu_1}{\sigma_1}\frac{x_2-\mu_2}{\sigma_2} + \frac{(x_2-\mu_2)^2}{\sigma_2^2}\right].$$

EXERCISES 5.4

1. A large mail-order retailer sells the same blouse in four different colors: $\mathcal{R}, \mathcal{B}, \mathcal{G},$ and $\mathcal{Y}$. From previous records, you know that the proportions of orders received for blouses of these colors are .2, .4, .3, .1, respectively. If orders for 100 of these blouses are received in a given week, what would you use as a model for $X_\mathcal{R}, X_\mathcal{B}, X_\mathcal{G}, X_\mathcal{Y}$, the numbers ordered for the different colors?

2. The joint *pdf* for X, Y is $f_{X,Y}(x,y) = 12xy(1-x)$, for $0 < x < 1, 0 < y < 1$.
   a. Find the marginal *pdf*s for X and Y.
   b. Are X and Y independent?

3. A toy store sells the same doll wearing three different colored dresses. On a given day they have 150 of these dolls in stock, with equal numbers of dolls wearing the three different colored dresses. Suppose that 30 of these dolls are sold on this day, and that the colors selected by the purchasers are randomly determined. What is the probability law for the numbers of dolls sold wearing the three different colored dresses? What is the probability that all 30 dolls sold wear the same color dress? What is the probability that the three colors each were chosen 10 times?

4. The conditional *pdf* for X, given $Y = y$, is $f_{X|Y}(x|y) = 1/y^2$ for $0 < x < y^2$, while the marginal *pdf* for Y is $f_Y(y) = 4y^3$ for $0 < y < 1$. What is the marginal *pdf* for X?

5. Model the maximum air temperature to be observed on two successive days at a weather station as a pair of random variables X_1, X_2, which are bivariate normal with parameters $\mu_1 = \mu_2 = 75, \sigma_1 = \sigma_2 = 8$, and $\rho = .9$.
   a. If the observed maximum temperature on the first day is 73, what is the probability law for X_2, the maximum temperature on the following day?
   b. Evaluate the probability the observed maximum temperature on the second day exceeds 73 given that $X_1 = 73$.

CHAPTER 6

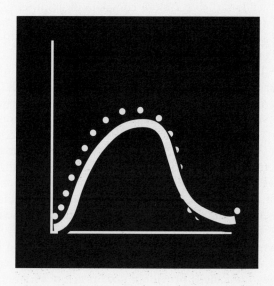

EXPECTATION, MOMENTS

We have discussed a number of different probability distributions for single random variables; the observed values for a single random variable X can be pictured as points on a line, a one-dimensional space. Thus a single random variable X can be called one-dimensional. The concept of expected value (or expectation) of a random variable, or of a function of a random variable, is useful in describing many aspects of the distribution of the random variable (for example, measures of the middle and the spread of the distribution).

A pair of random variables X, Y is called two-dimensional since their observed values occur together, and can be pictured as points in a two-dimensional space (similarly $X_1, X_2 \ldots, X_n$ is called n-dimensional). The concept of expectation again proves indispensable in describing aspects of the probability distribution for two- and higher-dimensional random variables. As is discussed in Section 6.1, expectation can again be used to measure the middle of the distribution, as well as its spread. The means of the individual random variables describe the center of gravity of the distribution; the individual variances, though, do not totally describe the variability of the distribution. A new quantity, called the covariance of two random variables, is needed to give a more complete description of the variability of the joint distribution of X, Y. This quantity is also the basis for the correlation of two random variables, a simple measure of how the random variables "move together" over repeated observed values.

The mean (a number describing the center of the distribution) and the variance (a number describing the spread of the distribution) of a random variable X are determined by $E[X]$ and $E[X^2]$, the averages of the first two powers of X. As discussed in Section 6.2, the moments of a random variable are given by $E[X^k]$, for $k = 1, 2, 3 \ldots$, the averages of positive integer powers of X. This sequence of moments can, in sufficiently regular cases, uniquely determine the distribution for X; in such cases, then, they provide an alternative way of defining the distribution for X. The moment generating function for X provides a succinct way of describing the moments of X and can prove useful in finding the distribution for certain functions of X.

The conditional distribution for Y, given $X = x$, can be different from the marginal distribution for Y. Expectation can be used with such conditional distributions, as discussed in Section 6.3; such conditional expectations (as they are called) provide a powerful tool in modeling and describing jointly distributed random variables.

6.1 | Expectation

Expected values of functions of single, or one-dimensional, random variables are useful for many purposes. This concept of expectation proves equally useful for higher-dimensional random variables as well and is defined in the same way. Recall that if X is discrete and has probability function $p_X(x)$, then the expected value for a function $w(X)$ is

$$E[w(X)] = \sum_{x \in R_X} w(x) p_X(x),$$

while if X is continuous with *pdf* $f_X(x)$, the expected value for $w(X)$ is

$$E[w(X)] = \int_{-\infty}^{\infty} w(x) f_X(x) \, dx.$$

In either of these two situations the possible observed values for the function $w(X)$ are "averaged" with respect to the probability law.

As with our discussion of joint probability laws, we shall consider the concept of expectation explicitly for two-dimensional random variables; the extension to three or more random variables involves more dimensions, but generally no fundamentally different concepts. Thus let $w(x_1, x_2)$ be a real-valued function of two variables, whose domain includes the range for the random variables X_1, X_2. Then the expected value for $w(X_1, X_2)$ is again defined to be the average of $w(x_1, x_2)$ with respect to the probability law, as given in the following definition.

DEFINITION 6.1

If $w(x_1, x_2)$ is a real-valued function of two variables, whose domain includes R_{X_1, X_2}, then the *expected value* of $w(X_1, X_2)$ is

$$
E[w(X_1, X_2)] = \begin{cases} \displaystyle\sum_{(x_1, x_2) \in R_{X_1, X_2}} w(x_1, x_2) p_{X_1, X_2}(x_1, x_2), & \text{if } X_1, X_2 \text{ is discrete,} \\[4mm] \displaystyle\iint\limits_{(x_1, x_2) \in R_{X_1, X_2}} w(x_1, x_2) f_{X_1, X_2}(x_1, x_2)\, dx_1\, dx_2, & \text{if } X_1, X_2 \text{ is continuous,} \end{cases}
$$

as long as the (multiple) sum or integral is absolutely convergent.

In the continuous case the range of integration for both variables can just as well be taken as $(-\infty, \infty)$ since $f_{X_1, X_2}(x_1, x_2) = 0$ for $(x_1, x_2) \notin R_{X_1, X_2}$ (as in the one-dimensional case). The following example illustrates the use of this definition.

Example 6.1

Assume that a new-car dealer sells automobiles made by two different major manufacturers (called 1 and 2); also assume that the (admittedly oversimplified) values in Table 6.1 give the joint probability function for X_1, X_2, where X_1 is the number of automobiles from manufacturer 1 sold on any given day and X_2 is the number sold from manufacturer 2. Further assume that the profit made by the dealership is \$1000 for each car sold of type 1 and is \$800 for each car sold of type 2. The profit per day from selling these two types of automobiles then is $w(X_1, X_2) = 1000X_1 + 800X_2$, a simple linear function of X_1 and X_2.

The expected profit per day is the sum of a total of 12 terms (since there are 12 pairs in R_{X_1, X_2}):

Table 6.1

$p_{X_1, X_2}(x_1, x_2)$, for Example 6.1

		$x_1 =$				
		0	**1**	**2**	**3**	**Total**
	0	.20	.15	.10	.05	.50
$x_2 =$	**1**	.15	.10	.05	.03	.33
	2	.07	.05	.03	.02	.17
Total		.42	.30	.18	.10	1.00

$$E[w(X_1, X_2)] = \sum_{x_1, x_2 \in R_{X_1, X_2}} (1000x_1 + 800x_2) p_{X_1, X_2}(x_1, x_2)$$

$$= (0 + 0)(.20) + (1000 + 0)(.15) + \cdots + (2000 + 1600)(.03)$$
$$+ (3000 + 1600)(.02)$$
$$= 1496.$$

As a continuous example, suppose you buy your favorite breakfast cereal in a box labeled "Net Contents 12 oz." The total weight of this package is the sum of the weight of the cereal plus the weight of the packaging used, both of which may vary slightly from package to package. Let Y_1, Y_2 represent the true weights of the contents and the packaging, respectively, for one of these boxes selected at random from a supermarket shelf. We shall assume that, for any given box, the amount of cereal used and the weight of the packaging employed are independent. (For example, an overweight amount of cereal should be just as likely with a box having thick cardboard as with one having thin cardboard.) Thus we need only specify the marginal probability laws for Y_1, Y_2, and the joint *pdf* is given by the product of the marginal *pdf*s. We shall take the marginal *pdf* for Y_1 to be $f_{Y_1}(y_1) = 50(y_1 - 11.9)$, for $11.9 < y_1 < 12.1$, a triangular *pdf*. As the *pdf* for Y_2 we shall use $f_{Y_2}(y_2) = (1000/9)(y_2 - 2)^2$, for $1.7 < y_2 < 2$; this *pdf* is a section of a parabola. Since Y_1 and Y_2 are independent the joint *pdf* for these two random variables is the product of the two marginal *pdf*s:

$$f_{Y_1, Y_2}(y_1, y_2) = \frac{5 \times 10^4}{9}(y_1 - 11.9)(y_2 - 2)^2,$$

for $11.9 < y_1 < 12.1$ and $1.7 < y_2 < 2$. Then $w(Y_1, Y_2) = Y_1 + Y_2$ gives the total weight of the package selected, and its expected value is

$$E[Y_1 + Y_2] = \int_{y_1=11.9}^{12.1} \int_{y_2=1.7}^{2} (y_1 + y_2) \frac{5 \times 10^4}{9}(y_1 - 11.9)(y_2 - 2)^2 \, dy_2 \, dy_1$$

$$= \int_{y_1=11.9}^{12.1} y_1 50(y_1 - 11.9) \, dy_1 + \int_{y_2=1.7}^{2} y_2 \frac{10^3}{9}(y_2 - 2)^2 \, dy_2$$

$$= 1.775 + 12.033 = 13.808 \text{ ounces.}$$

Integration or summation of two- or higher-dimensional functions is required in evaluating the expected values of functions of several random variables; however, as in the case of one-dimensional random variables, the expected value of a constant is itself (since the constant factors through the summation or integration and integrating the *pdf* or summing the probability function over the full range must give 1). In addition, the expected value of a sum of terms must be equal to the sum of the expectations (and again constants factor through expected values). To see that this must be the case, let a_1, a_2 be constants, let X_1, X_2 be jointly distributed random variables, and let $w_1(X_1, X_2), w_2(X_1, X_2)$ be two arbitrary functions of X_1, X_2 (whose expectations exist). Define $w(X_1, X_2) = a_1 w_1(X_1, X_2) + a_2 w_2(X_1, X_2)$. Then, if X_1, X_2

are continuous random variables with joint *pdf* $f_{X_1,X_2}(x_1,x_2)$, the expected value of $w(X_1,X_2)$ is

$$E[w(X_1,X_2)] = \int_{-\infty}^{\infty}\int_{-\infty}^{\infty}(a_1w_1(x_1,x_2) + a_2w_2(x_1,x_2))f_{X_1,X_2}(x_1,x_2)\,dx_2\,dx_1$$

$$= a_1\int_{-\infty}^{\infty}\int_{-\infty}^{\infty}w_1(x_1,x_2)f_{X_1,X_2}(x_1,x_2)\,dx_2\,dx_1$$

$$+ a_2\int_{-\infty}^{\infty}\int_{-\infty}^{\infty}w_2(x_1,x_2)f_{X_1,X_2}(x_1,x_2)\,dx_2\,dx_1$$

$$= a_1\,E[w_1(X_1,X_2)] + a_2\,E[w_2(X_1,X_2)].$$

Exactly the same reasoning is appropriate if X_1, X_2 are discrete, with summation replacing integration. Thus, in either the discrete or continuous case we can write

$$E[10 + 5X_1^2 + 3X_2^2 - 7X_1X_2] = 10 + 5E[X_1^2] + 3E[X_2^2] - 7E[X_1X_2],$$

for example. The next two examples illustrate some of this reasoning.

Example 6.2

Suppose a projectile is fired at a target in the two-dimensional (x_1,x_2)-plane; the target is located at $(0,0)$, the origin of the coordinate system. We assume the impact point of the projectile is (X_1,X_2), where X_1 is the horizontal coordinate of the impact point and X_2 is the vertical coordinate of the impact point. If the aiming mechanism is unbiased we would then assume that $E[X_1] = E[X_2] = 0$ (i.e., the probability laws for X_1 and X_2 are centered at 0, in the sense that the means are 0). The square of the distance from the impact point to the target is then $R^2 = w(X_1,X_2) = X_1^2 + X_2^2$, and the expected value for the square of this radial distance is simply $E[R^2] = E[X_1^2 + X_2^2] = E[X_1^2] + E[X_2^2] = \sigma_1^2 + \sigma_2^2$, the sum of the two variances of the random variables (granted their means are both 0).

Example 6.3

Recall that a binomial random variable X with parameters n and p gives the number of successes observed in n independent Bernoulli trials, each with parameter p. It was pointed out earlier that we could also define n Bernoulli random variables ($Y_i = 1$, if a success occurs on trial i and $Y_i = 0$, if not) and then the value for the binomial random variable X is in fact given by the *sum* of these n Bernoulli random variables; that is, $X = \sum_{i=1}^{n} Y_i$. Thus, by using the fact that the expected value of a sum of terms must equal the sum of their individual expectations, we get

$$E[X] = E\left[\sum_{i=1}^{n} Y_i\right] = \sum_{i=1}^{n} E[Y_i] = np,$$

since $E[Y_i] = p$ for $i = 1, 2, \ldots, n$. This is of course the same value derived earlier for the mean of X; in some senses this procedure for deriving the result is simpler than using the original definition. In the same way, this approach gives an easy way of deriving the mean for the negative binomial probability law (as a sum of geometrics) and the mean of the Erlang (as a sum of exponentials).

As another example of the additivity of expectations, suppose we have a standard, well-shuffled 52-card deck of cards and turn them face up one at a time. What is the expected number of cards turned over before the first ace occurs? The answer to this query can be easily gotten by considering the position of each of the 48 non-ace cards in the deck relative to the positions of the four aces. More specifically, let the 48 non-aces be numbered from 1 to 48, and consider the positions of card 1 and the four aces; let $Y_1 = 1$ if card 1 *precedes all* four aces, and let $Y_1 = 0$ otherwise. Then Y_1 is a Bernoulli random variable with parameter $p = \frac{1}{5}$, since the probability that card 1 will be the first of five cards is $\frac{1}{5}$. Similarly, let $Y_2 = 1$ if card 2 *precedes all* four aces and let $Y_2 = 0$ otherwise. We continue in this way to define a Bernoulli variable for each non-ace card, giving 48 random variables $Y_1, Y_2, \ldots, Y_{48}$. The expected value for each Y_i is then simply $p = \frac{1}{5}$, the probability that card i will be first in a group of five cards (itself plus the aces). The total number of non-aces to precede the first ace in the well-shuffled deck is simply $\sum_{i=1}^{48} Y_i$, the total number of 1's, and the expected number of cards before the first ace is $E[\sum_{i=1}^{48} Y_i] = \sum_{i=1}^{48} E[Y_i] = 48(\frac{1}{5}) = 9.6$.

Linear functions of random variables are used a great deal in many applied problems (as in Examples 6.1 and 6.3). A linear function of two or more random variables can be thought of as a new random variable (and is sometimes called a *linear combination* of the variables involved). It is straightforward to find the mean and variance of this new random variable, in terms of the means and variances of the underlying random variables. Suppose X_1, X_2 are jointly distributed random variables with means μ_1, μ_2, variances σ_1^2, σ_2^2. Let a_0, a_1, a_2 be constants and define a new random variable $Y = a_0 + a_1 X_1 + a_2 X_2$. Since the value for Y is determined by the values for X_1, X_2, we can find the mean and variance for Y from knowledge of the probability law for X_1, X_2. The mean for Y is

$$\mu_Y = E[Y] = E[a_0 + a_1 X_1 + a_2 X_2] = a_0 + a_1 \mu_1 + a_2 \mu_2,$$

which is the same linear function of the means of X_1, X_2; that is, the observed value for y is $a_0 + a_1 y_1 + a_2 y_2 = g(y_1, y_2)$ and the mean for Y is $a_0 + a_1 \mu_1 + a_2 \mu_2 = g(\mu_1, \mu_2)$. The variance of Y, however, depends on more than just the individual variances of X_1 and X_2; it is given by

$$
\begin{aligned}
\sigma_Y^2 &= E[(Y - \mu_Y)^2] = E[(a_0 + a_1 X_1 + a_2 X_2 - a_0 - a_1 \mu_1 - a_2 \mu_2)^2] \\
&= E[a_1^2 (X_1 - \mu_1)^2 + a_2^2 (X_2 - \mu_2)^2 + 2a_1 a_2 (X_1 - \mu_1)(X_2 - \mu_2)] \\
&= a_1^2 E[(X_1 - \mu_1)^2] + a_2^2 E[(X_2 - \mu_2)^2] + 2a_1 a_2 E[(X_1 - \mu_1)(X_2 - \mu_2)] \\
&= a_1^2 \sigma_1^2 + a_2^2 \sigma_2^2 + 2a_1 a_2 \sigma_{12}.
\end{aligned}
\tag{1}
$$

The quantity σ_{12} is called the covariance of X_1, X_2, also denoted by

$Cov[X_1, X_2]$. This quantity occurs in considering the variances of linear functions of random variables and is formally defined as follows.

DEFINITION 6.2

If X_1, X_2 are jointly distributed random variables, their *covariance* is defined by

$$\text{Cov}[X_1, X_2] = \sigma_{12} = \text{E}[(X_1 - \text{E}[X_1])(X_2 - \text{E}[X_2])].\tag{2}$$

The covariance of the two random variables gets its name from its "variance-like" structure, with both X_1 and X_2 involved. Indeed, note that if X_2 is replaced by X_1, the covariance of X_1, X_1 simply reduces to the variance of X_1. The covariance of two random variables affects the variance of a linear combination of the two; it also describes a certain aspect of the joint behavior of the two random variables.

The covariance of two random variables provides a measure of linear association between the two. The value for the covariance is not easily comprehended from a plot of the (two-dimensional) probability law (nor are the individual variances easily comprehended from the plot of a one-dimensional law). The way in which the covariance measures the strength of the linear association of the two random variables is fairly transparent, in one sense. From Eq. (2) we have

$$\begin{aligned}
\text{Cov}[X_1, X_2] &= \text{E}[X_1 X_2 - X_1\,\text{E}[X_2] - \text{E}[X_1]X_2 + \text{E}[X_1]\,\text{E}[X_2]] \\
&= \text{E}[X_1 X_2] - \text{E}[X_1]\,\text{E}[X_2] - \text{E}[X_1]\,\text{E}[X_2] + \text{E}[X_1]\,\text{E}[X_2] \\
&= \text{E}[X_1 X_2] - \text{E}[X_1]\,\text{E}[X_2].
\end{aligned}\tag{3}$$

Thus, the covariance of two random variables is simply the difference between the average of their product and the product of their averages. The covariance can be negative or positive, and indeed in many important cases its value is zero. Equation (3) is an effective computational formula for the covariance of X_1 and X_2.

Example 6.4

Table 6.2 reproduces the assumed joint probability function for the numbers of cars of two different makes that a dealer sells daily, discussed in Example 6.1.

Table 6.2

$p_{X_1,X_2}(x_1, x_2)$, for Example 6.4

		$x_1 =$				
		0	**1**	**2**	**3**	**Total**
	0	.20	.15	.10	.05	.50
$x_2 =$	**1**	.15	.10	.05	.03	.33
	2	.07	.05	.03	.02	.17
Total		.42	.30	.18	.10	1.00

Using the two marginal probability laws we have

$$E[X_1] = 0(.42) + 1(.30) + 2(.18) + 3(.10) = .96,$$
$$E[X_2] = 0(.50) + 1(.33) + 2(.17) = .67.$$

Since the product $x_1 x_2 = 0$ if *either* $x_1 = 0$ or $x_2 = 0$, the expected value for $X_1 X_2$ is the sum of six terms:

$$E[X_1 X_2] = 1(1)(.10) + 1(2)(.05) + 1(3)(.03) + 2(1)(.05)$$
$$+ 2(2)(.03) + 2(3)(.02) = .63,$$

so the covariance of X_1, X_2 is $.63 - (.96)(.67) = -.0132$. The average of the products of these two variables is smaller than the product of their averages. Using the two marginal probability laws, it is also easy to see that the variances of X_1, X_2 are $\sigma_1^2 = .9984$ and $\sigma_2^2 = .5611$. Since X_1, X_2 give the numbers of cars sold per day by this dealer, the total number of cars he will sell per day is then $Y = X_1 + X_2$, their sum; the expected number of cars to be sold in a given day is then

$$\mu_Y = \mu_{X_1} + \mu_{X_2} = .96 + .67 = 1.63,$$

the sum of the two means. The variance of the number of cars sold in a given day is

$$\sigma_Y^2 = \sigma_1^2 + \sigma_2^2 + 2\,\text{Cov}[X_1, X_2] = .9984 + .5611 + 2(-.0132)$$
$$= 1.5331,$$

which is *smaller* than the sum of the two individual variances because of the negative covariance.

In Example 6.1 we also discussed a model for the total weight $Y_1 + Y_2$ of a cereal package, where Y_1 represents the weight of the cereal, Y_2 represents the weight of the packaging used, and the two variables were assumed independent. Because of this assumed independence,

$$E[Y_1 Y_2] = \int_{-\infty}^{\infty} \int_{-\infty}^{\infty} y_1 y_2 f_{Y_1}(y_1) f_{Y_2}(y_2)\, dy_1\, dy_2$$
$$= \int_{-\infty}^{\infty} y_1 f_{Y_1}(y_1)\, dy_1 \int_{-\infty}^{\infty} y_2 f_{Y_2}(y_2)\, dy_2$$
$$= E[Y_1]\, E[Y_2],$$

from which it follows that the covariance of these two random variables is necessarily 0. The variance of $Y_1 + Y_2$ then is simply the sum of their two individual variances, or $.010244 + .003375 = .0136$.

These results about the mean and variance of a linear function generalize very easily to linear functions of n random variables, where n can be any positive integer. Assume $X_1, X_2, \ldots, X_n$ are jointly distributed random variables with means $\mu_1, \mu_2, \ldots, \mu_n$ and variances $\sigma_1^2, \sigma_2^2, \ldots, \sigma_n^2$, respectively. Let $a_1, a_2, \ldots, a_n$ be constants, and define the linear function $Y =$

$a_1X_1 + \cdots + a_nX_n$. Immediately the mean for Y is $\mu_Y = a_1\mu_1 + \cdots + a_n\mu_n$, and its variance is

$$
\begin{aligned}
\sigma_Y^2 &= \mathrm{E}[(Y - \mu_Y)^2] \\
&= \mathrm{E}\left[\left(\sum_{i=1}^{n} a_i(X_i - \mu_i)\right)^2\right] \\
&= \sum_{i=1}^{n}\sum_{j=1}^{n} \mathrm{E}[a_i a_j(X_i - \mu_i)(X_j - \mu_j)] \\
&= \sum_{i=1}^{n} a_i^2 \,\mathrm{Var}[X_i] + 2\sum_{i<j}\sum a_i a_j \,\mathrm{Cov}[X_i, X_j],
\end{aligned}
$$

since every possible cross-product occurs two times. Indeed, if $b_1, b_2, \ldots, b_n$ are also constants and we have a second linear function of $X_1, X_2, \ldots, X_n$ defined by $Z = b_1 X_1 + \cdots + b_n X_n$, then it is easy to see that the covariance between Y and Z is

$$
\begin{aligned}
\mathrm{Cov}[Y, Z] &= \mathrm{E}[(Y - \mu_Y)(Z - \mu_Z)] \\
&= \mathrm{E}\left[\left(\sum_{i=1}^{n} a_i(X_i - \mu_i)\right)\left(\sum_{j=1}^{n} b_j(X_j - \mu_j)\right)\right] \\
&= \sum_{i=1}^{n}\sum_{j=1}^{n} a_i b_j \,\mathrm{E}[(X_i - \mu_i)(X_j - \mu_j)] \\
&= \sum_{i=1}^{n} a_i b_i \,\mathrm{Var}[X_i] + \sum_{\substack{i=1 \\ i \neq j}}^{n}\sum_{j=1}^{n} a_i b_j \,\mathrm{Cov}[X_i, X_j],
\end{aligned}
$$

since $\mathrm{Cov}[X_i, X_i] = \mathrm{Var}[X_i]$. These results are summarized in the following theorem.

THEOREM 6.1 Let $a_1, a_2, \ldots, a_n$ and $b_1, b_2, \ldots, b_n$ be constants and let $X_1, X_2, \ldots, X_n$ be jointly distributed random variables. If we define two new random variables by $Y = \sum_{i=1}^{n} a_i X_i$ and $Z = \sum_{i=1}^{n} b_i X_i$, then

$$
\mathrm{E}[Y] = \sum_{i=1}^{n} a_i \,\mathrm{E}[X_i],
$$

$$
\mathrm{Var}[Y] = \sum_{i=1}^{n} a_i^2 \,\mathrm{Var}[X_i] + 2\sum_{\substack{i=1 \\ i<j}}^{n}\sum_{j=1}^{n} a_i a_j \,\mathrm{Cov}[X_i, X_j],
$$

$$
\mathrm{Cov}[Y, Z] = \sum_{i=1}^{n} a_i b_i \,\mathrm{Var}[X_i] + \sum_{\substack{i=1 \\ i \neq j}}^{n}\sum_{j=1}^{n} a_i b_j \,\mathrm{Cov}[X_i, X_j].
$$

Example 6.5

To illustrate Theorem 6.1, let X_1, X_2 represent jointly distributed random variables with means $\mu_1 = 5, \mu_2 = 10$, variances $\sigma_1^2 = 4, \sigma_2^2 = 2$, and $\text{Cov}[X_1, X_2] = \sigma_{12} = -1$. Then the random variable $Y = 3X_1 + X_2$ has mean $\mu_Y = 3(5) + 1(10) = 25$ and variance $\sigma_Y^2 = 3^2(4) + 1^2(2) + 2(3)(1)(-1) = 32$. If $Z = -2X_1 + 7X_2$ is a second linear function of X_1, X_2, its mean is $\mu_Z = -2(5) + 7(10) = 60$, while its variance is $\sigma_Z^2 = (-2)^2(4) + 7^2(2) + 2(-2)(7)(-1) = 142$; the covariance of Y and Z is $\text{Cov}[Y, Z] = (3)(-2)(4) + (1)(7)(2) + ((3)(7) + (-2)(1))(-1) = -29$.

In some types of problems, it is of interest to choose the multipliers in a linear function of random variables to accomplish certain goals. Let X_1, X_2 have the means, variances, and covariance just specified. Can we find a linear function $Y = a_1 X_1 + a_2 X_2$ whose mean is 0? Since for any constants a_1, a_2 we have $\mu_Y = 5a_1 + 10a_2$, this query amounts to finding a_1, a_2 so that $5a_1 + 10a_2 = 0$; that is, any choice with $a_1 = -2a_2$ will give a random variable Y with mean 0, so $Y = -2a_2 X_1 + a_2 X_2$ has mean 0. (Or in general, $Y = a_1 X_1 + a_2 X_2$ will have mean 0 as long as $a_1 = -a_2 \mu_2 / \mu_1$, where μ_1, μ_2 are the means for X_1, X_2.) The variance of Y is then $\sigma_Y^2 = (-2a_2)^2(4) + a_2^2(2) + 2(-2a_2)(a_2)(-1) = 22a_2^2$. If we further wanted Y to have variance 1 (in addition to $\mu_Y = 0$), we can accomplish this by selecting $a_2 = 1/\sqrt{22}$ and $a_1 = -2a_2 = -2/\sqrt{22}$.

Theorem 6.1 states the general result that holds for linear functions of the elements of the same random variables. If the covariances between the individual random variables in $X_1, X_2, \ldots, X_n$ are all 0 (i.e., if $\text{Cov}[X_i, X_j] = 0$ for all $i \neq j$), then the results for the variances and covariance of two linear functions are much simpler, given as the following corollary.

COROLLARY 6.1 Let $X_1, X_2, \ldots, X_n$ be jointly distributed, such that the covariance between any two of them is 0, and let $a_1, a_2, \ldots, a_n$ and $b_1, b_2, \ldots, b_n$ be constants. Then the random variables $Y = \sum_{i=1}^n a_i X_i$ and $Z = \sum_{i=1}^n b_i X_i$ have variances given by

$$\text{Var}[Y] = \sum_{i=1}^n a_i^2 \, \text{Var}[X_i], \qquad \text{Var}[Z] = \sum_{i=1}^n b_i^2 \, \text{Var}[X_i],$$

and covariance given by

$$\text{Cov}[Y, Z] = \sum_{i=1}^n a_i b_i \, \text{Var}[X_i].$$

The "co-variability" of two random variables is measured by their covariance; the magnitude of this number is affected by the scales of measurement employed for the two random variables (which is also true of the mean and the variance of a random variable). For example, suppose T_1, T_2 are jointly distributed with $\text{Cov}[T_1, T_2] = 1$. If we define two new random variables, $S_1 = 2T_1, S_2 = 3T_2$, say, then $\text{E}[S_1] = 2\,\text{E}[T_1], \text{E}[S_2] = 3\,\text{E}[T_2]$, and

$E[S_1 S_2] = 6\,E[T_1 T_2]$ giving $\text{Cov}[S_1, S_2] = 6\,\text{Cov}[T_1, T_2]$. Multiplying random variables by constants alters (by expanding, shrinking, or even reversing) the corresponding scales. This may result in different values for the covariance depending on the constants used.

The correlation between two random variables is designed to address this problem. This coefficient measures the degree of linear association between the two variables, and it remains unchanged by such linear transformations. It thus allows comparisons of degrees of linear association between pairs of random variables on a dimensionless scale. Formally the correlation coefficient is defined as follows.

DEFINITION 6.3

The *correlation* between X_1 and X_2 is defined by

$$\rho = \frac{\text{Cov}[X_1, X_2]}{\sqrt{\text{Var}[X_1]\,\text{Var}[X_2]}} = \frac{\sigma_{12}}{\sigma_1 \sigma_2},$$

where σ_1, σ_2 are the standard deviations of X_1, X_2, and ρ is called the correlation coefficient.

Example 6.6

Suppose $X_1, X_2, \ldots, X_n$ are uncorrelated random variables, which means the correlation is 0 for each possible pair, and each variable has mean μ and variance σ^2. The *average* of these n random variables then is commonly denoted by $\bar{X}$, that is,

$$\bar{X} = \frac{1}{n} \sum_{i=1}^{n} X_i,$$

a linear function of $X_1, X_2, \ldots, X_n$ with each multiplier equal to $a_i = 1/n$, the reciprocal of the number of random variables involved. The mean of $\bar{X}$ is then immediately

$$\sum_{i=1}^{n} \frac{1}{n}\mu = \mu,$$

the same as the common mean of the X_i's and the variance of $\bar{X}$ is

$$\sum_{i=1}^{n} \left(\frac{1}{n}\right)^2 \sigma^2 = \frac{\sigma^2}{n},$$

smaller than the common variance of the X_i's by a factor of $1/n$. If the common variance for the X_i values is 5, say, and $\bar{X}$ is the average of $n = 10$ such uncorrelated random variables, then the variance for $\bar{X}$ is $\frac{5}{10} = .5$. This means that $\bar{X}$ varies less about μ than do the original X_i values, a fact that has important consequences in many statistical problems.

The sign of the correlation coefficient ρ is always identical to the sign of the covariance of the two variables: If $\mathrm{Cov}[X_1, X_2] < 0$, then X_1 and X_2 are *negatively correlated*, while if $\mathrm{Cov}[X_1, X_2] > 0$, then X_1 and X_2 are *positively correlated*. The correlation coefficient is 0, and the variables are said to be *uncorrelated*, whenever their covariance is 0.

Example 6.7

Recall that if we begin observing events in a Poisson process at time $t = 0$ and let X_1, X_2 represent the times of the first two events to occur, their joint *pdf* is $f_{X_1,X_2}(x_1, x_2) = \lambda^2 e^{-\lambda x_2}$ for $0 < x_1 < x_2$, where λ is the parameter of the process. We have seen that X_1 is exponential with parameter λ while X_2 is Erlang with parameters $r = 2, \lambda$ (marginal probability laws). It follows then that their means and variances are $\mu_1 = 1/\lambda, \mu_2 = 2/\lambda, \sigma_1^2 = 1/\lambda^2, \sigma_2^2 = 2/\lambda^2$. To evaluate the covariance between these two random variables we require

$$\mathrm{E}[X_1 X_2] = \int_0^\infty \int_{x_1}^\infty x_1 x_2 \lambda^2 e^{-\lambda x_2} \, dx_2 \, dx_1 = \frac{3}{\lambda^2}. \tag{4}$$

From this we immediately have $\mathrm{Cov}[X_1, X_2] = 1/\lambda^2$ and thus the correlation between X_1 and X_2 is

$$\rho = \frac{\mathrm{Cov}[X_1, X_2]}{\sigma_1 \sigma_2} = \frac{1}{\sqrt{2}} = .707.$$

This correlation coefficient is dimensionless and indeed has the same value regardless of the value for λ, the parameter of the process. These two random variables are positively correlated.

Since the covariance of two variables involves their product, it is measured in units given by the product of the units of X_1 and X_2, as is the product of their standard deviations; their correlation coefficient is then dimensionless and in fact is a *bounded* measure. As will be shown, the correlation between any two random variables must lie between -1 and 1, inclusive, with the two endpoints indicating that extremely special circumstances hold for the joint probability law of the two variables. The reasoning we shall employ is based on a special case of the *Cauchy–Schwarz inequality*.

Let X, Y be jointly distributed random variables with means equal to 0 and variances equal to 1 (so $\mathrm{E}[X^2] = \mathrm{E}[Y^2] = 1$). Since $(X - Y)^2$ is a non-negative random variable (it never equals a negative value) its expected value must also be nonnegative (as is also true of $(X + Y)^2$). That is,

$$\mathrm{E}[(X - Y)^2] = \mathrm{E}[X^2] - 2\,\mathrm{E}[XY] + \mathrm{E}[Y^2] = 2(1 - \mathrm{E}[XY]) \geq 0,$$

which then says that $\mathrm{E}[XY] \leq 1$. In the same way, $\mathrm{E}[(X + Y)^2] = 2(1 + \mathrm{E}[XY]) \geq 0$ implies that $\mathrm{E}[XY] \geq -1$; thus in any case we have $|\mathrm{E}[XY]| \leq 1$, for any two random variables whose means equal 0 and whose variances equal 1.

Now suppose V_1, V_2 are jointly distributed random variables with means μ_1, μ_2 and variances σ_1^2, σ_2^2, respectively, and let X, Y be their standard forms, that is,

$$X = \frac{V_1 - \mu_1}{\sigma_1}, \qquad Y = \frac{V_2 - \mu_2}{\sigma_2}.$$

Then we certainly have $E[X] = E[Y] = 0$ and $E[X^2] = E[Y^2] = 1$ so

$$E[XY] = \frac{E[(V_1 - \mu_1)(V_2 - \mu_2)]}{\sigma_1 \sigma_2} = \rho,$$

from which it follows that $|\rho| \leq 1$ for *any* two random variables V_1, V_2. This establishes the following theorem.

THEOREM 6.2 If V_1, V_2 are jointly distributed random variables, then, $|\rho| \leq 1$, where ρ is the correlation between V_1 and V_2 (so long as the expectations exist).

The two extreme values for ρ, -1 and 1, only occur for degenerate distributions. By examining the reasoning employed above, you can see that $E[XY] = 1$ only if $E[(X - Y)^2] = 0$. As discussed earlier, $E[(X - Y)^2] = 0$ implies that $P(X - Y = 0) = 1$, in other words, that $X = Y$ with probability 1. This means that *all* the probability in the (x, y)-plane is concentrated on the line where $y = x$ (i.e., only points on this line contain any probability). That is, knowledge of the observed value for X gives the observed value for Y with probability 1 (and vice versa). In terms of V_1, V_2, this says that

$$P\left(V_2 = \mu_2 - \frac{\sigma_2}{\sigma_1}\mu_1 + \frac{\sigma_2}{\sigma_1}V_1\right) = 1.$$

Thus all the probability in the (v_1, v_2)-plane lies on a straight line with slope σ_2/σ_1. Such a distribution is called degenerate since it is not truly two-dimensional, with all the probability actually lying on a straight line (with a positive slope if $\rho = 1$).

The value for ρ will be -1 only if $E[(X + Y)^2] = 0$, which occurs only with $P(X + Y = 0) = 1$; in this case,

$$P\left(V_2 = \mu_2 + \frac{\sigma_2}{\sigma_1}\mu_1 - \frac{\sigma_2}{\sigma_1}V_1\right) = 1.$$

All of the probability in the (v_1, v_2)-plane is again concentrated on a straight line, this time with negative slope $-\sigma_2/\sigma_1$. This is another example of a degenerate two-dimensional probability law. (See Fig. 6.1.)

An example of this situation is given by letting V_1 be the number of successes to occur in n independent Bernoulli trials with parameter p, with V_2 the number of failures to occur on the same trials. It is easy to see that $\rho = -1$ for this case (by actually evaluating it from the joint probability law for V_1, V_2) and to see that all of the probability is in fact concentrated on the line $V_2 = n - V_1$, with slope -1. If we have observed the value for V_1, the value for V_2 is also completely specified. For example, knowing 18 successes occurred in 30 trials also means that 12 failures occurred.

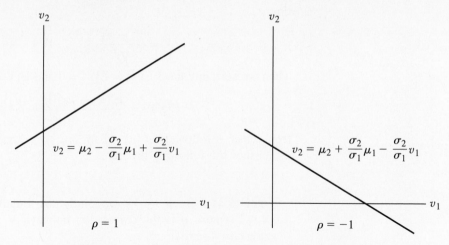

Figure 6.1 With $\rho = \pm 1$, probability is concentrated on a line

If $\rho \neq \pm 1$, then the joint probability distribution for V_1, V_2 is not concentrated on a line in the (v_1, v_2)-plane. When $0 < \rho < 1$, then $E[V_1 V_2] > E[V_1] E[V_2]$; the average of the product is greater than the product of the two averages. This is frequently interpreted to mean that there is a "tendency" for larger values of V_1 to be accompanied by larger values of V_2; the relationship between the two random variables, while not perfect, is direct, or increasing.

Similarly, if $-1 < \rho < 0$, then necessarily $E[V_1 V_2] < E[V_1] E[V_2]$; the average of the product is smaller than the product of the averages. This is frequently interpreted to mean that there is a "tendency" for larger values of V_1 to be accompanied by smaller values of V_2 (and vice versa); the relationship between the two random variables, while not perfect, is indirect, or decreasing.

For the case in which $\rho = 0$, $E[V_1 V_2] = E[V_1] E[V_2]$; the average of the product is identical with the product of the averages. The degree of *linear* association between the two variables is 0; this does not mean that they must be totally unrelated, as the following example should help clarify.

Example 6.8

Suppose we number three slips of paper 1, 2, 3, and then place them in a hat. One of these slips is then selected at random; let X be the number on the slip selected. Then we easily see that X is discrete uniform with $n = 3$, and thus $\mu_X = 2$, the middle value in the discrete range. Now let us define two new random variables, V_1, V_2, as follows: $V_1 = X - 2$ and $V_2 = V_1^2 = (X - 2)^2$. V_1 and V_2 then are certainly related to one another since V_2 is the square of V_1; we would have perfect knowledge of the value for V_2 if the value for V_1 were known. The joint range for these two random variables is simply $R_{V_1, V_2} = \{(-1, 1), (0, 0), (1, 1)\}$ (since X can only equal 1 or 2 or 3), and their joint probability function equals $\frac{1}{3}$ at each of these points. From this, it is easy to see that $E[V_1] = 0$ (so the product $E[V_1] E[V_2] = 0$ as well) and that $E[V_1 V_2] = 0$. Thus we have $\text{Cov}[V_1, V_2] = \rho = 0$, so V_1, V_2 are uncorrelated, in spite of the fact that knowing the value for V_1 also gives the value for V_2. These two variables have correlation 0 but are perfectly related (in a nonlinear manner, since every observed value for V_1, V_2 lies on the parabola $v_2 = v_1^2$).

We have seen that the expected value of a linear function of n random variables is the same linear function of the expectations. Is the same true for multiplicative functions? That is, if X_1, X_2 are jointly distributed and $w(X_1, X_2) = w_1(X_1)w_2(X_2)$, is it true that $E[w_1(X_1)w_2(X_2)] = E[w_1(X_1)] E[w_2(X_2)]$? In general, the answer to this query is no, but in the special case that X_1, X_2 are *independent* random variables the answer is yes.

Recall that if X_1, X_2 are independent, then their joint probability law is given by the product of their marginal probability functions (or marginal *pdf*s) for all elements in the range. Thus, for the continuous case, if $w(X_1, X_2) = w_1(X_2)w_2(X_2)$ and X_1, X_2 are independent, then

$$E[w(X_1, X_2)] = \int_{-\infty}^{\infty} \int_{-\infty}^{\infty} w_1(x_1)w_2(x_2)f_{X_1}(x_1)f_{X_2}(x_2)\,dx_1\,dx_2,$$

$$= \int_{-\infty}^{\infty} w_1(x_1)f_{X_1}(x_1)\,dx_1 \int_{-\infty}^{\infty} w_2(x_2)f_{X_2}(x_2)\,dx_2$$

$$= E[w_1(X_1)] E[w_2(X_2)].$$

Exactly the same reasoning holds if X_1, X_2 are discrete. Indeed, for an arbitrary number of independent random variables $X_1, X_2, \ldots, X_n$, the same conclusion holds, as given in the following theorem.

THEOREM 6.3 If $X_1, X_2, \ldots, X_n$ are independent random variables, then

$$E\left[\prod_{i=1}^{n} w_i(X_i)\right] = \prod_{i=1}^{n} E[w_i(X_i)],$$

for any functions $w_i(\cdot)$, $i = 1, 2, \ldots, n$, whose expectations exist.

Example 6.9

Suppose a fair die is rolled two times and X_1, X_2 represent the two numbers that result; also suppose that you will be paid $D = X_1 X_2$ dollars, the product of the numbers that occur on the two rolls. How much should you pay to make this a fair game? Recall that a fair game is one in which your expectation is 0. Since $E[D] = E[X_1 X_2] = E[X_1] E[X_2] = (3.5)^2 = 12.25$ (because X_1, X_2 are independent), your expected gain would be 0 if you were to pay $12.25 to play this game. The probability law for D is far from symmetric. It is not hard to see that $P(D < 12.25) = \frac{23}{36} = .639$ so you would expect to lose money on about 64% of the games you play, while still averaging out to a zero expectation. This is directly caused by the range for D, the amount you win; you are assured to win at least $1 each time you play, and (with small probability) could win as much as $36 from one play.

If X_1, X_2 are independent random variables, then $E[X_1 X_2] = E[X_1]\,E[X_2]$ and the covariance (and correlation) between X_1 and X_2 is necessarily 0. It is important to recognize that independence implies a covariance of zero, but that zero covariance for two random variables does not by itself imply they must be independent.

EXERCISES 6.1

1. Suppose a rectangle is constructed in the (x_1, x_2)-plane with base length x_1 and height x_2, where x_1 is the observed value of a random variable X_1 that is uniform on the interval $(1, 2)$, while x_2 is the observed value of a random variable X_2 that, given $X_1 = x_1$, is uniform on the interval $(0, x_1)$. What is the expected area of the rectangle?

2. A fair die is rolled twice; let X_1, X_2 be the numbers of spots on the top face for these two rolls. Evaluate $E[X_1 + X_2]$, $E[X_1 X_2]$, and $E[X_1/X_2]$.

3. Let $Y_1, Y_2, \ldots, Y_k$ be independent binomial random variables with parameters n_i, p. (Note that the n_i parameters may be different, but the p parameters are equal.) Evaluate the mean and variance of $\sum_{i=1}^{k} Y_i$.

4. Let $Y_1, Y_2, \ldots, Y_n$ be independent Poisson random variables, with μ_i the parameter for Y_i; evaluate the mean and variance of $\sum_{i=1}^{n} Y_i$.

5. The width of a rectangle is given by the observed value of a random variable X_1 whose *pdf* is

$$f_{X_1}(x_1) = 2x_1 - 1, \qquad \text{for } \tfrac{1}{2} < x_1 < \tfrac{3}{2}.$$

The height of the rectangle is given by X_2, which is uniform on the interval $(x_1 + \tfrac{1}{2}, x_1 + \tfrac{3}{2})$. What is the expected area of this rectangle?

6. The expected values for the random variables X and Y are 5 and -5, respectively, while their variances are 4 and 9, and their covariance is -5. Define $W = X + Y$ and $V = X - Y$, and evaluate the means, variances, and covariance for W, V.

7. Suppose X_1, X_2 are random variables with the same mean μ and the same variance σ^2, while their covariance is $\rho\sigma^2$. What is the covariance of $U = X_1 + X_2$ and $V = X_1 - X_2$?

8. Let $X_1, X_2, \ldots, X_n$ be independent random variables, each with mean μ and variance σ^2, and define $U = \sum_{i=1}^{n} a_i X_i$, where $a_1, a_2, \ldots, a_n$ are constants satisfying $\sum_{i=1}^{n} a_i = 1$.
   a. Show that the expected value for U is μ, the common expected value for the X_i.
   b. Show that the variance of U is $\sigma^2 \sum_{i=1}^{n} a_i^2$, and that this variance is minimized with $a_i = 1/n$, for $i = 1, 2, \ldots, n$.

9. Suppose X_1, X_2 are independent random variables, each with the same mean μ. If a is any constant, then $E[aX_1 + (1 - a)X_2] = \mu$; also assume that the variance for X_1 is $\sigma_1^2 = k\sigma_2^2$, where k is a known constant and $\mathrm{Var}[X_2] = \sigma_2^2$. With $U = aX_1 + (1 - a)X_2$, find the value of a that minimizes σ_U^2 (which will be a function of k).

10. What is the expected position number of the *last* ace in a well-shuffled 52-card deck?

11. Show that the magnitude of the correlation between two random vari-

ables is invariant with linear transformations. That is, if X, Y are jointly distributed random variables with correlation ρ, and we define $U = a + bX$, $V = c + dY$, where $a, b \neq 0$ and $c, d \neq 0$ are constants, then the magnitude of the correlation between U and V is also $|\rho|$.

12. Let T_1, T_2 be jointly distributed with *pdf*

$$f_{T_1, T_2}(t_1, t_2) = t_1 e^{-t_2}, \qquad \text{for } 0 < t_1 < t_2.$$

Evaluate $\text{Cov}[T_1, T_2]$ and the correlation ρ between T_1 and T_2.

13. If X_1 is a discrete random variable and $X_2 = X_1^2$, then X_1, X_2 may be uncorrelated, as shown in the text. However, they may also be highly correlated.
    a. Show that if the range of X_1 contains only two points, and the probability function for X_1 has value $\frac{1}{2}$ for each of them, and $\mu_{X_1} \neq 0$, then X_1 and $X_2 = X_1^2$ are perfectly correlated. (Consider the geometry of this situation.)
    b. What happens in part a if $\mu_{X_1} = 0$?

14. Suppose X_1, X_2 are jointly distributed random variables such that $\text{E}[X_1] = \text{E}[X_2] = 3$, $\text{Var}[X_1] = \text{Var}[X_2] = 2$, and $\text{Cov}[X_1, X_2] = 1$.
    a. Find a_1, a_2 such that $Y = a_1 X_1 + a_2 X_2$ has mean 0 and variance 1.
    b. Find b_1, b_2 such that $\text{Cov}[Y, Z] = 0$, where $Z = b_1 X_1 + b_2 X_2$. What are the values of μ_Z and σ_Z?

6.2 | Moments and Generating Functions

Random variables are used across a wide variety of applications; they are used to model numeric values of many different types: results of sample surveys, planned experiments, testing of the quality of a lot of produced items, and so forth. We have seen that a model for a discrete random variable can be described by its probability function, or if one wants, by the *cdf*; similarly, the model for a continuous random variable can be described by its *pdf* (or again the *cdf*). As we shall see in this section, additional descriptors for the probability law of a random variable are provided by its moments or by what is called the moment generating function for the probability law.

The material in this section applies equally well to either the discrete or the continuous case, and proves especially useful for certain theoretical developments in describing relationships between probability laws. Let us start by defining the moments of a probability law.

DEFINITION 6.4

Assume X is a random variable with a given probability law. Then the kth *moment of X* is $m_k = \text{E}[X^k]$, for $k = 1, 2, 3 \ldots$, as long as the expectation exists.

The first moment for a random variable X is thus $m_1 = \mathrm{E}[X^1] = \mu_X$, the mean value, a measure of the middle of the probability law for X. The second moment $m_2 = \mathrm{E}[X^2] = \sigma_X^2 + \mu_X^2$ is related to the variability, or spread, of the probability law for X. The third moment for a random variable, $m_3 = \mathrm{E}[X^3]$, has been suggested as a measure of *skewness* of the *pdf* or probability function, a comparison of the distribution of the probability mass to the left and to the right of the mean value, m_1. The fourth moment, $m_4 = \mathrm{E}[X^4]$, has been used as a measure of *kurtosis*, the relative flatness, or peakedness, of the probability distribution in the neighborhood of the mean, m_1.

For sufficiently regular probability laws, the knowledge of the full sequence of moments $m_1, m_2, m_3, \ldots$ is tantamount to knowledge of the probability law itself; that is, the moments of a probability law can provide another way of specifying the probability law. In fact, certain types of constructs involving random variables are most easily comprehended through the use of the moments, or as will be developed, the moment generating function for the probability law.

Example 6.10

We have seen that $\mathrm{E}[X^k] = \Gamma(k+r)/\lambda^k (r-1)!$, if X is an Erlang random variable with parameters r, λ. Thus with $r = 1$ (so X is exponential with parameter λ) the sequence of moments for the probability law is given by $1/\lambda, 2/\lambda^2, 6/\lambda^3, \ldots$, while with $r = 2$ the sequence of moments of X is $2/\lambda, 6/\lambda^2, 24/\lambda^3, \ldots$.

If X is Bernoulli with parameter p, the sequence of moments is especially simple: $m_k = \mathrm{E}[X^k] = 0^k(1-p) + 1^k p = p$, for all k. If X is uniform on the interval (a, b), then

$$m_k = \mathrm{E}[X^k] = \int_a^b x^k \frac{1}{b-a}\, dx = \frac{b^{k+1} - a^{k+1}}{(k+1)(b-a)}.$$

As given by the original definition, and illustrated in Example 6.10, the kth moment of X is the expected value for $w(X)$, where $w(X) = X^k$; thus finding m_k requires evaluating a sum if X is discrete or an integral if X is continuous. Interestingly, it is possible to find a function whose *derivatives* give the moments, for either the discrete or the continuous case, as long as a certain expectation exists.

Suppose X is discrete with range R_X and t is a constant. Define

$$m_X(t) = \mathrm{E}\left[e^{tX}\right] = \sum_{x \in R_X} e^{tx} p_X(x),$$

and let us assume that this expectation exists for *all* t in some interval that includes 0. Note that $m_X(0) = \mathrm{E}[e^{0 \cdot X}] = \mathrm{E}[1] = 1$, for any X. Then the derivative of $m_X(t)$ with respect to t is

$$m_X'(t) = \frac{d}{dt} \sum_{x \in R_X} e^{tx} p_X(x) = \sum_{x \in R_X} \frac{d}{dt} e^{tx} p_X(x)$$

$$= \sum_{x \in R_X} x e^{tx} p_X(x),$$

since the derivative of the sum is the sum of the derivatives. If we now set $t = 0$, we have

$$m'_X(0) = \sum_{x \in R_X} x e^{0 \cdot x} p_X(x) = \sum_{x \in R_X} x p_X(x) = \mathrm{E}[X] = m_1,$$

the first moment of X, since $e^{0 \cdot x} = e^0 = 1$, for any x. It is easy to see that the second and third derivatives in turn are

$$m''_X(t) = \sum_{x \in R_X} x^2 e^{tx} p_X(x), \tag{1}$$

$$m'''_X(t) = \sum_{x \in R_X} x^3 e^{tx} p_X(x). \tag{2}$$

Setting $t = 0$ in Eqs. (1) and (2) gives $m''_X(0) = m_2$ and $m'''_X(0) = m_3$; in short, the kth derivative of $m_X(t)$ evaluated at $t = 0$, which we denote $m_X^{(k)}(0)$, gives m_k, the kth moment of X.

This discussion assumes a discrete X, where expectations are given by sums. If X is continuous, the same operations are appropriate: Again the kth derivative of $m_X(t) = \mathrm{E}[e^{tX}]$, evaluated at $t = 0$, gives the kth moment of X. This function is called the moment generating function for X, as given in the following definition.

DEFINITION 6.5

If X is a random variable, and the expected value

$$m_X(t) = \mathrm{E}[e^{tX}]$$

exists for all t in an interval that includes 0, $m_X(t)$ is called the *moment generating function* for X (or for the probability law for X).

The following example evaluates the moment generating functions for some familiar probability laws.

Example 6.11

Assume X is an Erlang random variable with parameters r, λ; the moment generating function for X is

$$
\begin{aligned}
m_X(t) &= \int_0^\infty e^{tx} \frac{\lambda^r x^{r-1}}{(r-1)!} e^{-\lambda x} \, dx \\
&= \int_0^\infty \frac{\lambda^r x^{r-1}}{(r-1)!} e^{-x(\lambda - t)} \, dx \\
&= \int_0^\infty \frac{\lambda^r}{(r-1)!} \left(\frac{u}{\lambda - t} \right)^{r-1} e^{-u} \frac{du}{\lambda - t} \\
&= \left(\frac{\lambda}{\lambda - t} \right)^r \frac{\Gamma(r)}{(r-1)!} = \left(\frac{\lambda}{\lambda - t} \right)^r,
\end{aligned}
\tag{3}
$$

for any $t < \lambda$. It is easy to verify that the kth derivative of $m_X(t)$, evaluated at $t = 0$, is

$$m_X^{(k)}(0) = \frac{r(r+1)\cdots(r+k-1)}{\lambda^k} = \frac{\Gamma(k+r)}{\lambda^k(r-1)!},$$

the kth moment of X. Recall that the Erlang probability law is a special case of the gamma probability law: The integer parameter r in the Erlang distribution is replaced in the gamma distribution by $n > 0$, which varies continuously. The integration in Eq. (3) is equally valid for the gamma probability law, for $n > 0$, so the moment generating function for a gamma random variable Y with parameters n, λ is

$$m_Y(t) = \left(\frac{\lambda}{\lambda - t}\right)^n.$$

If X is Bernoulli with parameter p, then

$$m_X(t) = e^{t(0)}(1-p) + e^{t(1)}p = (1-p) + pe^t,$$

so $m_X^{(k)}(0) = p$, the kth moment of X. If U is uniform on the interval (a, b), then

$$m_U(t) = \int_a^b \frac{e^{tx}}{b-a}\,dx = \frac{e^{tb} - e^{ta}}{t(b-a)};$$

the derivatives of $m_U(t)$, evaluated at $t = 0$ (using L'Hôpital's rule) give the moments of U.

If Z is a standard normal random variable, its moment generating function is

$$
\begin{aligned}
m_Z(t) = E[e^{tZ}] &= \int_{-\infty}^{\infty} e^{tz} \frac{1}{\sqrt{2\pi}} e^{-z^2/2}\,dz \\
&= \int_{-\infty}^{\infty} \frac{1}{\sqrt{2\pi}} e^{-(z^2 - 2tz + t^2 - t^2)/2}\,dz \\
&= e^{t^2/2} \int_{-\infty}^{\infty} \frac{1}{\sqrt{2\pi}} e^{-(z-t)^2/2}\,dz = e^{t^2/2}.
\end{aligned}
\tag{4}
$$

The last integral in Eq. (4) gives the area under a normal *pdf*, with mean t and variance 1, which equals 1 for any value of t.

Suppose X, Y are continuous random variables with moment generating functions $m_X(t), m_Y(t)$, respectively; let us also assume that $m_X(t) = m_Y(t)$ for all values of t in an interval that includes 0. From these assumptions we would conclude that the moments for X and the moments for Y are identical, since the equality of the moment generating functions in a neighborhood of 0 also says their derivatives at 0 (which give the moments) must be equal. This equality of the moment generating functions gives

$$\int_{-\infty}^{\infty} e^{tx} f_X(x)\,dx = \int_{-\infty}^{\infty} e^{ty} f_Y(y)\,dy,$$

for all values of t in an interval containing 0. Taking the difference of these two integrals (and using x as the variable of integration for both) then gives

$$\int_{-\infty}^{\infty} e^{tx}\big(f_X(x) - f_Y(x)\big)\,dx = 0, \tag{5}$$

for all values of t in an interval including 0. Equation (5) is consistent with the statement $f_X(x) = f_Y(x)$ for all values of x, which implies that X and Y have identical probability laws. The proof for this conclusion is beyond the level of this text, so it is simply stated in the following theorem.

THEOREM 6.4 If the moment generating functions for X and Y are identical for all values of t in some interval that includes 0, then the probability laws for X and Y are identical.

This result proves powerful in establishing (or identifying) probability laws for random variables. If we are able to evaluate the moment generating function for X and find $m_X(t) = e^{t^2/2}$, say, this is sufficient to conclude that the probability law for X is the standard normal (since this is the standard normal moment generating function).

The moment generating function for a linear function of a random variable is simply related to the moment generating function of the random variable itself. Indeed, let X be a random variable with moment generating function $m_X(t)$ and define $Y = a + bX$, where a, and b are constants. Then

$$m_Y(t) = \mathrm{E}[e^{tY}] = \mathrm{E}[e^{t(a+bX)}] = e^{ta}\,\mathrm{E}[e^{tbX}] = e^{ta}m_X(tb);$$

the moment generating function for Y is the constant e^{ta} times the moment generating function for X with argument tb rather than t. This establishes the following theorem.

THEOREM 6.5 If X is a random variable with moment generating function $m_X(t)$, and a, b are real constants such that $Y = a + bX$, then the moment generating function for Y is

$$m_Y(t) = e^{ta}m_X(tb).$$

Example 6.12

Let Z be a standard normal random variable, and define $X = \mu_X + \sigma_X Z$. Then the moment generating function for Z is $m_Z(t) = e^{t^2/2}$, and the moment generating function for X is then

$$m_X(t) = e^{t\mu_X}m_Z(t\sigma_X) = e^{t\mu_X + t^2\sigma_X^2/2}.$$

Recalling that X itself then is normal with parameters μ_X, σ_X^2, this gives the moment generating function for an arbitrary normal random variable (one with an arbitrary mean and variance).

The moment generating function for $-Z$, the negative of a standard normal random variable, is then $m_Z(-t) = e^{(-t)^2/2} = e^{t^2/2} = m_Z(t)$; that is, the negative of a standard normal random variable also has the standard normal probability law. If U is uniform on $(0, 1)$ and we define $V = 1 - U$, we have seen that the probability law for V is again uniform on the same interval. That this is the case can also be derived from the moment generating function since

$$m_V(t) = e^t m_U(-t) = e^t \left(\frac{e^{-t} - 1}{-t} \right) = \frac{e^t - 1}{t} = m_U(t).$$

The moments of a probability law can be used to characterize the probability distribution; there are other, related sequences that can equally well be used for the same purpose. One of these sequences is the sequence of moments about the mean.

DEFINITION 6.6

Let X be a random variable with mean μ_X. The kth *moment of X about the mean* is defined to be

$$\mu_k = \mathrm{E}[(X - \mu_X)^k],$$

for $k = 1, 2, 3, \ldots$, as long as the expectation exists.

The first moment about the mean is always 0 since

$$\mu_1 = \mathrm{E}[X - \mu_X] = \mathrm{E}[X] - \mu_X = m_1 - m_1 = 0, \tag{6}$$

for any probability law. The second moment about the mean is the variance of the probability law:

$$\mu_2 = \mathrm{E}[(X - \mu_X)^2] = \sigma_X^2 = m_2 - m_1^2. \tag{7}$$

The third moment about the mean is

$$\mu_3 = \mathrm{E}[(X - \mu_X)^3] = \mathrm{E}[X^3] - 3\mu_X \mathrm{E}[X^2] + 3\mu_X^2 \mathrm{E}[X] - \mu_X^3$$
$$= m_3 - 3m_1 m_2 + 2m_1^3. \tag{8}$$

The kth moment about the mean is clearly a function of the first k moments of X, as illustrated in Eqs. (7) and (8) for $k = 2, 3$. Knowledge of the moments $m_1, m_2, \ldots$ (frequently referred to as moments about 0 to distinguish them from moments about the mean) allows us to evaluate the sequence of moments about the mean $\mu_1, \mu_2, \ldots$ for the same random variable, and vice versa.

Since $Y = X - \mu_X$ is a linear function of X, the moment generating function for Y is $m_Y(t) = e^{-t\mu_X} m_X(t)$. However, we can also see that $\mathrm{E}[Y^k] = \mathrm{E}[(X - \mu_X)^k]$, so the kth moment (about 0) for Y is identical to μ_k, the kth moment about the mean for X. Thus the kth derivative of $e^{-t\mu_X} m_X(t)$, evaluated at $t = 0$, gives μ_k, the kth moment of X about its mean; that is, the function $e^{-t\mu_X}m_X(t)$ generates moments about the mean for X.

A number of other generating functions have also proved useful for various purposes. Suppose X has moment generating function $m_X(t)$. Since $e^{tx} > 0$ for all real t and x, $m_X(t) > 0$ for all possible values of t and we have no difficulty in defining $c_X(t) = \ln m_X(t)$, the cumulant generating function for X.

DEFINITION 6.7

Granted X is a random variable with moment generating function $m_X(t)$, the *cumulant generating function* for X is $c_X(t) = \ln m_X(t)$.

One reason this function is of interest is given by examining the first two derivatives of $c_X(t)$:

$$c_X'(t) = \frac{m_X'(t)}{m_X(t)} \tag{9}$$

$$c_X''(t) = \frac{m_X''(t)m_X(t) - (m_X'(t))^2}{m_X^2(t)}. \tag{10}$$

Setting $t = 0$ in Eqs. (9) and (10) gives

$$c_X'(0) = \frac{m_1}{1} = \mu_X \tag{11}$$

$$c_X''(0) = \frac{m_2 \cdot 1 - m_1^2}{1^2} = \sigma_X^2, \tag{12}$$

since $m_X(0) = 1, m_X'(0) = m_1$, and $m_X''(0) = m_2$. That is, the first and second derivatives of $c_X(t)$ evaluated at $t = 0$ directly give the mean and variance for X (these are also called the first two *cumulants*).

Example 6.13

We have seen that the moment generating function for a gamma random variable is $m_Y(t) = (\lambda/(\lambda - t))^n$, so its cumulant generating function is $c_Y(t) = n \ln \lambda - n \ln(\lambda - t)$; thus $c_Y'(0) = n/\lambda = \mu_Y$, and $c_Y''(0) = n/\lambda^2 = \sigma_Y^2$. The moment generating function for a normal random variable with parameters μ_X, σ_X^2 is $m_X(t) = e^{t\mu_X + t^2\sigma_X^2/2}$, so its cumulant generating function is $c_X(t) = t\mu_X + t^2\sigma_X^2/2$, and the first two cumulants are $c_X'(0) = \mu_X$ and $c_X''(0) = \sigma_X^2$. Since all further derivatives of this function are 0, the third, fourth, and higher cumulants are 0 for a normal random variable.

In evaluating the variance for discrete random variables, we saw that it is frequently more straightforward to evaluate the sum defining $E[X(X - 1)]$ than to evaluate $E[X^2]$ directly. The quantity $E[X(X - 1)]$ is called the second factorial moment for X because it is the product of two terms and has a factorial-like structure; the kth *factorial moment* is the expected value of the product of k terms and is defined to be $E[X(X-1)\cdots(X-(k-1))]$ for $k = 1, 2, 3, \ldots$. These quantities are generated by the derivatives of the factorial moment generating function for discrete X.

DEFINITION 6.8

If X is a discrete random variable with probability function $p_X(x)$, then the factorial moment generating function for X is

$$\psi_X(t) = E[t^X],$$

as long as the expectation exists for all values of t in an interval that includes 1.

To see how this function generates the factorial moments, note that the first derivative of $\psi_X(t)$ is

$$\psi_X'(t) = \frac{d}{dt} \sum_{R_X} t^x p_X(x) = \sum_{R_X} x t^{x-1} p_X(x).$$

Subsequent derivatives "bring down" succeeding values of the exponent of t. The kth derivative then is

$$\psi_X^{(k)}(t) = \sum_{R_X} x(x-1)\cdots(x-(k-1)) t^{x-k} p_X(x). \tag{13}$$

If we evaluate this derivative at $t = 1$ (not 0), we have

$$\psi_X^{(k)}(1) = \sum_{R_X} x(x-1)\cdots(x-(k-1)) p_X(x)$$
$$= E[X(X-1)\cdots(X-(k-1))],$$

the kth factorial moment for X.

Example 6.14

Suppose X is a binomial random variable with parameters n and p. Its factorial moment generating function is then

$$\psi_X(t) = \sum_{x=0}^{n} t^x \binom{n}{x} p^x q^{n-x} = \sum_{x=0}^{n} \binom{n}{x} (pt)^x q^{n-x} = (pt+q)^n,$$

from the binomial theorem. The first two derivatives of $\psi_X(t)$ are $\psi_X'(t) = np(pt+q)^{n-1}$ and $\psi_X''(t) = n(n-1)p^2(pt+q)^{n-2}$, from which we find $\psi_X'(1) = np = E[X]$ and $\psi_X''(1) = n(n-1)p^2 = E[X(X-1)]$, the first two factorial moments.

If Y is Poisson with parameter μ_Y, its factorial moment generating function is

$$\psi_Y(t) = \sum_{y=0}^{\infty} t^y \frac{\mu_Y^y}{y!} e^{-\mu_Y} = \sum_{y=0}^{\infty} \frac{(\mu_Y t)^y}{y!} e^{-\mu_Y} = e^{t\mu_Y} e^{-\mu_Y}$$

$$= e^{(t-1)\mu_Y}.$$

Thus $\psi_Y'(t) = \mu_Y e^{(t-1)\mu_Y}$ and $\psi_Y''(t) = \mu_Y^2 e^{(t-1)\mu_Y}$. The first two factorial moments of Y are then $\psi_Y'(1) = \mu_Y = \mathrm{E}[Y]$, $\psi_Y''(1) = \mu_Y^2 = \mathrm{E}[Y(Y-1)]$.

The factorial moment generating function is given by a logarithmic translation of the argument of the moment generating function. It is true that $t^X \equiv e^{X \ln t}$ for any $t > 0$, as is easily verified by taking the logarithm of both sides of this equation; thus $\psi_X(t) = \mathrm{E}[t^X] \equiv \mathrm{E}[e^{X \ln t}] = m_X(\ln t)$, and of course, $m_X(t) = \mathrm{E}[e^{tX}] \equiv \mathrm{E}[(e^t)^X] = \psi_X(e^t)$. Knowing the factorial moment generating function for a discrete X immediately allows us to also find the moment generating function for the same random variable.

Example 6.15

Suppose that a binomial random variable has factorial moment generating function $\psi_X(t) = (.7t + .3)^{13}$; its moment generating function is then $\psi_X(e^t) = m_X(t) = (.7e^t + .3)^{13}$, and its cumulant generating function is $c_X(t) = \ln m_X(t) = 13\ln(.7e^t + .3)$. A Poisson random variable with parameter 10 has factorial moment generating function $\psi_Y(t) = e^{10(t-1)}$; its moment generating function is $m_Y(t) = \psi_Y(e^t) = e^{10(e^t-1)}$, and its cumulant generating function is $c_Y(t) = \ln m_X(t) = 10(e^t - 1)$.

The factorial moment generating function, $\psi_X(t) = \mathrm{E}[t^X]$, can be used to find the factorial moments of a discrete random variable by evaluating its derivatives at $t = 1$. This same function has another use (and another name) for certain types of discrete random variable: Assume that X is discrete with range $R_X \subset \{0, 1, 2, \ldots\}$, the nonnegative integers, and probability function $p_X(x)$ for $x \in R_X$. As we saw in Eq. (13), the kth derivative of $\psi_X(t)$ is

$$\psi_X^{(k)}(t) = \sum_{x=0}^{\infty} x(x-1)\cdots(x-(k-1))\, t^{x-k} p_X(x)$$

$$= \sum_{x=k}^{\infty} x(x-1)\cdots(x-(k-1))\, t^{x-k} p_X(x),$$

since $x(x-1)\cdots(x-(k-1)) = 0$ for $x = 0, 1, 2, \ldots, k-1$. Thus

$$\psi_X^{(k)}(t) = k! p_X(k) + \frac{(k+1)!}{1!} t p_X(k+1) + \frac{(k+2)!}{2!} t^2 p_X(k+2) + \cdots. \tag{14}$$

Every term in Eq. (14), except the first, is multiplied by t raised to some integral power; thus, if we were to set $t = 0$ we are left with only the first term. That is, $\psi_X^{(k)}(0) = k! p_X(k)$; equivalently, we can evaluate the probability function for X from the value of the kth derivative evaluated at $t = 0$:

$p_X(k) = \psi_X^{(k)}(0)/k!$ for $k = 0, 1, 2, \ldots$. Because of this, $\psi_X(t)$ is also called the probability generating function for a discrete random variable whose range is a subset of the nonnegative integers. This discussion of $\psi_X(t)$ is summarized in the following theorem.

THEOREM 6.6 Let X be a discrete random variable with range R_X, and define $\psi_X(t) = \mathrm{E}[t^X]$, as long as the expectation exists for all t in an interval including 1. Then

$$\psi_X^{(k)}(1) = \mathrm{E}[X(X - 1) \cdots (X - (k - 1))],$$

for $k = 1, 2, 3, \ldots$, and $\psi_X(t)$ is called the factorial moment generating function for X.

If additionally the range for X is a subset of the nonnegative integers, then

$$\psi_X^{(k)}(0) = k! p_X(k),$$

for $k = 0, 1, 2, \ldots$, and $\psi_X(t)$ is also called the *probability generating function* for X.

Example 6.16

The probability generating function for a Poisson random variable Y is $\psi_Y(t) = e^{\mu_Y(t-1)}$. The kth derivative of this function with respect to t is $\psi_Y^{(k)}(t) = \mu_Y^k e^{\mu_Y(t-1)}$; evaluating this derivative at $t = 0$ gives $\psi_Y^{(k)}(0) = \mu_Y^k e^{-\mu_Y} = k! p_Y(k)$, as of course it must.

Suppose you are given the function $\psi_W(t) = 3t/(4 - t)$ and are told that it is the probability generating function for a discrete random variable W whose range is a subset of the nonnegative integers. It is straightforward to see that the kth derivative of $\psi_W(t)$ is $4 \cdot 3 \cdot k!/(4 - t)^{k+1}$, so $\psi_W^{(k)}(0) = k!(\frac{3}{4})(\frac{1}{4})^{k-1}$ for $k = 1, 2, 3, \ldots$; the probability function for W is $p_W(k) = \psi_W^{(k)}(0)/k! = \frac{3}{4}(\frac{1}{4})^{k-1}$. That is, W is a geometric random variable with $p = \frac{3}{4}$.

Another way to look at the probability generating function

$$\psi_X(t) = t^0 p_X(0) + t p_X(1) + t^2 p_X(2) + t^3 p_X(3) + \cdots$$

is simply to realize that if we write $\psi_X(t)$ as a sum of nonnegative powers of t, then

1. The powers of t that occur identify the elements of the range R_X; for example, if t^3 does *not* occur in this series, it is because the probability at $x = 3$ is 0.
2. The multipliers of all the powers of t must sum to 1, and these multipliers themselves give the values of the probability function at the respective element of R_X.

The following example illustrates this reasoning.

Example 6.17

Suppose a random variable Y has probability generating function

$$\psi_Y(t) = \frac{t^2}{2} + \frac{t^4}{4} + \frac{t^8}{8} + \frac{t^{16}}{8}.$$

Then the range for Y is necessarily $R_Y = \{2, 4, 8, 16\}$, since these are the only powers of t to occur. Said another way, $p_Y(5) = 0$; since t^5 is *not* represented in this series, its multiplier is 0. The probabilities of occurrence of these elements listed in R_Y are $\frac{1}{2}, \frac{1}{4}, \frac{1}{8}, \frac{1}{8}$, respectively.

Table 6.3 displays the moment generating functions for some standard continuous probability laws discussed earlier. The kth derivative evaluated at 0 gives the kth moment for the corresponding probability law. The function $e^{-t\mu_X} m_X(t)$ generates moments about the mean, and the cumulant generating function is $\ln m_X(t)$.

Table 6.4 displays the probability generating functions for some discrete probability laws discussed earlier. The kth derivative of these functions, eval-

Table 6.3

Moment Generating Functions for Some Continuous Laws

Probability Law	Moment Generating Function
Uniform	$\dfrac{e^{tb} - e^{ta}}{t(b-a)}$
Exponential	$\dfrac{\lambda}{\lambda - t}$
Gamma	$\dfrac{\lambda^n}{(\lambda - t)^n}$
Normal	$e^{t\mu + t^2\sigma^2/2}$

Table 6.4

Probability Generating Functions for Some Discrete Laws

Probability Law	Probability Generating Function
Discrete uniform	$\dfrac{t(1 - t^N)}{N(1 - t)}$
Bernoulli	$q + pt$
Binomial	$(q + pt)^n$
Geometric	$\dfrac{pt}{1 - qt}$
Negative binomial	$\dfrac{(pt)^r}{(1 - qt)^r}$
Poisson	$e^{\mu(t-1)}$

uated at $t = 0$, gives $k!p_X(k)$. Recall that probability generating functions are also called factorial moment generating functions, since the kth derivative, evaluated at $t = 1$, gives the kth factorial moment.

Suppose $X_1, X_2, \ldots, X_n$ are independent random variables and we define $Y = \sum_{i=1}^{n} a_i X_i$, where $a_1, a_2, \ldots, a_n$ are constants, so Y is a linear function of $X_1, X_2, \ldots, X_n$. Then the mean for Y is $\mu_Y = \sum_{i=1}^{n} a_i \mu_i$ and the variance for Y is $\sigma_Y^2 = \sum_{i=1}^{n} a_i^2 \sigma_i^2$, where μ_i and σ_i^2 are the means and variances respectively of X_i, $i = 1, 2, \ldots, n$. We can also easily evaluate the moment generating function for Y in terms of the moment generating functions for $X_1, X_2, \ldots, X_n$. If we recognize the resulting function as being the moment generating function for a known probability law, Y has that probability law.

Recall that the expected value of a product of functions of independent random variables is simply given by the product of the individual expectations. Thus the moment generating function for $Y = \sum_{i=1}^{n} a_i X_i$ is

$$
\begin{aligned}
m_Y(t) = \mathrm{E}[e^{tY}] &= \mathrm{E}\left[e^{t(a_1 X_1 + a_2 X_2 + \cdots + a_n X_n)}\right] \\
&= \mathrm{E}\left[e^{ta_1 X_1} e^{ta_2 X_2} \cdots e^{ta_n X_n}\right] \\
&= \mathrm{E}\left[e^{ta_1 X_1}\right] \mathrm{E}\left[e^{ta_2 X_2}\right] \cdots \mathrm{E}\left[e^{ta_n X_n}\right] \\
&= m_{X_1}(ta_1) m_{X_2}(ta_2) \cdots m_{X_n}(ta_n),
\end{aligned}
\tag{15}
$$

since the expectation of a product is the product of the expectations, granted that $X_1, X_2, \ldots, X_n$ are independent. Equation (15) gives the following theorem.

THEOREM 6.7 If $X_1, X_2, \ldots, X_n$ are independent random variables with moment generating functions $m_{X_1}(t), m_{X_2}(t), \ldots, m_{X_n}(t)$ and if $a_1, a_2, \ldots, a_n$ are constants, then the moment generating function for $Y = \sum_{i=1}^{n} a_i X_i$ is given by

$$m_Y(t) = m_{X_1}(ta_1) m_{X_2}(ta_2) \cdots m_{X_n}(ta_n).$$

Theorem 6.7 proves quite powerful in deriving distributions for linear functions of independent random variables. The following example mentions several results that follow from Theorem 6.7.

Example 6.18

Suppose Y_1, Y_2 are independent binomial random variables with parameters n_1, p_1 and n_2, p_2, respectively. Then the moment generating function for $W = Y_1 + Y_2$ is

$$m_W(t) = m_{Y_1}(t) m_{Y_2}(t) = (q_1 + p_1 e^t)^{n_1} (q_2 + p_2 e^t)^{n_2}.$$

If $p_1 = p_2 = p$ (and then $q_1 = q_2 = 1 - p = q$ as well), the moment generating function for $W = Y_1 + Y_2$ is

$$m_W(t) = (q + pe^t)^{n_1 + n_2}.
\tag{16}$$

Since Eq. (16) is the moment generating function of a binomial random variable, W is itself binomial with parameters $n_1 + n_2, p$, regardless of whether $n_1 = n_2$; it is necessary that $p_1 = p_2$ for $Y_1 + Y_2$ to follow the binomial probability law, since the two bases in the product $m_{Y_1}(t) m_{Y_2}(t)$ are equal (for all values of t) only for this case.

The moment generating function for a gamma random variable with parameters n, λ can be written $(1 - t/\lambda)^{-n}$ for $t < \lambda$. Thus the moment generating function for a sum of two independent gamma random variables X_1, X_2, with parameters n_1, λ_1 and n_2, λ_2, respectively, is given by $(1 - t/\lambda_1)^{-n_1} (1 - t/\lambda_2)^{-n_2}$. If $\lambda_1 = \lambda_2 = \lambda$, then the moment generating function for $X_1 + X_2$ is simply $(1 - t/\lambda)^{-(n_1 + n_2)}$, so $X_1 + X_2$ is itself a gamma random variable with parameters $n_1 + n_2, \lambda$ for this case. The binomial and gamma probability laws "reproduce" themselves (with certain parameter restrictions), in the sense that sums of independent random variables with the same probability law also follow that probability law. In Exercises 6.2 you will be asked to determine the reproductive properties of the negative binomial, Poisson, Erlang, and normal probability laws, all of which are easily explored using the moment generating function approach.

Recall that the square of a standard normal random variable has the χ^2 (chi-square) distribution with one degree of freedom, and that this distribution is a special gamma random variable with parameters $n = \lambda = \frac{1}{2}$. If $Z_1, Z_2, \ldots, Z_k$ are independent standard normal random variables, what is the probability law for $W = \sum_{i=1}^{k} Z_i^2$, the sum of their squares? Letting $U_i = Z_i^2$ for $i = 1, 2, \ldots, k$, then, we know that each U_i is a χ^2 random variable, and $W = \sum_{i=1}^{k} U_i$. The fact that $Z_1, Z_2, \ldots, Z_k$ are independent implies that $U_1, U_2, \ldots, U_k$ are independent as well. The moment generating function for the χ^2 distribution with one degree of freedom is

$$m_{U_i}(t) = \frac{1}{(1 - 2t)^{1/2}},$$

so the moment generating function for W is simply given by the product

$$m_W(t) = \prod_{i=1}^{k} m_{U_i}(t) = \frac{1}{(1 - 2t)^{k/2}};$$

W is called a chi-square random variable with k degrees of freedom (since it is the sum of squares of k independent standard normal random variables), its only parameter. This moment generating function for W is that of a gamma random variable with parameters $n = \frac{k}{2}$ and $\lambda = \frac{1}{2}$; that is, the χ^2 distribution with k degrees of freedom is also a special gamma random variable. It follows immediately that the mean of the χ^2 distribution is $n/\lambda = k$, its degrees of freedom, and its variance is $n/\lambda^2 = 2k$, twice its degrees of freedom. The *pdf* for the χ^2 distribution with k degrees of freedom is thus

$$f_W(t) = \frac{t^{\frac{k}{2} - 1}}{2^k \Gamma(\frac{k}{2})} e^{-t/2}, \qquad t > 0,$$

where the parameter $k = 1, 2, 3, \ldots$ and is called the "degrees of freedom" for W. Much of this discussion is summarized in the following theorem.

THEOREM 6.8   Let $Z_1, Z_2, \ldots, Z_k$ be independent standard normal random variables; then $W = \sum_{i=1}^{k} Z_i^2$ has the χ^2 distribution with k degrees of freedom and $E[W] = k$, $\text{Var}[W] = 2k$.

Example 6.19

The χ^2 distribution with $k = 2$ degrees of freedom is identical to the exponential probability law with $\lambda = \frac{1}{2}$ (the gamma with $n = 1$ and $\lambda = \frac{1}{2}$). This allows easy evaluation of probability statements for this special case. For example, if X_1, X_2 are independent normal random variables, each with mean 0 and variance 4, then $X_1/2, X_2/2$ are independent standard normal. $W = (X_1^2 + X_2^2)/4$ then has the χ^2 distribution with two degrees of freedom (that is, W is exponential with $\lambda = \frac{1}{2}$) and

$$P(X_1^2 + X_2^2 \leq 2) = P(W \leq \tfrac{2}{4}) = 1 - e^{-\frac{1}{2}(\frac{2}{4})} = .221,$$

while

$$P(X_1^2 + X_2^2 > 4) = P(W > \tfrac{4}{4}) = e^{(-\frac{1}{2})(\frac{4}{4})} = .607.$$

Indeed, for any *even* number of degrees of freedom k, W is an Erlang random variable with $r = k/2$, $\lambda = \frac{1}{2}$, so the *cdf* for W can be written as a sum of Poisson probabilities:

$$F_W(t) = 1 - \sum_{i=0}^{\frac{k}{2}-1} P(X_{t/2} = i),$$

where $X_{t/2}$ is Poisson with parameter $t/2$. For odd degrees of freedom, numerical integration is called for to evaluate $F_W(t)$.

EXERCISES 6.2

1. Evaluate the moment generating function for a binomial random variable with $n = 2$ and p. Use this function to evaluate the full series of moments for $k = 1, 2, 3, \ldots$.

2. Let X be discrete uniform with parameter n; evaluate m_1 and m_2, the first two moments for X.

3. A random variable U has moment generating function $m_U(t) = (7/(7 - t))^3$. Evaluate the mean and variance for U. What is the probability law for U?

4. Let Y be a geometric random variable with parameter p; find the moment generating function and the factorial moment generating function for Y.

5. A random variable Y has moment generating function $m_Y(t) = e^{t(5+t)}$; find the mean and variance for Y. What is the probability law for Y?

6. The moment generating function for a discrete random variable G is

$$m_G(t) = \cosh t = \frac{e^t + e^{-t}}{2}.$$

What is the probability law for G?

7. A discrete random variable W, whose range is a subset of the non-negative integers, has probability generating function

$$\psi(t) = \frac{t^2}{3} + \frac{t^4}{2} + \frac{t^8}{6}.$$

What is the probability function for W?

8. Find the cumulant generating function for the negative binomial probability law with parameters r, p, and use it to re-evaluate the mean and variance of the probability law.

9. If X is a random variable with cumulant generating function $c_X(t)$, find the cumulant generating function for $Y = a + bX$, where a, b are constants. Use this to show (again) that $\mu_Y = a + b\mu_X$ and that $\sigma_Y^2 = b^2 \sigma_X^2$.

10. Suppose X is a discrete random variable with factorial moment generating function $\psi_X(t)$. Let $Y = a + bX$, where a, b are constants, and express the factorial moment generating function for Y in terms of $\psi_X(t)$.

11. If the first three moments of a random variable V are 5, 27, and 155, evaluate the second and third moments of V about its mean.

12. If X is a random variable whose mean is μ_X and whose second and third moments about the mean are μ_2 and μ_3, express the second and third moments of X (about 0) in terms of μ_X, μ_2, μ_3 (the moments about the mean). (*Hint:* $x^3 \equiv ((x - \mu_X) + \mu_X)^3$.)

13. A random variable W has mean 5, while its second and third moments about the mean are 2 and 4; what are its first three moments m_1, m_2, m_3?

14. Granted $\psi_X(t) = t^{100}$ is the probability generating function for X, what is the probability law for X?

15. Suppose for a random variable X the first two cumulants are, say, 1 and 5, and all successive cumulants are 0. What must be the probability law for X?

16. *For the mathematically inclined* The kth moment of a random variable exists if and only if $E[|X|^k] < \infty$. Show that if the kth moment exists, then so does the mth moment, where $m < k$. (*Hint:* If $|X| < 1$, then $|X|^m < 1$, so $|X|^m < |X|^k + 1$. If $|X| > 1$, then $|X|^m < |X|^k$ for any $m < k$. Thus $|X|^m < |X|^k + 1$.)

17. If X has mean μ_X and variance σ_X^2, express the moment generating function for $Y = (X - \mu_X)/\sigma_X$, the standard form for X, in terms of $m_X(t)$.

18. Let $Y_1, Y_2, \ldots, Y_r$ be independent, geometric random variables, each with the same parameter p. Find the moment generating function for $X = \sum_{i=1}^r Y_i$. What is the probability law for X?

19. If $X_1, X_2, \ldots, X_n$ are independent Poisson random variables, with parameters $\mu_1, \mu_2, \ldots, \mu_n$, respectively, find the moment generating function for $W = \sum_{i=1}^n a_i X_i$, where $a_1, a_2, \ldots, a_n$ are constants, and identify its probability law.

20. Suppose $Y_1, Y_2, \ldots, Y_n$ are independent normal random variables, where the mean and variance for Y_i are μ_i, σ_i^2; find the moment generating function for $V = \sum_{j=1}^n a_j Y_j$, where $a_1, a_2, \ldots, a_n$ are constants, and identify its probability law.

21. Let $X_1, X_2, \ldots, X_n$ be independent Erlang random variables, where the parameters for X_i are r_i, λ_i. What must be true of the parameters for $\sum_{i=1}^{n} X_i$ to follow the Erlang probability law?

22. Suppose $X_1, X_2, \ldots, X_n$ are independent negative binomial random variables where the parameters for X_i are r_i, p_i. What must be true of these parameters for $\sum_{i=1}^{n} X_i$ to follow the negative binomial probability law?

23. If $U_1, U_2, \ldots, U_n$ are independent χ^2 random variables, where U_i has ν_i degrees of freedom, show that $V = \sum_{i=1}^{n} U_i$ has the χ^2 distribution with $\sum_{i=1}^{n} \nu_i$ degrees of freedom.

24. Let $U_1, U_2, \ldots, U_n$ be independent uniform $(0, 1)$ random variables, and find the probability law for $V = \ln(1/U_1^2 U_2^2 \cdots U_n^2)$. (*Hint:* Write $V = -2 \sum_{i=1}^{n} \ln U_i$. What is the probability law for $-2 \ln U_i$?)

6.3 | Conditional Expectation

Conditional probability is a very useful tool in modeling many real-world phenomena; closely allied with conditional probability is the idea of conditional expectation, averages taken with respect to a conditional probability law. Conditional expectation and some of its applications will be discussed in this section. The following example introduces the concept.

Example 6.20

Suppose we observe events whose occurrences through time satisfy the assumptions for a Poisson process with parameter λ. If we begin our observation at time $t = 0$ and let T_1 be the time the first event occurs and T_2 the time the second event occurs, then the marginal probability law for T_1 is exponential with mean $1/\lambda$. The conditional probability law for T_1, given $T_2 = t_2$, is uniform on $(0, t_2)$ as we saw earlier. The mean (or balance point) of this conditional *pdf* is $t_2/2$. This latter value is called the conditional expectation of T_1, given that $T_2 = t_2$; we shall use the stylus ($|$) with our expected value notation to indicate conditional expectation, as opposed to marginal, or unconditional, expectation. Thus for this case we shall write $E[T_1 \mid t_2] = \mu_{T_1 \mid t_2} = t_2/2$, to stress the fact that conditional probability is involved. For this same case we know that the marginal probability law for T_2 is Erlang with parameters $r = 2$ and λ, so the (unconditional) expected value for T_2 is $E[T_2] = 2/\lambda$. Recall also that the conditional probability law for T_2, given $T_1 = t_1$, is the shifted exponential with *pdf* $f_{T_2 \mid T_1}(t_2 \mid t_1) = \lambda e^{-\lambda(t_2 - t_1)}$ for $t_2 > t_1$; the mean of this *pdf* is $t_1 + 1/\lambda$. Thus $E[T_2 \mid t_1] = \mu_{T_2 \mid t_1} = t_1 + 1/\lambda$ using our conditional expectation notation. Notice that these conditional expectations do not equal the marginal expectations; this happens because T_1, T_2 are not independent random variables.

Our definition of conditional expectation follows.

DEFINITION 6.9 ───────────────

Let X_1, X_2 be jointly distributed random variables. The *conditional expectation* for a function $w(X_1, X_2)$, given $X_2 = x_2$, is

$$E[w(X_1, X_2) \mid x_2] = E[w(X_1, x_2) \mid x_2]$$

$$= \begin{cases} \displaystyle\sum_{x_1 \in R_{X_1}} w(x_1, x_2) p_{X_1 \mid X_2}(x_1 \mid x_2), & \text{if } X_1, X_2 \text{ is discrete,} \\[2ex] \displaystyle\int_{-\infty}^{\infty} w(x_1, x_2) f_{X_1 \mid X_2}(x_1 \mid x_2)\, dx_1, & \text{if } X_1, X_2 \text{ is continuous.} \end{cases}$$

The reason we have written $E[w(X_1, X_2) \mid x_2] = E[w(X_1, x_2) \mid x_2]$ in this definition is that any variable conditioned on becomes a fixed constant when finding the conditional expected value for any function $w(X_1, X_2)$. This point can be a little subtle; it says, for example, that $E[XY \mid x] \equiv x\, E[Y \mid x]$, or indeed, that $E[g(X)h(Y) \mid x] \equiv g(x)\, E[h(Y) \mid x]$ for any functions $g(\cdot)$ and $h(\cdot)$, since constants factor through expectations.

If X_1, X_2 are two jointly distributed random variables, the conditional expectation of X_1, given $X_2 = x_2$, is called the *conditional mean of X_1, given $X_2 = x_2$*: If X_1, X_2 are continuous, then

$$E[X_1 \mid x_2] = \mu_{X_1 \mid x_2} = \int_{-\infty}^{\infty} x_1 f_{X_1 \mid X_2}(x_1 \mid x_2)\, dx_1,$$

and measures the balance point for the conditional *pdf*. If X_1, X_2 are independent, then the marginal *pdf* for X_1 is identical to the conditional *pdf* for X_1, given *any* observed value x_2 for X_2, which then implies that the conditional mean for X_1 is the marginal mean μ_{X_1}. This equivalence may also occur even if X_1, X_2 are *not* independent; that is, the conditional expectation for X_1 may be μ_{X_1}, even though X_1, X_2 are not independent, as illustrated in the following example.

Example 6.21

In Chapter 5 we modeled the impact point for a dart thrown at a circular dart board with radius 1, using the uniform *pdf* $f_{X_1, X_2}(x_1, x_2) = 1/\pi$, for $x_1^2 + x_2^2 \leq 1$. The marginal *pdf* for X_1 is

$$f_{X_1}(x_1) = \frac{2}{\pi}\sqrt{1 - x_1^2}, \qquad \text{for } -1 < x_1 < 1.$$

It is easy to verify that $E[X_1] = 0$, since this *pdf* is symmetric about $x_1 = 0$. As verified earlier, X_1, X_2 are not independent and the conditional *pdf* for X_1, given $X_2 = x_2$, is uniform,

$$f_{X_1 \mid X_2}(x_1 \mid x_2) = 1/2\sqrt{1 - x_2^2}, \qquad \text{for } -\sqrt{1 - x_2^2} < x_1 < \sqrt{1 - x_2^2},$$

and is symmetric about $x_1 = 0$, for any $-1 < x_2 < 1$. The conditional mean for X_1 is then $0 = \mu_{X_1}$, no matter what the given value for X_2. X_1, X_2 are not independent, but $\mu_{X_1 \mid x_2} = \mu_{X_1}$ for any value of x_2.

As we have seen in earlier chapters, conditional probability proves useful and powerful in modeling many practical problems; the same is true for conditional expectation. The theorem of total probability (Theorem 1.7) states that the unconditional probability of any event A can be written

$$P(A) = \sum_{i=1}^{k} P(A \cap B_i),$$

if the events $B_1, B_2, \ldots, B_k$ partition the sample space S; we also know that $P(A \cap B_i) = P(B_i)P(A \mid B_i)$. Thus, as was shown in Chapter 1,

$$P(A) = \sum_{i=1}^{k} P(A \mid B_i)P(B_i); \tag{1}$$

that is, the unconditional probability of A occurring can be expressed as an average of the conditional probabilities of A occurring. (This is an averaging process since $P(B_i) \geq 0$, for all i, and $\sum_{i=1}^{k} P(B_i) = 1$ granted that $S = B_1 \cup B_2 \cup \cdots \cup B_k$.) This result has an exact counterpart in conditional expectations of functions, which holds for either discrete or continuous random variables; we shall explicitly discuss it only for the discrete case.

As given in Definition 6.9, the conditional expected value for $w(X_1, X_2)$, given $X_2 = x_2$, is $\mathrm{E}[w(X_1, x_2) \mid x_2] = g(x_2)$, which is a (possibly constant) function of x_2. We could then further evaluate the expected (average) value of this function, namely $\mathrm{E}[g(X_2)]$; this gives

$$\mathrm{E}[g(X_2)] = \mathrm{E}[\mathrm{E}[w(X_1, X_2) \mid X_2]] = \sum_{x_2 \in R_{X_2}} g(x_2) p_{X_2}(x_2)$$

$$= \sum_{x_2 \in R_{X_2}} \{\mathrm{E}[w(X_1, x_2) \mid x_2]\} p_{X_2}(x_2)$$

$$= \sum_{x_2 \in R_{X_2}} \left[\sum_{x_1 \in R_{X_1}} w(x_1, x_2) p_{X_1 \mid X_2}(x_1 \mid x_2) \right] p_{X_2}(x_2)$$

$$= \sum_{x_2 \in R_{X_2}} \left[\sum_{x_1 \in R_{X_1}} w(x_1, x_2) \frac{p_{X_1, X_2}(x_1, x_2)}{p_{X_2}(x_2)} \right] p_{X_2}(x_2)$$

$$= \sum_{x_2 \in R_{X_2}} \sum_{x_1 \in R_{X_1}} w(x_1, x_2) p_{X_1, X_2}(x_1, x_2) = \mathrm{E}[w(X_1, X_2)].$$

Thus the average value of a conditional expectation is the unconditional expectation, just as the unconditional probability of an event A occurring is the average of its conditional probabilities; this is summarized in the following theorem.

THEOREM 6.9 If X_1 and X_2 are jointly distributed random variables and $g(x_2) = \mathrm{E}[w(X_1, x_2) \mid x_2]$ is the conditional expectation of $w(X_1, X_2)$ given $X_2 = x_2$, then

$$\mathrm{E}[g(X_2)] = \mathrm{E}\left[\mathrm{E}[w(X_1, X_2) \mid X_2]\right] = \mathrm{E}[w(X_1, X_2)].$$

This result proves extremely useful in evaluating expected values, in any case in which it is easy to get the conditional expectation followed by the expected value of the resulting function. In particular, if X_1, X_2 are two jointly distributed random variables and $\mathrm{E}[X_1 \mid x_2]$ is the conditional mean of X_1, then the unconditional expectation of X_1 is given by $\mathrm{E}[\mathrm{E}[X_1 \mid X_2]] = \mathrm{E}[X_1]$; the expected value for the conditional mean is the unconditional mean of X_1. Notice that, if $\mathrm{E}[X_1 \mid x_2] = c$, a constant for any given x_2, it follows that $c = \mu_{X_1}$; that is, the value for this constant conditional mean must be the unconditional mean.

Example 6.22

Suppose the probability a witness to a certain type of crime will in fact report it to the police is .2; assume these instances (reporting of crime witnessed) are well modeled as independent Bernoulli trials with probability of success given by $p = .2$. Then the number of such crimes reported to the police is a binomial random variable Y. If on the average the police receive two such reports per month ($\mu_Y = 2$), what is the expected number of such crimes committed per month? If we let X be the number of crimes actually committed per month, then $\mathrm{E}[Y \mid x] = .2x$. Since $\mathrm{E}[\mathrm{E}[Y \mid X]] = \mathrm{E}[Y]$, it then follows that $\mathrm{E}[.2X] = .2\,\mathrm{E}[X] = \mathrm{E}[Y] = 2$; that is, $\mathrm{E}[X] = 10$.

The bivariate normal probability law was introduced earlier, in Section 5.3. Recall that we assumed X to be standard normal while the conditional probability law for Y given $X = x$ was normal with mean ρx and variance $1 - \rho^2$. Then $\mathrm{E}[Y \mid x] = \rho x$ implies that $\mathrm{E}[XY] = \mathrm{E}[X\,\mathrm{E}[Y \mid X]] = \mathrm{E}[\rho X^2] = \rho$ is the covariance between X and Y for this probability law. Since both marginals are standard normal (and their variances are thus 1), ρ is also the correlation between X and Y, as mentioned in Section 5.3.

Recall the covariance of two random variables has computational formula $\mathrm{Cov}[X_1, X_2] = \mathrm{E}[X_1 X_2] - \mathrm{E}[X_1]\,\mathrm{E}[X_2]$; since

$$\mathrm{E}[X_1 X_2] = \mathrm{E}[X_1\,\mathrm{E}[X_2 \mid X_1]] = \mathrm{E}[X_2\,\mathrm{E}[X_1 \mid X_2]],$$

$\mathrm{Cov}[X_1, X_2] = 0$ if *either* $\mathrm{E}[X_2 \mid x_1]$ or $\mathrm{E}[X_1 \mid x_2]$ is a constant (and as already noted, that constant value will then be the marginal mean). Thus the covariance (and correlation) between two random variables equals 0 if they are independent or (a weaker condition) if *either* of the conditional means is constant. As you will see in Exercises 6.3, it is quite easy to construct examples in which one of the two conditional means is constant but the other is not. This relationship is not symmetric; that is, it may be that the conditional mean for X_1, given $X_2 = x_2$, is constant but that the conditional mean for X_2, given $X_1 = x_1$, is not.

Suppose $X_1, X_2, \ldots, X_k$ is multinomial with parameters n, $p_1, p_2, \ldots, p_k$. Let us use conditional probability to evaluate the covariance of any two of these random variables X_i, X_j, where $i \neq j$. For convenience we shall let $j = k$ (so that the value for X_k is given), but the reasoning is easily seen to hold if we condition on any single random variable in $X_1, X_2, \ldots, X_k$. Recall that the marginal probability law for X_i is binomial with parameters n, p_i, for $i = 1, 2, \ldots, k$, so the variance of X_i is $np_i(1 - p_i)$. We saw earlier that the *conditional* probability law for $X_1, X_2, \ldots, X_{k-1}$, given $X_k = x_k$, is again multinomial, with parameters $n - x_k$, $p'_1, p'_2, \ldots, p'_{k-1}$, where $p'_i = p_i/(1 - p_k)$ (so $\sum_i p'_i = 1$). From this it follows that the conditional probability law for X_i, given $X_k = x_k$, is binomial with parameters $n - x_k$, and $p_i/(1 - p_k)$, and thus $\mathrm{E}[X_i \mid x_k] = (n - x_k)p_i/(1 - p_k)$. We then have $\mathrm{E}[X_i x_k \mid x_k] = x_k \, \mathrm{E}[X_i \mid x_k] = x_k(n - x_k)p_i/(1 - p_k)$, and the unconditional expected value is

$$\mathrm{E}[X_i X_k] = \mathrm{E}\left[X_k(n - X_k)\frac{p_i}{1 - p_k} \right]$$

$$= \frac{p_i}{1 - p_k} n(n - 1)p_k(1 - p_k)$$

$$= n(n - 1)p_i p_k.$$

Thus $\mathrm{Cov}[X_i, X_k] = n(n - 1)p_i p_k - (np_i)(np_k) = -np_i p_k$ and the correlation between X_i, X_k is then

$$\rho = \frac{-np_i p_k}{\sqrt{np_i(1 - p_i)np_k(1 - p_k)}} = -\sqrt{\frac{p_i p_k}{(1 - p_i)(1 - p_k)}}.$$

This negative correlation (and covariance) indicates an inverse or decreasing relation between X_i and X_k, which is not surprising. The larger the value of X_k, the more restricted the range of X_i. Notice, then, that if we perform n Bernoulli trials, where X_1 is the number of successes and $X_2 = n - X_1$ is the number of failures, then the covariance of X_1 and X_2 is $-np_1 p_2 = -np(1 - p)$, and the correlation coefficient is -1. Exercises 6.3 ask you to evaluate these same quantities for the multi-hypergeometric random vector.

The mean of the conditional probability law for Y, given $X = x$, measures the balance point of this conditional probability law and may or may not depend on the given value x. As we have seen, $\mathrm{E}[\mathrm{E}[Y \mid X]] = \mathrm{E}[Y]$, the expected value of the conditional mean is the unconditional mean value for Y. The *variability* of a conditional probability law is measured by its conditional variance:

$$\sigma^2_{Y \mid x} = \mathrm{Var}[Y \mid x] = \mathrm{E}[(Y - \mathrm{E}[Y \mid x])^2 \mid x].$$

Might one expect the average of the conditional variance to equal the unconditional variance, the property found for the conditional mean? Let us examine this question.

If we add and subtract the same constant a within the parentheses defining the conditional variance, its value remains unchanged:

$$\mathrm{Var}[Y \mid x] = \mathrm{E}[(Y - a + a - \mathrm{E}[Y \mid x])^2 \mid x]$$

$$= \mathrm{E}[(Y - a)^2 + 2(a - \mathrm{E}[Y \mid x])(Y - a) + (a - \mathrm{E}[Y \mid x])^2 \mid x]$$

$$= \mathrm{E}[(Y - a)^2 \mid x] - (a - \mathrm{E}[Y \mid x])^2, \tag{2}$$

since $(a - E[Y \mid x])$ is a constant, which factors through the expectation, and $E[Y - a \mid x] = -(a - E[Y \mid x])$. In particular, the constant a used in Eq. (2) could itself be the mean of Y (i.e., $a = \mu_Y$), which would result in the equation

$$\text{Var}[Y \mid x] = E[(Y - \mu_Y)^2 \mid x] - (\mu_Y - E[Y \mid x])^2. \qquad (3)$$

Note that the last term on the right in Eq. (3) is the square of the difference between the conditional mean and its own expectation, the unconditional mean. The expected value of this term is called the *variance of the conditional mean* and is denoted by $\text{Var}[E[Y \mid X]]$. Taking the expectation of Eq. (3) then gives

$$E[\text{Var}[Y \mid X]] = \text{Var}[Y] - \text{Var}[E[Y \mid X]]$$

or

$$\text{Var}[Y] = E[\text{Var}[Y \mid X]] + \text{Var}[E[Y \mid X]]. \qquad (4)$$

If Y and X are jointly distributed random variables, then the (marginal) variance of Y is given by the sum of the expected value of the conditional variance for Y plus the variance of the conditional expected value $E[Y \mid X]$, as summarized in the following theorem.

> **THEOREM 6.10** If X and Y are jointly distributed random variables, then
>
> $$E[Y] = E[E[Y \mid X]],$$
> $$\text{Var}[Y] = E[\text{Var}[Y \mid X]] + \text{Var}[E[Y \mid X]],$$
>
> as long as the expectations exist.

The following example illustrates these computations.

Example 6.23

The number of customers to arrive for teller service at a bank during a 1-hour period is a Poisson random variable X with parameter μ. Each customer either does or does not want to cash a personal check, independently, with probability p. Let X be the number of customers arriving for service, and let Y be the number wanting to cash a personal check. Then the probability law for Y, given $X = x$, is binomial with parameters x, p. The conditional expectation for Y, given $X = x$, is xp, and the conditional variance for Y is $xp(1 - p)$. From Theorem 6.10 the unconditional expectation for Y is

$$E[Y] = E[E[Y \mid X]] = E[Xp] = p\,E[X] = p\mu. \qquad (5)$$

Since

$$\text{Var}[Y] = \text{Var}[E[Y \mid X]] + E[\text{Var}[Y \mid X]], \qquad (6)$$

we also have

$$\text{Var}[Y] = \text{Var}[Xp] + E[Xp(1 - p)]$$
$$= p^2\,\text{Var}[X] + p(1 - p)\,E[X] = p\mu.$$

Note that Theorem 6.10 says that

$$E\left[\,\text{Var}[\,Y\mid X\,]\,\right] = \text{Var}[Y] - \text{Var}\left[\,E[\,Y\mid X\,]\,\right];$$

that is, the expected value of the conditional variance is the marginal variance minus the variance of the conditional mean. Thus the expected value of the conditional variance can be no larger than the marginal variance. If we let $E[\,Y\mid x\,] = g(x)$, then the variance of the conditional mean is $\sigma^2_{g(X)}$, the variance of the function $g(X)$; if $g(X)$ is constant, then the variance of $g(X)$ is 0 and the expected value of the conditional variance is the same as the marginal variance. If $g(X)$ is not a constant function but the conditional variance is a constant $c \geq 0$, then its value is given by

$$c = \sigma_Y^2 - \sigma^2_{g(X)} = \sigma_Y^2\left(1 - \frac{\sigma^2_{g(X)}}{\sigma_Y^2}\right).$$

Note that the conditional variance is smaller than the marginal variance by the factor $1 - \sigma^2_{g(X)}/\sigma_Y^2$, a fact with important implications in signal processing and methods of statistical inference.

Example 6.24

Suppose X_1, X_2 are jointly distributed random variables with identical marginal distributions; each has mean μ and variance σ^2, and the correlation between them is ρ (so *neither* conditional mean is constant if $\rho \neq 0$). Let us also assume that the conditional mean for X_2, given $X_1 = x_1$, is a *linear* function of x_1, that $E[X_2\mid x_1] = a + bx_1 = h(x_1)$, say. Certainly, the constants a, b must be related to the parameters of this joint probability law; let us find what their values are.

Taking the expected value of $a + bX_1$ gives $a + b\mu$, and the expected value of the conditional expectation is $E[X_2] = \mu$, so we must have $\mu = a + b\mu$, or $a = \mu(1 - b)$. Since $E[X_1X_2] - \mu^2 = \rho\sigma^2$ and

$$E[X_1X_2] = E[X_1\,h(X_1)] = E[X_1(a + bX_1)] = a\mu + b(\sigma^2 + \mu^2),$$

it follows that we must have $\mu^2 + \rho\sigma^2 = a\mu + b(\sigma^2 + \mu^2)$. Substituting the value for a into this equation and solving for b gives $b = \rho$, so $a = \mu(1 - \rho)$, and the conditional expectation for X_2 must be $h(x_1) = \mu(1 - \rho) + \rho x_1 = \mu + \rho(x_1 - \mu)$. The variance of the conditional mean is

$$\text{Var}\left[\,E[X_2\mid X_1]\,\right] = \text{Var}[h(X_1)] = \text{Var}[\mu(1 - \rho) + \rho X_1] = \rho^2\sigma^2,$$

so the expected value for the conditional variance is $\sigma^2 - \rho^2\sigma^2$. If this conditional variance is a constant, then its value must be $\sigma^2(1 - \rho^2)$. Note that this is exactly the structure for the bivariate normal where both means and variances are equal.

Example 6.25

Theorem 6.10 is very useful in modeling the mean and variance of a sum of a random number of random variables. Suppose we want to model the total dollar amount of sales made by a retail store on a given day; the number N

of such sales is clearly not a constant from day to day. Let X_i represent the dollar amount of sale i on this day. Granted N sales are made, the total dollar amount of the sales for this day is then $Y = \sum_{i=1}^{N} X_i$, which gives the total of a random number of random variables. To keep this model simple, let us assume that the individual dollar amounts, the X_i's, while random, each have the same mean μ and the same variance σ^2, and that the covariance between any two X_i's is 0 no matter what the value of N. The expected dollar total for the sales made on this day is then

$$E[Y] = E\left[\sum_{i=1}^{N} X_i\right].$$

If $N = n$, then the conditional mean of the total is

$$E[Y \mid n] = E\left[\sum_{i=1}^{n} X_i \,\middle|\, n\right] = \sum_{i=1}^{n} E[X_i] = n\mu;$$

the unconditional expected total of the sales for this day is then the expected value of this conditional expectation:

$$E[Y] = E[E[Y \mid N]] = E[N\mu] = \mu\,E[N].$$

Not surprisingly, this is the product of the expected number of sales to be made times the expected amount for each sale. The conditional variance of Y, given $N = n$, is

$$Var[Y \mid n] = E[(Y - n\mu)^2 \mid n] = E\left[\left[\sum_{i=1}^{n}(X_i - \mu)\right]^2 \,\middle|\, n\right]$$

$$= \sum_{i=1}^{n}\sum_{j=1}^{n} E[(X_i - \mu)(X_j - \mu) \mid n]. \tag{7}$$

Because of our assumption that the covariance of X_i and X_j, given $N = n$, is 0 for all $i \neq j$, only the terms in this sum for which $i = j$ will be nonzero. Thus Eq. (7) reduces to $Var[Y \mid n] = n\sigma^2$, where σ^2 is the common variance of the X_i's. The (marginal or unconditional) variance for Y is

$$Var[Y] = E[Var[Y \mid N]] + Var[E[Y \mid N]]$$
$$= E[N\sigma^2] + Var[N\mu] = \sigma^2\,E[N] + \mu^2\,Var[N]. \tag{8}$$

Notice that the variance of the total sales is larger because of the randomness of N; that is, if the number of sales were a constant n each day, then the variance of Y would simply be $n\sigma^2$ (because then $E[n] = n$ and $Var[n] = 0$). The second term in Eq. (8) reflects the inflation in the variance of Y caused by the randomness or variability in the number of sales.

 **EXERCISES 6.3**

1. Let X, Y be jointly distributed random variables, and assume that $E[Y \mid x] = a + bx$, where a and b are constants. Show that $\mu_Y = a + b\mu_X$.

2. A computer simulation requires knowledge of the times to failure of cer-

tain types of gear; when a failure time is required, the computer selects an observed value x from the probability law whose *pdf* is $f_X(x) = 2x$ for $0 < x < 1$. The reciprocal of this value, $1/x$, is then used as the parameter for the exponential distribution of Y, the time to failure; thus the conditional *pdf* for Y, given $X = x$, is

$$f_{Y|X}(y \mid x) = \frac{1}{x} e^{-y/x}, \qquad \text{for } y > 0.$$

The actual failure time employed is then selected from this (conditional) probability law for Y. Evaluate the expected value and the variance for Y.

3. An enthusiastic lottery player tosses a fair coin until the first head occurs; the (random) flip number N on which this first head occurs then gives the number of tickets he will buy. Assume that the tickets he buys are independent Bernoulli trials with parameter p (success meaning the ticket wins some prize), and let X be the number of winning tickets he buys. Evaluate the mean and variance for X.

4. The marginal *pdf* for X is

$$f_X(x) = bx, \qquad \text{for } 0 < x < \sqrt{\frac{2}{b}},$$

and the conditional *pdf* for Y, given $X = x$, is Erlang with parameters r and $\lambda = 1/x$. Thus the joint *pdf* for X, Y is

$$f_{X,Y}(x, y) = \frac{b y^{r-1}}{x^{r-1} \Gamma(r)} e^{-y/x}, \qquad \text{for } 0 < x < \sqrt{\frac{2}{b}}, \ y > 0,$$

where $r > 0$, and $b > 0$.
a. Evaluate $E[Y \mid x]$, and use this to find $E[XY]$.
b. Find the correlation between Y and X.

5. Let X, Y be jointly distributed random variables, and assume that $E[Y \mid x] = bx$, where $b \neq 0$ is a constant and $\mu_X \neq 0$.
a. Show that $b = \mu_Y / \mu_X$. b. Show that $E[Y/X] = E[Y]/E[X]$.

6. Assume that X, Y are jointly distributed random variables, and that $E[Y \mid x] = bx$, where $b \neq 0$ is a constant and $\mu_X \neq 0$. Show that the correlation between X and Y is $\rho = \mu_Y \sigma_X / \mu_X \sigma_Y$. (*Hint:* Consider $\text{Cov}[X, Y] = E[(X - \mu_X)Y]$.)

7. In a computer video game, you are to annihilate as many invaders as possible. Assume that the number of invaders presented to you per minute is a Poisson random variable X_1 with parameter μ; you have one opportunity to hit each one presented. Also assume that the number you hit, given x_1 are presented, is a binomial random variable X_2 with parameters x_1, p.
a. Evaluate $E[X_2]$, the number of invaders you hit (in a 1-minute game).
b. If you play a 3-minute game, evaluate the expected number you hit.

8. Let X_1 have density $2(1 - x_1)$, for $0 < x_1 < 1$, and let the conditional *pdf* for X_2, given $X_1 = x_1$, be uniform for $-1 + x_1 < x_2 < 1 - x_1$. Evaluate $E[X_2 \mid x_1], E[X_1 \mid x_2]$, and $\text{Cov}[X_1, X_2]$. This is a case for which $E[X_2 \mid x_1] = E[X_2]$ but $E[X_1 \mid x_2] \neq E[X_1]$.

9. Let $X_1, X_2, \ldots, X_k$ be multi-hypergeometric random variables with parameters m, $r_1, r_2, \ldots, r_k$, n. Evaluate the covariance and correlation between X_i, X_k, the ith and kth components of $X_1, X_2, \ldots, X_k$.

10. Suppose X, Y are jointly distributed random variables, with $E[X \mid y] = c$, where $c \neq 0$ is constant, and $E[Y \mid x] = g(x)$, a function of x. Show that

$$E[Y] = \frac{E[XY]}{E[X]} = E[g(X)].$$

11. Suppose X_1, X_2 are jointly distributed, each with mean μ, variance σ^2, correlation ρ. Also suppose that the conditional mean for X_2, given $X_1 = x_1$, is linear in x_1. Let $Y = X_1 + X_2$, and find the conditional mean for Y given $X_1 = x_1$. Also evaluate the expected value of the conditional variance for Y given $X_1 = x_1$.

12. Let X_1, X_2 be independent, each with mean μ and variance σ^2. Define $Y = X_1 X_2$, and find the conditional mean for Y, given $X_1 = x_1$. Also evaluate the expected value of the conditional variance of Y, given $X_1 = x_1$.

13. Let X_1, X_2, X_3 be identically distributed random variables, each with mean μ and variance σ^2; the correlation between each pair is ρ, and the conditional expectation of X_i, given $X_j = x_j$, is linear in x_j for all $i \neq j$. Define $Y = X_1 + X_2 + X_3$. Evaluate the conditional mean for Y, given $X_1 = x_1$, and find the expected value of the conditional variance for Y, given $X_1 = x_1$. Given $Y = y$, evaluate the conditional expectation for X_1, and find the expected value of the conditional variance for X_1.

14. Let X_1, X_2 be jointly distributed random variables, each with mean μ and variance σ^2; the correlation between them is ρ, and the conditional mean of X_2, given $X_1 = x_1$, is linear in x_1. Define $Y = X_1 - X_2$. Evaluate the conditional mean for Y, given $X_1 = x_1$, and find the expected value of the conditional variance for Y, given $X_1 = x_1$.

15. Show that if X, Y are jointly distributed, then

$$\text{Var}[Y] \geq \min(E[\text{Var}[Y \mid X]], \text{Var}[E[Y \mid X]]).$$

16. The number of sales made by a retail store per business day is a random variable N with mean 200 and standard deviation 20. For any given value of N, the dollar amounts of the individual sales are uncorrelated, each with mean 50 and standard deviation 10. What is the expected total dollar amount of the sales made by this store in 1 day? What is the standard deviation of the daily total dollar amounts of sales?

6.4 | Summary

The expected value of a function of two random variables is given by the following:

EXPECTED VALUE, DISCRETE

$$E[w(X_1, X_2)] = \sum_{(x_1, x_2) \in R_{X_1, X_2}} w(x_1, x_2) p_{X_1, X_2}(x_1, x_2)$$

EXPECTED VALUE, CONTINUOUS

$$E[w(X_1, X_2)] = \iint\limits_{(x_1, x_2) \in R_{X_1, X_2}} w(x_1, x_2) f_{X_1, X_2}(x_1, x_2)\, dx_1\, dx_2$$

The covariance of two random variables is $\mathrm{Cov}[X_1, X_2] = E[X_1 X_2] - E[X_1]\,E[X_2]$, a measure of how the two vary together. The correlation between X_1 and X_2 is $\rho = \mathrm{Cov}[X_1, X_2]/\sigma_1 \sigma_2$, which is dimensionless and lies between -1 and 1.

If a_i and b_i, where $i = 1, 2, \ldots, n$ are constants and $X_1, X_2, \ldots, X_n$ are jointly distributed random variables, and if we define two new random variables $Y = \sum_{i=1}^{n} a_i X_i$ and $Z = \sum_{i=1}^{n} b_i X_i$, then

---

$$E[Y] = \sum_{i=1}^{n} a_i\, E[X_i],$$

$$\mathrm{Var}[Y] = \sum_{i=1}^{n} a_i^2\, \mathrm{Var}[X_i] + 2 \sum_{i=1}^{n} \sum_{\substack{j=1 \\ i<j}}^{n} a_i a_j\, \mathrm{Cov}[X_i, X_j],$$

$$\mathrm{Cov}[Y, Z] = \sum_{i=1}^{n} a_i b_i\, \mathrm{Var}[X_i] + \sum_{i=1}^{n} \sum_{\substack{j=1 \\ i \neq j}}^{n} a_i b_j\, \mathrm{Cov}[X_i, X_j].$$

The expected value of a product of functions of independent random variables is equal to the product of their individual expectations. The covariance and correlation between independent random variables is necessarily 0.

The moment generating function for a random variable X is $m_X(t) = E[e^{tX}]$. The kth derivative of $m_X(t)$, evaluated at $t = 0$, gives the kth moment for X:

$$m_k = E[X^k] = m_X^{(k)}(t)|_{t=0}.$$

The moment generating function for $Y = a + bX$ is $m_Y(t) = e^{ta} m_X(tb)$. The probability generating function for a discrete random variable X with $R_X \subset \{0, 1, 2, \ldots\}$ is $\psi_X(t) = E[t^X]$; this function is also called the factorial moment generating function. Derivatives of $\psi_X(t)$ can be used to find the probability function for X and the factorial moments of X.

$$\psi_X^{(k)}(t)|_{t=0} = k! p_X(k), \qquad \psi_X^{(k)}(t)|_{t=1} = E[X(X-1)\cdots(X-k+1)],$$

The factorial moment generating function shares a simple relationship with the moment generating function.

$$\psi_X(t) = m_X(\ln t), \qquad \psi_X(e^t) = m_X(t).$$

The moment generating function for a linear function of independent random variables, $Y = \sum_{i=1}^{n} a_i X_i$, is $m_Y(t) = \prod_{i=1}^{n} m_{X_i}(a_i t)$.

The conditional expectation of $w(X_1, X_2)$, given $X_2 = x_2$, is its average with respect to the conditional probability law for X_1 given $X_2 = x_2$; this conditional expectation defines a function of the given variable x_2. The expectation of the conditional expectation is the unconditional expectation:

$$E[E[w(X_1, X_2)| X_2]] = E[w(X_1, X_2)].$$

If X_1, X_2 are jointly distributed, then $E[E[X_2 \,|\, X_1]] = E[X_2]$. The expected value of the conditional variance is given by the difference between the marginal variance and the variance of the conditional mean. The covariance and correlation between X_1, X_2 is 0 if *either* conditional mean is constant.

If X, Y are jointly distributed random variables, then

$$E[Y] = E[E[Y \,|\, X]]$$
$$\mathrm{Var}[Y] = E[\mathrm{Var}[Y \,|\, X]] + \mathrm{Var}[E[Y \,|\, X]],$$

as long as the expectations exist.

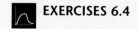 **EXERCISES 6.4**

1. X_1, X_2, X_3 are independent normal random variables, each with mean 5 and variance 25. Let $\bar{X}$ be their average.
   a. What is the probability law for $\bar{X}$?
   b. Evaluate $P(X_1 < 3)$ and $P(\bar{X} < 3)$.

2. X_1, X_2, X_3 are independent exponential random variables, each with mean 5. Let $\bar{X}$ be their average.
   a. What is the probability law for $3\bar{X}$?
   b. Evaluate $P(X_1 < 3)$ and $P(\bar{X} < 3)$.

3. An insurance company that insures automobiles receives N collision insurance claims in a given month; assume N is Poisson with parameter 200. The dollar amount of the ith claim is X_i, for $i = 1, 2, \ldots, N$. Assume that the X_i values are uncorrelated, each with mean 1000 and standard deviation 200, no matter what the value of N. The dollar total of the claims submitted in this month is then $T = \sum_{i=1}^{N} X_i$. Evaluate the mean and standard deviation of T.

4. Model the number of two-car head-on automobile collisions, on an overcrowded stretch of two-lane highway in a given year, as a Poisson random variable with $\mu = 3$. Also assume that the total number of persons in the two cars involved is a random variable with mean 3.5 for each collision, and that the persons in the cars are killed independently with probability .3. What is the expected number of fatalities for this stretch of road in a year?

5. The amount of annual rainfall, recorded in inches at a tree farm, is a random variable R with *pdf*

$$f_R(r) = \begin{cases} \dfrac{r-10}{75}, & \text{for } 10 < r < 20, \\ \dfrac{25-r}{37.5}, & \text{for } 20 < r < 25, \\ 0, & \text{otherwise.} \end{cases}$$

The increase in height of a "typical" tree on this farm (in inches), given $R = r$ inches of rainfall, is a random variable G whose mean is $E[G \mid r] = 3 + \frac{3}{2}r$ and whose variance is $\text{Var}[G \mid r] = r(60 - r)/100$. Evaluate $E[G]$, and $\text{Var}[G]$, the mean and variance of the "typical" increase in height.

6. The probability generating function for X is

$$\psi_X(t) = \frac{t(1 - t^6)}{6(1 - t)}.$$

What is the probability function for X?

7. Let X have the shifted exponential distribution with *pdf* $f_X(x) = \lambda e^{-\lambda(x-a)}$ for $x > a$, where a is a constant. Find the moment generating function for X.

8. The range for the alternative negative binomial random variable is the set of nonnegative integers, counting the number of failures *before* the rth success. Find the factorial moment generating function for this probability law.

9. The conditional probability function for X, given $N = n$, is binomial with parameters n, p, and the probability law for N is Poisson with parameter μ. Find the conditional factorial moment generating function for X, given $N = n$, and use this to evaluate $\psi_X(t)$, the factorial moment generating function for X.

10. *Poisson sampling* Each automobile crossing a bridge is classified as being red in color, or blue, or neither of these. Assume that the number of red automobiles to cross this bridge in the time interval $(0, t]$ is a Poisson random variable X_1 with parameter $\mu_1 t$; similarly, assume the numbers of automobiles (X_2, X_3) in the other two classes (blue or neither) to cross this bridge in the same time period are also Poisson random variables with parameters $\mu_2 t, \mu_3 t$, respectively. X_1, X_2, X_3 are independent.
   a. What is the joint probability function for X_1, X_2, X_3?
   b. What is the probability distribution for $Y = X_1 + X_2 + X_3$?
   c. Given that $Y = n$ automobiles crossed this bridge in this time period, what is the conditional probability function for X_1, X_2, X_3?

CHAPTER 7

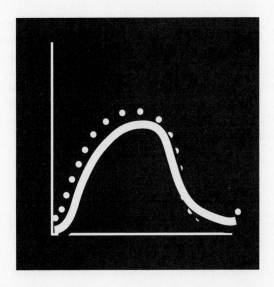

TRANSFORMATIONS
AND LIMIT THEOREMS

In Section 4.4 we considered the problem of finding the probability law for $g(X)$, a function of X, granted the distribution for X was known. This same type of problem will be revisited in Sections 7.1 and 7.2, in the context of two or more variables. For example, if $X_1, X_2, \ldots, X_5$ are independent uniform $(0, 1)$ random variables, what is the probability law for R, the difference between the largest and smallest of these five random variables? Or if Y_1, Y_2, Y_3 are independent exponential random variables, each with parameter 1, what is the probability law for the middle-ranked value? These questions can be investigated through the order statistics (ranked values) of the random variables; the *cdf* proves to be a useful tool in such problems. These order statistics will be explored in Section 7.1.

We have seen some results about the means and variances for linear functions of $X_1, X_2, \ldots, X_n$. If we observe X_1, X_2, say, we might want the probability law for their sum, or their difference, or their ratio. Section 7.2 will show how the *pdf* for the sum of two independent random variables is related to the individual *pdf*s of the two; the *pdf* for the ratio of two random variables is also derived. These results are useful in determining the probability laws for some commonly employed statistical techniques; both Student's T and Snedecor's F distributions are derived in this section.

Sequences of random variables and some limiting results will be discussed in Section 7.3. The Law of Large Numbers describes the limiting behavior of certain sequences of probability statements and is useful in justifying the relative frequency interpretation of probabilities. The Central Limit Theorem shows that the probability distribution of certain sequences of random variables converges to the standard normal. This behavior is the basis for using the normal to approximate the binomial and Poisson distributions as discussed in Chapter 4.

7.1 | Order Statistics

In Section 4.3 we introduced the idea of observing independent uniform $(0, 1)$ random variables and ranking these values from smallest to largest, denoted $U_{(1)}, U_{(2)}, \ldots, U_{(n)}$. These values are called *order statistics*, to reflect the fact that they are ranked values. We found that $U_{(i)}$ is a beta random variable with parameters $i, n - i + 1$.

We can extend the concept of order statistics to random variables that are not uniform: Let $X_1, X_2, \ldots, X_n$ be independent, continuous random variables, with the same *cdf* $F_X(t)$. Then the ranked values from smallest to largest, $X_{(1)}, X_{(2)}, \ldots, X_{(n)}$, are called the order statistics for a random sample of values from the probability law specified by $F_X(t)$. What is the probability law for $X_{(j)}$, the jth ranked value? We can derive this probability law from our knowledge of the distribution for $U_{(j)}$, the jth ranked value of n uniform $(0, 1)$ random variables.

Recall the probability integral transform from Section 4.4: If X is any continuous random variable with *cdf* $F_X(t)$, then $U = F_X(X)$ is uniform $(0, 1)$. If $X_1, X_2, \ldots, X_n$ are independent, continuous, then $U_1 = F_X(X_1), U_2 = F_X(X_2), \ldots, U_n = F_X(X_n)$ are independent uniform $(0, 1)$. Moreover, $U_{(j)} = F_X(X_{(j)})$; that is, the jth ordered uniform variable is given by $F_X(t)$ evaluated at $X_{(j)}$, the jth ordered value of $X_1, X_2, \ldots, X_n$. Conversely, $X_{(j)} = F_X^{-1}(U_{(j)})$, so we can derive the *cdf* for $X_{(j)}$ from the *cdf* for $U_{(j)}$. In fact, using the reasoning in Chapter 4,

$$F_{X_{(j)}}(t) = P(X_{(j)} \le t) = P(F_X^{-1}(U_{(j)}) \le t)$$
$$= P\left(U_{(j)} \le F_X(t)\right) = F_{U_{(j)}}\left(F_X(t)\right).$$

As we saw earlier, the *cdf* for $U_{(j)}$ is

$$F_{U_{(j)}}(t) = \sum_{i=j}^{n} \binom{n}{i} t^i (1 - t)^{n-i}, \qquad \text{for } 0 < t < 1,$$

so the *cdf* for $X_{(j)}$ is

$$F_{X_{(j)}}(t) = \sum_{i=j}^{n} \binom{n}{i} (F_X(t))^i (1 - F_X(t))^{n-i}, \tag{1}$$

for $0 < F_X(t) < 1$, that is for $t \in R_X$. The derivative of Eq. (1) gives the *pdf* for $X_{(j)}$:

$$\frac{d}{dt} F_{X_{(j)}}(t) = f_{X_{(j)}}(t) = n \binom{n-1}{j-1} f_X(t) (F_X(t))^{j-1} (1 - F_X(t))^{n-j}. \tag{2}$$

Note that the same term-to-term cancellation in taking the derivative occurs as in Section 4.3. This discussion is summarized in the following theorem.

THEOREM 7.1 If $X_1, X_2, \ldots, X_n$ are independent continuous random variables with the same *cdf* $F_X(t)$ and *pdf* $f_X(t)$, while $X_{(1)}, X_{(2)}, \ldots, X_{(n)}$ represent their ordered values, then the *cdf* for $X_{(j)}$ is

$$F_{X_{(j)}}(t) = \sum_{i=j}^{n} \binom{n}{i} (F_X(t))^i (1 - F_X(t))^{n-i},$$

for $j = 1, 2, \ldots, n$. The *pdf* for $X_{(j)}$ is given by

$$f_{X_{(j)}}(t) = n f_X(t) \binom{n-1}{j-1} (F_X(t))^{j-1} (1 - F_X(t))^{n-j}.$$

Example 7.1

Suppose $X_1, X_2, \ldots, X_n$ are independent random variables, each exponential with parameter λ; $X_{(1)}$ is the smallest or minimum of these random variables, and its *cdf* is given by

$$F_{X_{(1)}}(t) = \sum_{i=1}^{n} \binom{n}{i} (F_X(t))^i (1 - F_X(t))^{n-i},$$
$$= 1 - (1 - F_X(t))^n = 1 - e^{-n\lambda t},$$

the *cdf* for an exponential random variable with parameter $n\lambda$. Thus, if $X_{(1)}$ is the smallest of, say, $n = 10$ independent exponential random variables each with parameter λ, its probability law is also exponential and has parameter 10λ. The observed value for $X_{(1)}$ behaves just like an observed value for the time of the first occurrence in a Poisson process with parameter 10λ (i.e., a Poisson process with events occurring at $n = 10$ times the rate of the process governing the individual observations $X_1, X_2, \ldots, X_{10}$). The *cdf* of the largest, or maximum, of n exponential random variables is

$$F_{X_{(n)}}(t) = (F_X(t))^n = (1 - e^{-\lambda t})^n,$$

which is *not* the *cdf* for an exponential random variable. The *pdf* for $X_{(n)}$ is given by the derivative of this *cdf*:

$$f_{X_{(n)}}(t) = n\lambda e^{-\lambda t} (1 - e^{-\lambda t})^{n-1}.$$

Suppose a computer is used to generate $n = 7$ observed values for independent normal random variables, each with parameters $\mu = 3$ and $\sigma = 2$; none of the order statistics from a group of independent normal random variables has a normal probability law, as is easy to verify. The probability that any single one of these variables will exceed 4 is

$$1 - F_X(4) = P(X > 4) = P(Z > .5) = .3085,$$

where Z is standard normal. The probability that the *smallest* number generated exceeds 4 is then $1 - F_{X_{(1)}}(4) = (.3085)^7 = .0003$, and the probability that the *largest* number generated exceeds 4 is $1 - F_{X_{(7)}}(4) = 1 - (.6915)^7 = .9244$. The probability the "middle" number generated, $X_{(4)}$, exceeds 4 is

$$P(X_{(4)} > 4) = 1 - F_{X_{(4)}}(4)$$

$$= \sum_{i=0}^{3} \binom{7}{i} (.6915)^i (.3085)^{7-i} = .1373,$$

while the probability the third smallest $(X_{(3)})$ *does not* exceed 4 is

$$F_{X_{(3)}}(4) = \sum_{i=3}^{7} \binom{7}{i} (.6915)^i (.3085)^{7-i} = .9675.$$

The *pdf* for the jth order statistic $X_{(j)}$ from $X_1, X_2, \ldots, X_n$ can also be gotten using a simple heuristic line of reasoning. (This approach allows relatively easy evaluation of the *pdf* for two or more order statistics.) Recall that the *pdf* for a continuous random variable at any value x is proportional to the probability of getting an observed value in the neighborhood of x:

$$P(x - \tfrac{1}{2}\Delta x < X_{(j)} < x + \tfrac{1}{2}\Delta x) \doteq f_{X_{(j)}}(x)\Delta x.$$

The jth order statistic will equal a value in a neighborhood of x if and only if we find, among the n observed values,

1. an observed value in this neighborhood, and
2. exactly $j - 1$ observed values less than x, and
3. $n - j$ observed values greater than x.

The observed value in the neighborhood of x could be any of the n values; any of the remaining $j - 1$ observed values could be the ones smaller than x with the remaining $n - j$ values exceeding x. With $f_X(x)$ and $F_X(t)$ the common *pdf* and *cdf* for the random variables $X_1, X_2, \ldots, X_n$, the probability of getting an observed value in the neighborhood of x is $f_X(x)\Delta x$, the probability any one of them is smaller than x is $F_X(x)$, and the probability any one of them is larger than x is $1 - F_X(x)$. That is, the *pdf* for $X_{(j)}$ should be approximately

$$f_{X_{(j)}}(x)\Delta x \doteq n f_X(x)\,\Delta x \binom{n-1}{j-1} (F_X(x))^{j-1} (1 - F_X(x))^{n-j}.$$

Note that this yields our previous result, Eq. (2), where $j = 1, 2, \ldots, n$. The factor n represents the number of ways to select one of the n values to fall in the neighborhood of x; $f_X(x)\Delta x$ represents the probability of getting an observation in this neighborhood; and the remaining terms are simply the binomial probability of finding $j - 1$ values to the left, and $n - j$ values to the right, of x, assuming all the X_i's are continuous with the same probability law.

Example 7.2

Let X be triangular with *cdf* $F_X(t) = t^2$ for $0 < t < 1$ and *pdf* $f_X(t) = 2t$ for $0 < t < 1$. If X_1, X_2, X_3 are independent with this common probability law, the *pdf* for the middle value $X_{(2)}$ (where $n = 3$ and $j = 2$) is $f_{X_{(2)}}(t) = 3(2)(2t)(t^2)(1 - t^2) = 12t^3(1 - t^2)$. The mean for this random variable is then $\frac{24}{35}$, slightly larger than $\mu_X = \frac{2}{3}$.

The same heuristic used for determining the *pdf* of a single order statistic is also useful in finding the joint *pdf* for two or more order statistics; in the latter case we employ a multinomial counting random variable rather than the binomial. To illustrate this type of reasoning, assume again that $X_1, X_2, \ldots, X_n$ are independent continuous random variables with *pdf* $f_X(x)$ and *cdf* $F_X(t)$. Now let t_1, t_2 be any two fixed values such that $t_1 < t_2$. Each random variable can be converted to a multinomial trial by defining the three classes $\{ X \leq t_1 \}, \{ t_1 < X \leq t_2 \}$, and $\{ t_2 < X \}$; each observed X_i will then provide an outcome in (exactly) one of these classes. The probabilities of these classes occurring are given by

$$p_1 = P(X \leq t_1) = F_X(t_1),$$
$$p_2 = P(t_1 < X \leq t_2) = F_X(t_2) - F_X(t_1),$$
$$p_3 = P(X > t_2) = 1 - F_X(t_2).$$

Now let i and j be two integers between 1 and n inclusive such that $i < j$, and consider the joint *pdf* for $X_{(i)}, X_{(j)}$, the ith and jth order statistics. Recall that

$$f_{X_{(i)}, X_{(j)}}(t_1, t_2)\Delta t_1 \Delta t_2$$

is approximately equal to the probability of finding the ith order statistic in a neighborhood of t_1 and the jth order statistic in a neighborhood of t_2. Using the same procedure as before, any of the n random variables could be selected as the one to fall in the neighborhood of t_1, and any of the remaining $n - 1$ variables could be the one to fall in the neighborhood of t_2. Then we must also have exactly $i - 1$ of the remaining $n - 2$ observed values to the left of t_1, exactly $j - i - 1$ in the interval (t_1, t_2), and the remaining $n - j$ to the right of t_2. These latter three counts then have the multinomial distribution with parameters $n - 2, p_1, p_2, p_3$. This counting problem is illustrated in Fig 7.1.

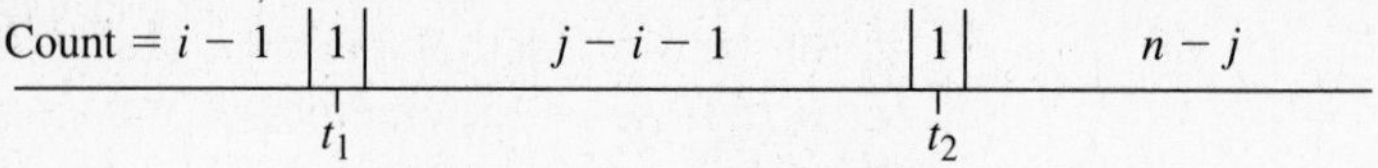

$$\text{Count} = i-1 \quad \boxed{1} \quad j-i-1 \quad \boxed{1} \quad n-j$$

$$t_1 \qquad\qquad\qquad\qquad t_2$$

Figure 7.1   Counts in classes

This then gives the joint *pdf* for $X_{(i)}, X_{(j)}$ to be

$$f_{X_{(i)},X_{(j)}}(t_1,t_2)\Delta t_1 \Delta t_2 = n f_X(t_1)\Delta t_1 (n-1) f_X(t_2) \Delta t_2$$
$$\times \begin{pmatrix} n-2 \\ i-1, j-i-1, n-j \end{pmatrix} (F_X(t_1))^{i-1} (F_X(t_2) - F_X(t_1))^{j-i-1} (1 - F_X(t_2))^{n-j}.$$

Example 7.3

Let $X_1, X_2, \ldots, X_n$ be independent uniform $(0,1)$ random variables. Then the joint *pdf* for $X_{(1)}$ and $X_{(n)}$, the smallest and largest of these random variables, is

$$f_{X_{(1)},X_{(n)}}(t_1,t_2) = n(n-1)1^2 \begin{pmatrix} n-2 \\ 0, n-2, 0 \end{pmatrix} t_1^0 (t_2 - t_1)^{n-2} (1 - t_2)^0$$
$$= n(n-1)(t_2 - t_1)^{n-2}, \qquad \text{for } 0 < t_1 < t_2 < 1. \qquad (3)$$

Then the probability the smallest random variable is no larger than .1 while the largest exceeds .9 is

$$P(X_{(1)} < .1, X_{(n)} > .9) = \int_0^{.1} \int_{.9}^1 n(n-1)(t_2 - t_1)^{n-2} \, dt_2 \, dt_1$$
$$= 1 + (.8)^n - 2(.9)^n.$$

With $n = 2$ this probability is .02, while with $n = 10$ it is .4100.

The range spanned by $X_1, X_2, \ldots, X_n$ is the difference between the largest and smallest values: $R = X_{(n)} - X_{(1)}$. This difference must be positive (since $X_{(n)} > X_{(1)}$). For $t > 0$, then, the *cdf* for R is

$$F_R(t) = P(R \le t) = P(X_{(n)} - X_{(1)} \le t) = P(X_{(n)} \le t + X_{(1)})$$
$$= \int_{-\infty}^{\infty} \int_{-\infty}^{t+t_1} f_{X_{(1)},X_{(n)}}(t_1,t_2) \, dt_2 \, dt_1. \qquad (4)$$

The *pdf* for R is then the derivative of Eq. (4):

$$f_R(t) = \int_{-\infty}^{\infty} f_{X_{(1)},X_{(n)}}(t_1, t + t_1) \, dt_1. \qquad (5)$$

This is illustrated below for n independent uniform $(0,1)$ random variables.

Example 7.4

Let $X_1, X_2, \ldots, X_n$ be independent uniform $(0,1)$. The joint *pdf* for the two extremes is then

$$f_{X_{(1)},X_{(n)}}(t_1,t_2) = n(n-1)(t_2 - t_1)^{n-2}, \qquad \text{for } 0 < t_1 < t_2 < 1,$$

from Eq. (3). The range is $R = X_{(n)} - X_{(1)}$, the difference between the largest and the smallest values. If the argument representing the smallest value, t_1 falls between 0 and $1 - t$, then the argument for the largest can vary from t_1 to $t + t_1$; if $1 - t < t_1 < 1$, the largest value can vary only from t_1 to 1 (for the joint *pdf* to remain positive). Thus, for $0 < t < 1$ the *cdf* for R, from Eq. (4), is

$$F_R(t) = P(R \le t) = P(X_{(n)} - X_{(1)} \le t) = P(X_{(n)} \le t + X_{(1)})$$

$$= \int_0^{1-t} \int_{t_1}^{t+t_1} n(n-1)(t_2 - t_1)^{n-2} \, dt_2 \, dt_1$$

$$+ \int_{1-t}^{1} \int_{t_1}^{1} n(n-1)(t_2 - t_1)^{n-2} \, dt_2 \, dt_1$$

$$= nt^{n-1}(1-t) + t^n,$$

and the *pdf* for R is given by the derivative

$$f_R(t) = \frac{dF_R(t)}{dt} = n(n-1)t^{n-2}(1-t), \qquad \text{for } 0 < t < 1.$$

Thus R is a beta random variable with $\alpha = n - 1$ and $\beta = 2$. Note that this probability law is the same as the probability law for $U_{(n-1)}$, the next-to-the-largest of n independent uniform $(0, 1)$ random variables. The expected value for the range is $\mu_R = (n-1)/(n+1)$, while its variance is $\sigma_R^2 = 2(n-1)/(n+1)^2(n+2)$.

EXERCISES 7.1

1. Let X be a random variable whose *pdf* is $f_X(t) = 2(1 - t)$ for $0 < t < 1$.
   a. What is the *cdf* for X?
   b. Find the *pdf* for $X_{(1)}$, the smallest of n independent random variables with this *pdf*.
   c. Find the *pdf* for $X_{(n)}$, the largest of n independent random variables with this *pdf*.

2. Assume that $R_1, R_2, \ldots, R_n$ are independent, each with *pdf*

$$f_R(r) = \begin{cases} re^{-r^2/2}, & \text{for } r > 0, \\ 0, & \text{for } r \le 0. \end{cases}$$

   a. Find the *pdf*s for the minimum value $R_{(1)}$ and the maximum value $R_{(n)}$ of these random variables.
   b. Evaluate the $100k$th quantiles for $R_{(1)}$ and $R_{(n)}$.
   c. Which of these two probability laws is the more variable, with $n = 10$, as measured by their interquartile ranges?

3. Assume $T_1, T_2, \ldots, T_n$ are independent random variables, each with *cdf*

$$F_T(t) = \begin{cases} 1 - \dfrac{1}{t}, & \text{for } t > 1, \\ 0, & \text{for } t \le 1. \end{cases}$$

   a. Find the *cdf*s for the minimum $T_{(1)}$ and the maximum $T_{(n)}$ of these random variables.

b. Evaluate the medians for these two extreme values, and their interquartile ranges.

4. Given that $R_1, R_2, \ldots, R_n$ are independent, each with *cdf*

$$F_R(t) = \begin{cases} 1 - e^{-t^2/2}, & \text{for } t > 0, \\ 0, & \text{for } t \le 0, \end{cases}$$

evaluate the probability that any single one of these random variables will lie between 1 and 2. Also determine the probability that the middle ranked value of $n = 5$ random variables will lie in this same interval.

5. If the random variable T has *pdf*

$$f_T(t) = \begin{cases} \dfrac{1}{t^2}, & \text{for } t > 1, \\ 0, & \text{for } t \le 1, \end{cases}$$

evaluate the probability that T lies between 4 and 5. Also find the probability that $T_{(4)}$ lies in this interval, given that $T_1, T_2, \ldots, T_6$ are independent with this same probability law.

6. A computer simulation, while running, generates the observed values for 11 independent exponential random variables, each with $\lambda = .1$. If these 11 numbers were ranked in order of magnitude, evaluate the probability that the middle value lies between 7 and 11.

7.2 | Functions of Random Variables

In Chapter 4 we examined some results useful in deriving the probability law of a function of a single random variable, and saw some applications of those ideas. There are also many applied problems requiring the probability distribution of a single function of n jointly distributed random variables $X_1, X_2, \ldots, X_n$, and some that require the joint distribution for several such functions of the same random variables. A number of techniques are useful in deriving such distributions, several of which will be discussed in this section.

If $X_1, X_2, \ldots, X_n$ are discrete, the probability law for any function of $X_1, X_2, \ldots, X_n$ again simply requires mapping the probability content at the individual points in $R_{X_1, X_2, \ldots, X_n}$ into the corresponding values of the function, totaling as necessary from the Theorem of Total Probability. The following example illustrates this procedure.

Example 7.5

A small chain of computer stores maintains four retail outlets and a single warehouse facility. Table 7.1 shows the number of orders the warehouse will receive from these outlets for a particular system, in a given week.

The warehouse's policy for deciding when it needs to replenish its supply of

Table 7.1

Probabilities of Orders from Four Outlets

	Probability of Ordering		
Outlet Number	*0*	*1*	*2*
1	.1	.8	.1
2	.2	.3	.5
3	.4	.2	.4
4	.3	.3	.4

these systems, to provide reliable and timely support of the stores, depends on the numbers of orders it receives from these outlets each week. Thus it would be of interest for the warehouse to know the probability law for the *total* number of orders it will receive per week for this system. If we formally let X_1, X_2, X_3, X_4 represent the numbers of orders the warehouse will receive for this type of system from the retail outlets in a given week, then $Y = X_1 + X_2 + X_3 + X_4$ gives the total number of orders for this system in the given week. If we further assume that X_1, X_2, X_3, X_4 are independent random variables, then their joint probability function is given by the product of the marginal probability functions for all possible observed values. The assumed information given in Table 7.1 specifies these marginal probability functions and allows us to find the probability law for Y using the theorem of total probability.

Clearly, with the given joint probability law for X_1, X_2, X_3, X_4, the range for $Y = \sum_{i=1}^{4} X_i$ is $R_Y = \{0, 1, 2, \ldots, 8\}$. The only way that Y can equal 0 is for *all* four X_i values to equal 0 (meaning that *none* of the four stores places an order), so $P(Y = 0) = (.1)(.2)(.4)(.3) = .0024$. Similarly, the only way that Y can equal 8 is for each of the X_i values to equal 2, giving $P(Y = 8) = (.1)(.5)(.4)(.4) = .008$. The other possible values for Y can occur in more than one way; their probabilities of occurrence are given by summing the probabilities of occurrence of all 4-tuples (x_1, x_2, x_3, x_4) for which $x_1 + x_2 + x_3 + x_4 = y \in R_Y$. Thus, for example,

$$P(Y = 1) = (.8)(.2)(.4)(.3) + (.1)(.3)(.4)(.3) + (.1)(.2)(.2)(.3) + (.1)(.2)(.4)(.3)$$
$$= .0264.$$

The total probability that $Y = 2$ is the sum of 10 terms, four of which are given by observing a single $x_i = 2$ with the others 0, and six of which have two of the x_i values equal to 1 with the other two equal to 0. Similarly, $P(Y = 3)$ is the sum of 16 terms, $P(Y = 4)$ of 19 terms, $P(Y = 5)$ of 16 terms, $P(Y = 6)$ of 10 terms, and $P(Y = 7)$ of 4 terms. The probability function for Y is given in Table 7.2.

This summation of probability values over the 4-tuples $(x_1, x_2, x_3, x_4) = y$ is a straightforward but tedious procedure.

Table 7.2

Probability Function for Y

$y =$	0	1	2	3	4	5	6	7	8
$P(Y = y) =$	.0024	.0264	.0782	.1760	.2300	.2476	.1526	.0788	.0080

The following example again illustrates the transfer of probability from a discrete vector-valued random variable to several functions of the elements of this vector; sometimes the summing of values for individual *n*-tuples can be avoided.

Example 7.6

Suppose a person is in the habit of purchasing five scratch-off lottery tickets each week; also assume the probability that each such ticket will win some prize is p and that individual tickets are well modeled as independent Bernoulli trials (with success meaning that *some* prize is won). If X_i is the number of winning tickets she purchases in week i, then this structure implies that X_i is binomial, with parameters 5 and p, for $i = 1, 2, \ldots, n$. The individual random variables $X_1, X_2, \ldots, X_n$ are then independent with joint probability function

$$p_{X_1, X_2, \ldots, X_n}(x_1, x_2, \ldots, x_n) = p_X(x_1) p_X(x_2) \cdots p_X(x_n), \tag{1}$$

where

$$p_X(x) = \binom{5}{x} p^x (1 - p)^{5-x}, \qquad \text{for } x = 0, 1, \ldots, 5 \tag{2}$$

is the binomial probability function. Now let us define the random variable Y_j to be the number of weeks (out of the n weeks) that this person buys j *winning* tickets, for $j = 0, 1, \ldots, 5$. Each of $Y_0, Y_1, \ldots, Y_5$ is then a function of $X_1, X_2, \ldots, X_n$. What is the joint probability law for $Y_0, Y_1, \ldots, Y_5$?

In a formal sense, the elements of $R_{X_1, X_2, \ldots, X_n}$ are n-tuples, each entry being one of the integers $0, 1, \ldots, 5$, locating 6^n separate points in an n-dimensional space. The amount of probability at each point is given by the product of binomial values, as specified by Eqs. (1) and (2). For each n-tuple in $R_{X_1, X_2, \ldots, X_n}$ then, $Y_0, Y_1, \ldots, Y_5$ simply gives the counts of the numbers of times each of these possible integers occurs. From the theorem of total probability we could then derive the joint probability function for $Y_0, Y_1, \ldots, Y_5$ from the joint probability function for $X_1, X_2, \ldots, X_n$ by summing the values of $p_{X_1, X_2, \ldots, X_n}(x_1, x_2, \ldots, x_n)$ over all the observed values $x_1, x_2, \ldots, x_n$ that give the same $y_0, y_1, \ldots, y_5$.

An easier approach to finding the probability function for $Y_0, Y_1, \ldots, Y_5$ is given by realizing that *each week* this person purchases five tickets defines a multinomial trial with $k = 6$ different outcomes, the number of winning tickets that she gets that week. Thus, from the binomial probability function, the probabilities of occurrence of the six different multinomial classes each week are given by

$$p_0 = (1-p)^5, \qquad p_1 = 5p(1-p)^4, \qquad p_2 = 10p^2(1-p)^3,$$
$$p_3 = 10p^3(1-p)^2, \qquad p_4 = 5p^4(1-p), \qquad p_5 = p^5.$$

$Y_0, Y_1, \ldots, Y_5$ is simply multinomial with parameters n (the number of weeks played) and $p_0, p_1, \ldots, p_5$:

$$p_{Y_0, Y_1, \ldots, Y_5}(y_0, y_1, \ldots, y_5) = \binom{n}{y_0, \ y_1, \ldots, y_5} p_0^{y_0} p_1^{y_1} \cdots p_5^{y_5},$$

for $y_i = 0, 1, \ldots, n$, where $i = 0, 1, \ldots, 5$, and $\sum_{i=0}^{5} y_i = n$.

If we assume the probability any ticket purchased will win some prize is $p = \frac{1}{9}$, and want to model the behavior for $Y_0, Y_1, Y_2, \ldots, Y_5$ for an $n = 8$-week period, then $Y_0, Y_1, Y_2, \ldots, Y_5$ has parameters $n = 8$ and p values given by .55493, .34683, .08671, .01084, .00068, .00002, for the different possible outcomes $x = 0, 1, \ldots, 5$ each week. The probability this person wins no prizes at all (so $Y_0 = 8$ and $Y_1 = Y_2 = \cdots = Y_5 = 0$) is then

$$p_{Y_0, Y_1, \ldots, Y_5}(8, 0, 0, 0, 0, 0) = \binom{8}{8, \ 0, \ 0, \ 0, \ 0, \ 0}(.55493)^8 = .00899,$$

while the probability she wins no prizes in 5 weeks and exactly one prize in each of the other 3 weeks is $\binom{8}{5, 3, 0, 0, 0, 0}(.55493)^5(.34683)^3 = .12295$.

For discrete $X_1, X_2, \ldots, X_n$ the more-or-less direct transfer of probability from the probability contents of the elements of $R_{X_1, X_2, \ldots, X_n}$ to the elements of the function(s) of interest is typically the simplest procedure to employ. Suppose X_1, X_2 are jointly continuous and we want to find the probability law for $Y = g(X_1, X_2)$, a function of the two random variables. One descriptor for the probability law for Y, as we know, is given by its *cdf* $F_Y(t) = P(Y \le t)$, for any real t. Thus we want to evaluate $P(Y \le t) = P(g(X_1, X_2) \le t)$ to find the probability law for Y. The statement $g(X_1, X_2) \le t$ defines a region in the (x_1, x_2)-plane for any fixed t, and we can then evaluate this probability by integrating the joint *pdf* for X_1, X_2 over this region.

One way of organizing (and thinking about) the evaluation of this probability is to use the conditional *pdf* for X_1, given $X_2 = x_2$ (or vice versa). That is, we might first evaluate the conditional *cdf* for Y given $X_2 = x_2$:

$$F_{Y|X_2}(t \mid x_2) = \int_{g(x_1, x_2) \le t} f_{X_1|X_2}(x_1 \mid x_2) \, dx_1. \tag{3}$$

This will in general be a function of the given value x_2. Then the expected value of Eq. (3) is

$$E[F_{Y|X_2}(t \mid X_2)] = \int_{-\infty}^{\infty} F_{Y|X_2}(t \mid x_2) f_{X_2}(x_2) \, dx_2$$
$$= \int_{-\infty}^{\infty} \left\{ \int_{g(x_1, x_2) \le t} f_{X_1, X_2}(x_1, x_2) \, dx_1 \right\} dx_2 = F_Y(t),$$

since the conditional *pdf* is the ratio of the joint *pdf* to the marginal *pdf*, and this marginal cancels in the double integral. As is the case for conditional expected values in general, taking expectations destroys the conditioning. This approach proves straightforward for many common problems.

Let us illustrate this approach in deriving the probability law for the sum of two independent continuous random variables, one of the most frequently employed transformations. Thus we desire the probability law for $Y = X_1 + X_2$, where X_1, X_2 are independent random variables, with *pdf* s $f_{X_1}(x_1)$ and $f_{X_2}(x_2)$. Since X_1, X_2 are independent, the conditional probability law for X_1 is identical with its marginal probability law, and the conditional *cdf* for Y, given $X_2 = x_2$, is

$$F_{Y \mid X_2}(t \mid x_2) = P(X_1 + x_2 \leq t) = P(X_1 \leq t - x_2) = F_{X_1}(t - x_2),$$

where $F_{X_1}(t)$ is the *cdf* for X_1. Thus the *cdf* for Y is

$$F_Y(t) = E[F_{X_1}(t - X_2)]$$
$$= \int_{-\infty}^{\infty} \left\{ \int_{-\infty}^{t-x_2} f_{X_1}(x_1)\, dx_1 \right\} f_{X_2}(x_2)\, dx_2.$$

The *pdf* for Y is given by the derivative of $F_Y(t)$:

$$f_Y(t) = \frac{dF_Y(t)}{dt} = \int_{-\infty}^{\infty} f_{X_1}(t - x_2) f_{X_2}(x_2)\, dx_2. \tag{4}$$

Equation (4) is called the *convolution formula*, and the *pdf* $f_Y(t)$ is called the *convolution* of the two functions $f_{X_1}(x_1)$ and $f_{X_2}(x_2)$; it plays a role in several areas of mathematics. The following example illustrates use of the convolution formula.

Example 7.7

If X_1, X_2 are independent standard normal random variables, then we can easily derive the *pdf* for their sum $Y = X_1 + X_2$ by using the convolution formula, Eq. (4):

$$f_Y(t) = \int_{-\infty}^{\infty} f_{X_1}(t - x_2) f_{X_2}(x_2)\, dx_2$$
$$= \int_{-\infty}^{\infty} \frac{1}{\sqrt{2\pi}} e^{-(t-x_2)^2/2} \frac{1}{\sqrt{2\pi}} e^{-x_2^2/2}\, dx_2$$
$$= \frac{1}{2\sqrt{\pi}} e^{-t^2/4} \int_{-\infty}^{\infty} \frac{1}{\sqrt{\pi}} e^{-(x_2 - t/2)^2}\, dx_2. \tag{5}$$

The last line results from completing the square in the exponent, giving

$$f_Y(t) = \frac{1}{2\sqrt{\pi}} e^{-t^2/4}. \tag{6}$$

Equation (6) is the density of a normal random variable with mean 0 and variance 2. The final integral in Eq. (5) equals 1 since it represents the total area under a normal *pdf* with mean $t/2$ and variance $\frac{1}{2}$.

You must exercise care in employing the convolution formula if the joint *pdf* is not positive over the whole (x_1, x_2)-plane. Suppose that X_1, X_2 are independent uniform random variables on the interval $(0, 1)$, and we again desire the probability law for $Y = X_1 + X_2$. The joint *pdf* for X_1, X_2 is the product of the two marginals, and thus equals 1 for $0 < x_1 < 1$ and $0 < x_2 < 1$. Thus the product $f_{X_1}(t - x_2)f_{X_2}(x_2)$ is positive only for

$$0 < t - x_2 < 1 \quad \text{and} \quad 0 < x_2 < 1. \tag{7}$$

The first inequality in Eq. (7) says $x_2 < t < 1 + x_2$, which implies the total limits for t must be the extremes for x_2 and $1 + x_2$, namely, 0 to 2 with $0 < x_2 < 1$. (This also follows from considering R_Y.) Equation (7) also says, then, that the product of the two densities equals 1 only for $0 < x_2 < t$ if $0 < t < 1$, and for $t - 1 < x_2 < 1$ if $1 < t < 2$. Thus the *pdf* for $Y = X_1 + X_2$ is

$$f_Y(t) = \begin{cases} \displaystyle\int_0^t dx_2 = t, & \text{if } 0 < t < 1, \\[2mm] \displaystyle\int_{t-1}^1 dx_2 = 2 - t, & \text{if } 1 < t < 2. \end{cases}$$

This *pdf* for the sum of two independent $(0, 1)$ random variables is triangular over the interval $(0, 2)$.

It is instructive to derive the *cdf* for $Y = X_1 + X_2$, where X_1, X_2 are independent uniform $(0, 1)$, by referring to Fig. 7.2. The event $\{Y \le t\} = \{X_1 + X_2 \le t\}$ defines the points in the triangle given on the left of the figure, for any $0 < t \le 1$. Thus the probability of this event is given by the volume under the joint *pdf* over this triangle; since the joint *pdf* is constant (with value 1 over the range), this volume in turn is simply the *area* of this triangle. That is, for $0 < t \le 1$, $F_Y(t) = P(Y \le t) = t^2/2$. For $1 < t < 2$, the region of interest is pictured on the right of the figure; the area of this shaded trapezoid is 1 minus the area of the upper (unshaded) triangle. Thus, for $1 < t < 2$, $F_Y(t) = 1 - (2 - t)^2/2$. You can easily see that the derivative of $F_Y(t)$ gives the above *pdf* found from the convolution formula.

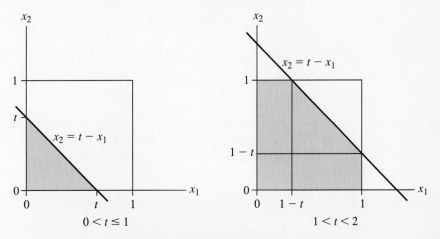

Figure 7.2 Regions for $F_Y(t)$

This conditional reasoning is equally simple in deriving the probability law for the ratio of two continuous random variables. Assume that X_1, X_2 are jointly distributed continuous random variables and that we would like to derive the probability law for $Y = g(X_1, X_2) = X_1/X_2$. Given that $X_2 = x_2$ (where $x_2 \neq 0$), the event $\{Y \leq t\}$ is the same as the event $\{X_1/x_2 \leq t\}$. The latter event is equivalent to $\{X_1 \leq tx_2\}$ if the given value is positive, and is equivalent to $\{X_1 \geq tx_2\}$ if the given value is negative. Thus the conditional *cdf* for Y, given $X_2 = x_2$, is

$$P(Y \leq t \mid x_2) = P\left(\frac{X_1}{x_2} \leq t \mid x_2\right)$$

$$= \begin{cases} P(X_1 \leq tx_2 \mid x_2), & \text{if } x_2 > 0, \\ P(X_1 \geq tx_2 \mid x_2), & \text{if } x_2 < 0. \end{cases} \tag{8}$$

If $x_2 > 0$, then Eq. (8) results in

$$P(Y \leq t \mid x_2) = F_{X_1 \mid X_2}(tx_2 \mid x_2),$$

so

$$\frac{d}{dt}P(Y \leq t \mid x_2) = x_2 f_{X_1 \mid X_2}(tx_2 \mid x_2).$$

If $x_2 < 0$, then Eq. (8) gives

$$P(Y \leq t \mid x_2) = 1 - F_{X_1 \mid X_2}(tx_2 \mid x_2),$$

so

$$\frac{d}{dt}P(Y \leq t \mid x_2) = -x_2 f_{X_1 \mid X_2}(tx_2 \mid x_2).$$

Note that, regardless of the value of x_2, we have

$$\frac{d}{dt}P(Y \leq t \mid x_2) = |x_2| f_{X_1 \mid X_2}(tx_2 \mid x_2). \tag{9}$$

The expected value of this conditional *pdf*, Eq. (9), gives the *pdf* for $Y = X_1/X_2$:

$$f_Y(t) = \frac{d}{dt}P(Y \leq t) = E\left[\frac{d}{dt}P(Y \leq t \mid X_2)\right]$$

$$= \int_{-\infty}^{\infty} |x_2| f_{X_1 \mid X_2}(tx_2 \mid x_2) f_{X_2}(x_2)\, dx_2$$

$$= \int_{-\infty}^{\infty} |x_2| f_{X_1, X_2}(tx_2, x_2)\, dx_2.$$

As with the convolution formula when evaluating $f_Y(t)$, consider carefully the region over which the joint *pdf* for X_1, X_2 is positive, since this region may depend on t.

Example 7.8

If Z_1, Z_2 are independent standard normal random variables, what is the probability law for their ratio $Y = Z_1/Z_2$? Because of the independence assumption, the joint *pdf* for Z_1, Z_2 is

$$f_{Z_1, Z_2}(z_1, z_2) = \frac{1}{2\pi}e^{-(z_1^2 + z_2^2)/2}$$

for all real (z_1, z_2), From the above discussion, the *pdf* for Y is then given by

$$f_Y(t) = \int_{-\infty}^{\infty} |z_2| \frac{1}{2\pi} e^{-(t^2+1)z_2^2/2} \, dz_2 \tag{10}$$

$$= \frac{1}{\pi} \int_0^{\infty} z_2 e^{-(t^2+1)z_2^2/2} \, dz_2$$

$$= -\frac{1}{\pi} \frac{1}{(t^2+1)} e^{-(t^2+1)z_2^2/2} \Big|_0^{\infty}$$

$$= \frac{1}{\pi(t^2+1)},$$

for any real t. Note that the integrand in Eq. (10) is a symmetric function. Thus the integral from $-\infty$ to ∞ is twice the value given by integrating from 0 to ∞. Thus $f_Y(t)$ is the *pdf* for a *Cauchy random variable*.

Now let U_1, U_2 be independent uniform random variables, each on the interval $(0, 1)$, and consider the *pdf* for their ratio $Y = U_1/U_2$. Clearly the range for Y is $R_Y = \{y : y > 0\}$, so $f_Y(t)$ is positive only for $t > 0$. The joint *pdf* for U_1, U_2 is positive only when both arguments lie between 0 and 1; that is, $f_{U_1, U_2}(tu_2, u_2) = 1$ only for $0 < tu_2 < 1$ *and* $0 < u_2 < 1$. The first of these inequalities requires that $u_2 < 1/t$, and the second requires that $u_2 < 1$, so this joint *pdf* is constant at 1 for $0 < u_2 < \min(1/t, 1)$. Thus, if $0 < t < 1$, then $1/t > 1$, and the joint *pdf* equals 1 for $0 < u_2 < 1$; if $t \geq 1$, then $f_{U_1, U_2}(tu_2, u_2) = 1$ only for $0 < u_2 < 1/t$. The *pdf* for $Y = U_1/U_2$, is then

$$f_Y(t) = \begin{cases} \displaystyle\int_0^1 u_2 \, du_2 = \frac{1}{2}, & \text{for } 0 < t < 1, \\ \displaystyle\int_0^{1/t} u_2 \, du_2 = \frac{1}{2t^2}, & \text{for } 1 \leq t. \end{cases} \tag{11}$$

The *pdf* for the ratio of two uniform random variables is itself constant (or uniform) for values between 0 and 1, and then declines like the reciprocal of $2t^2$.

As with the convolution (sum) of two independent uniform random variables, the *cdf* for the ratio $Y = U_1/U_2$ can be gotten quite simply by directly integrating the joint *pdf* for U_1, U_2 over the appropriate regions; the derivative then gives the *pdf* for Y. Since each of the uniform random variables ranges over the interval $(0, 1)$, the range for Y is $R_Y = \{y > 0\}$, the positive half of the real line. For $0 < t \leq 1$ the event $\{Y \leq t\}$ is equivalent to $\{U_1/U_2 \leq t\}$, which in turn is the region of points for which $\{U_1 \leq tU_2\}$. This region (and the joint range for U_1, U_2) is shaded in the left-hand plot in Fig. 7.3. Thus, for $0 < t \leq 1$, $F_Y(t)$ is given by the area of the shaded triangle (since the joint *pdf* is equal to 1 only for points in the square); that is, $F_Y(t) = t/2$. For $t > 1$, the corresponding region is shaded in the right-hand plot; again the *cdf* is the shaded area, or 1 minus the unshaded area, so $F_Y(t) = 1 - 1/2t$. You can easily verify that the derivative of $F_Y(t)$ gives Eq. (11).

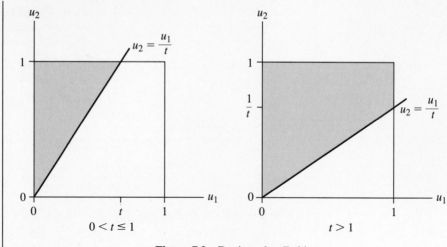

Figure 7.3   Regions for $F_Y(t)$

Several common procedures of statistical inference employ ratios of observed values of random variables; quantiles of the probability laws of such ratios are used in making inferences about the population of interest. One of the most frequently occurring of these ratios is a standard normal random variable divided by the square root of an independent χ^2 random variable over its degrees of freedom. Let us now derive the probability law for such a ratio. That is, if Z is a standard normal random variable, and W is an independent χ^2 random variable with k degrees of freedom, what is the probability law for

$$T = \frac{Z}{\sqrt{W/k}} = \frac{Z}{V},$$

where $V = \sqrt{W/k}$? First, since V is a monotonic increasing function of W, its *cdf* is given by $F_V(v) = F_W(kv^2)$ for $v > 0$. Thus the *pdf* for V is $f_V(v) = 2kvf_W(kv^2)$. Granted Z and W are independent, it follows that Z and V are as well. The joint *pdf* for Z and V is then simply

$$f_{Z,V}(z,v) = f_Z(z)f_V(v) = 2kvf_Z(z)f_W(kv^2),$$

for $-\infty < z < \infty, v > 0$. The *pdf* for $T = Z/V$ then is given by

$$f_T(t) = \int_0^\infty |v|2kvf_Z(tv)f_W(kv^2)\,dv$$

$$= \frac{\Gamma\left(\dfrac{k+1}{2}\right)}{\Gamma\left(\dfrac{k}{2}\right)\sqrt{k\pi}}\frac{1}{\left(1+\dfrac{t^2}{k}\right)^{(k+1)/2}}, \tag{12}$$

for any real t.

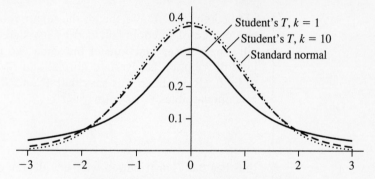

Figure 7.4 Student's T and normal *pdf*s

Equation 12 is called *Student's T density with k degrees of freedom*, the same parameter as the χ^2 random variable in the denominator; it is also a particular case of the Cauchy probability law (evaluate this density for $k = 1$). The graph of this *pdf* is a symmetric bell-shaped curve, centered at $t = 0$, that has "thicker tails" than does the standard normal, for any value of k. Figure 7.4 plots the standard normal *pdf* together with the T *pdf*s for $k = 1$ and $k = 10$ degrees of freedom. As $k \to \infty$, Student's T probability law converges to the standard normal. Since Z and W are independent, the expected value of T^m is

$$\mathrm{E}[T^m] = \left(\sqrt{k}\right)^m \mathrm{E}[Z^m] \, \mathrm{E}\left[\frac{1}{\sqrt{W^m}}\right] = k^{m/2} \, \mathrm{E}[Z^m] \, \mathrm{E}[W^{-m/2}], \qquad (13)$$

as long as the expectations exist. Since W is a χ^2 random variable with k degrees of freedom, it is also gamma with $n = k/2$ and $\lambda = \frac{1}{2}$. As mentioned earlier, then $\mathrm{E}[W^{-m/2}]$ exists as long as $-m/2 > -k/2$ (i.e., for powers m such that $m < k$). Thus, if T has one degree of freedom, $\mathrm{E}[T]$ does not exist (nor do any higher-order moments), and if T has two degrees of freedom, $\mathrm{E}[T^2]$ does not exist (nor does $\mathrm{Var}[T]$). With two or more degrees of freedom, the mean of T is 0 (since $\mathrm{E}[Z] = 0$). With three or more degrees of freedom, $\mathrm{Var}[T]$ exists and equals $k/(k-2) > 1$; this variance is larger than the standard normal variance because of the thicker tails of the T *pdf*.

Example 7.9

This T distribution is used frequently in constructing *confidence intervals* and *testing hypotheses* in problems of statistical inference. Selected quantiles of the T distributions are commonly published with texts on statistical inference. Suppose Z_1, Z_2 are independent standard normal random variables, and we let Y be their sum and let X be their difference. What is the probability that $|Y| < |X|$? This probability can be evaluated using the T distribution with $k = 1$ degrees of freedom, recognizing the fact that the event $\{|Y| < |X|\}$ is equivalent to the event $\{|Y|/|X| < 1\}$. We know that $Y = Z_1 + Z_2$ is normal with mean 0 and variance 2, so $Y/\sqrt{2}$ is also standard normal; $X = Z_1 - Z_2$ is also normal with mean 0 and variance 2, and $X/\sqrt{2}$ is standard normal. It follows then that $X^2/2$ is a χ^2 random variable with $k = 1$ degree of freedom, and $|X|/\sqrt{2}$ then is the square root of a χ^2 random variable (with one degree of freedom) divided by its degrees of freedom. Since Y and X are simple linear functions of the same two normal random variables, their joint distribution is bivariate normal and their correlation is easily verified to be 0. (Remember, this says that Y and X are then independent normal random variables.) The ratio

$$T = \frac{Y/\sqrt{2}}{|X|/\sqrt{2}} = \frac{Y}{|X|}$$

thus has the T distribution with one degree of freedom. From Eq. (12), the *pdf* for T is

$$f_T(t) = \frac{1}{\pi}\frac{1}{(1+t^2)}, \qquad \text{for } -\infty < t < \infty. \tag{14}$$

Equation (14) is directly integrable, and its *cdf* is

$$F_T(t) = \frac{1}{2} + \frac{1}{\pi}\tan^{-1}t,$$

the same as the *pdf* for the ratio of two standard normal random variables discussed in Example 7.8. (Can you see why?) Thus we have

$$P(|Y| < |X|) = P\left(\frac{|Y|}{|X|} < 1\right) = P\left(-1 < Y/|X| < 1\right) = P(-1 < T < 1)$$

$$= F_T(1) - F_T(-1) = \frac{1}{\pi}(\tan^{-1}(1) - \tan^{-1}(-1))$$

$$= \frac{1}{\pi}\left(\frac{\pi}{4} - \left(-\frac{\pi}{4}\right)\right) = \frac{1}{2}.$$

An additional ratio frequently employed in certain types of statistical applications involves independent χ^2 random variables. The probability law for such a ratio is called *Snedecor's F distribution*, so named by G. W. Snedecor, who helped popularize the use of many statistical procedures in the United States, in honor of Sir R. A. Fisher, the originator of much of the currently used statistical methodology. Let W_1, W_2 be independent χ^2 random variables with degrees of freeedom k_1, k_2, respectively, and define the ratio

$$\mathcal{F} = \frac{W_1/k_1}{W_2/k_2}.$$

This random variable has the F distribution with parameters k_1, k_2, again called degrees of freedom; its *pdf* is easily derived using our earlier discussion for a ratio. Let $f_{W_1}(w_1), f_{W_2}(w_2)$ represent the χ^2 *pdf*s. If we define $V_i = W_i/k_i$, a simple linear function of W_i, it is easy to see that the *pdf* for V_i is then $f_{V_i}(v_i) = k_i f_{W_i}(k_i v_i)$ for $i = 1, 2$, and $\mathcal{F} = V_1/V_2$. The *pdf* for $\mathcal{F}$ is then

$$f_{\mathcal{F}}(t) = \int_0^\infty v_2 k_1 k_2 f_{W_1}(tk_1 v_2) f_{W_2}(k_2 v_2)\, dv_2$$

$$= \frac{k_1}{k_2} \frac{\Gamma\left(\dfrac{k_1 + k_2}{2}\right)}{\Gamma\left(\dfrac{k_1}{2}\right)\Gamma\left(\dfrac{k_2}{2}\right)} \frac{\left(\dfrac{k_1}{k_2} t\right)^{(k_1 - 2)/2}}{\left(1 + \dfrac{k_1}{k_2} t\right)^{(k_1 + k_2)/2}}, \tag{15}$$

for any $t > 0$.

Example 7.10

The most frequent usage of this F distribution involves certain sums of squares of independent standard normal random variables. (Recall that such sums of squares follow χ^2 probability laws.) The exponential probability law, with $\lambda = \frac{1}{2}$, is the same as the χ^2 probability law with two degrees of freedom. It follows that if X_1, X_2 are independent, exponential, each with $\lambda = \frac{1}{2}$, they are also independent χ^2 random variables, each with two degrees of freedom. The ratio $Y = X_1/X_2$ then has the F distribution with degrees of freedom $k_1 = 2$ and $k_2 = 2$. (Note that because these degrees of freedom are equal, they cancel in forming the ratio.) That is, the *pdf* for Y is simply $f_Y(t) = 1/(1 + t)^2$ for $t > 0$, the same as the *pdf* for F given by Eq. (15) with $k_1 = k_2 = 2$. Then

$$P(X_1 > 3X_2) = P(Y > 3) = \int_3^\infty \frac{1}{(1 + t)^2} \, dt = \frac{1}{4}.$$

An $\mathcal{F}$ random variable is the ratio of two independent χ^2 random variables, each divided by its degrees of freedom, so we have $\mathcal{F} = k_2 W_1 / k_1 W_2$, where k_1, k_2 are the degrees of freedom for the numerator and denominator random variables, respectively. Because of the independence, it follows that

$$E[\mathcal{F}^m] = \frac{k_1^m}{k_2^m} E[W_1^m] E\left[\frac{1}{W_2^m}\right] = \frac{k_1^m}{k_2^m} E[W_1^m] E[W_2^{-m}],$$

as long as the expectations exist. As mentioned earlier in discussing Eq. (13) with the random variable T, $E[W_2^{-m}]$ exists as long as $-m > -k_2/2$. Thus, the mean $E[\mathcal{F}]$ exists as long as $1 < k_2/2$, that is, provided that $k_2 > 2$; this expected value is $\mu_{\mathcal{F}} = k_2/(k_2 - 2)$. The variance of $\mathcal{F}$ exists as long as $k_2 > 4$ and is given by

$$\sigma_{\mathcal{F}}^2 = 2k_2^2(k_1 + k_2 - 2)/k_1(k_2 - 2)^2(k_2 - 4).$$

For $k_1 > 2$ the *pdf* for $\mathcal{F}$ goes through a maximum at $(k_1 - 2)k_2/k_1(k_2 + 2)$, a value that is always less than 1, so the "most likely" observed value for an $\mathcal{F}$ random variable is smaller than 1, while its mean is always greater than 1 (when it exists).

It is easy to see that if $\mathcal{F}$ has the F distribution with degrees of freedom k_1, k_2, then its reciprocal $1/\mathcal{F}$ again has the F distribution, now with k_2, k_1 degrees of freedom. This result follows immediately from the fact that $1/\mathcal{F}$ again is the ratio of two independent χ^2 random variables over their degrees

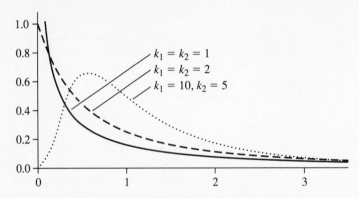

Figure 7.5 *pdf*s for Snedecor's F distribution

of freedom, except that now the numerator and denominator have been interchanged. (Note that the order of listing of the two degrees of freedom is important for this distribution; the first degree of freedom listed comes from the χ^2 random variable in the numerator.)

The *square* of Student's T random variable with k degrees of freedom,

$$T^2 = \frac{Z^2}{W/k},$$

has the F distribution with $k_1 = 1$, $k_2 = k$ degrees of freedom. This follows from the facts that Z^2 (the square of a standard normal random variable) has the χ^2 distribution with one degree of freedom, and Z^2 and W are independent random variables.

If $\mathcal{F}$ has the F distribution with k_1, k_2 degrees of freedom, then

$$X = \frac{k_1 \mathcal{F}}{k_2 + k_1 \mathcal{F}}$$

has the beta distribution with parameters $\alpha = k_1/2$ and $\beta = k_2/2$; that is, the beta probability law is a simple transform of the F distribution, and vice versa. You will be asked to verify this statement in Exercises 7.2. Figure 7.5 presents graphs of the $\mathcal{F}$ *pdf* for certain combinations of degrees of freedom.

EXERCISES 7.2

1. Suppose a pair of friends play the following game: Each of the two flip a fair coin simultaneously, one time after another. The game ends when the first head occurs; if only one of the two has a head on this flip, he is the winner. If both of them get their first head on the same flip, then the game ends in a tie.
   a. What is the probability law for the occurrence of the first head (perhaps simultaneously for both flippers)?
   b. Evaluate the probability that this game ends in a tie (i.e., that both get their first head on the same flip).

2. On any given day, a small business has three copies of item 1 for sale and four copies of item 2 for sale. There are thus $4 \times 5 = 20$ possibilities

for the numbers of sales that could be made each day (the elements of $\{0,1,2,3\} \times \{0,1,2,3,4\}$). Assume that these possibilities are all equally likely to occur each day. Also assume that the profit made on each sale of item 1 is \$10, while the profit made on each item of type 2 is \$30; let P be the amount of profit made on these two items per day. Find the probability function for P.

3. Assume that solar flares of a specified magnitude during a period of high activity occur like events in a Poisson process at a rate of 10 per 24-hour day; let $Y_1, Y_2, \ldots, Y_{14}$ be the daily counts to occur over a 2-week period.
   a. What is the joint probability law for $Y_1, Y_2, \ldots, Y_{14}$?
   b. Let X_{10} be the number of these days with exactly 10 solar flares. What is the probability law for X_{10}?
   c. Now let X_j be the number of these days with exactly j solar flares for $j = 8, 10, 12$. What is the joint probability law for X_8, X_{10}, X_{12}?

4. Let X_1, X_2 be independent exponential random variables, each with the same parameter λ. Evaluate the *pdf* for $Y = X_1 + X_2$ as the convolution of their *pdf*s.

5. The convolution of $f_{X_1}(x_1), f_{X_2}(x_2)$ gives the *pdf* for $Y = X_1 + X_2$, as mentioned in the text. How must the integral defining $f_Y(t)$ be changed if X_1, X_2 are not independent?

6. Assume that X_1, X_2 are independent lognormal random variables, and that $\ln X_i$ has mean μ_i, and variance σ_i^2 for $i = 1, 2$. What is the probability law for the product $Y = X_1 X_2$? (*Hint*: Recall that $\ln X_i$ is normal.)

7. Find the probability law for the product $U_1 U_2$ of two independent random variables, each uniform on the interval $(0, 1)$.

8. *Only for those who like integration challenges* Let U_1, U_2, U_3 be independent uniform on the interval $(0, 1)$, and find the *pdf* for $X = U_1 + U_2 + U_3$. (*Hint*: This can be written $X = Y + U_3$ where $Y = U_1 + U_2$. Consider evaluating $F_X(t)$.)

9. The designs of many types of equipment include redundancy of parts or systems to extend the usable lifetime of the item. One type of redundancy is described by placing parts in *parallel*, as pictured here for a device with two components; the device remains operable as long as *either* of the two parts (labeled 1 and 2) is still working.

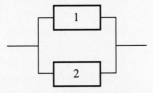

Now suppose that the times to failure for components 1 and 2 are exponential random variables T_1 and T_2 with parameters $\lambda = 1$ and $\lambda = .5$, respectively, and that T_1, T_2 are independent. Let D be the time to failure for the device and find the probability law for D.

10. Many devices are constructed with *serial* rather than parallel connections, as pictured here for a device with two components:

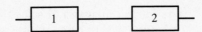

This device will fail when *either* of the two components fails, that is, when the first failure occurs. Assume the time to failure for each of these two components is exponential with parameter λ, and that their times to failure are independent. What is the probability law for D, the time to failure for the device?

11. Suppose X_1 is the time of occurrence of the first event in a Poisson process with parameter λ, starting from time $t = 0$, and X_2 is the time of occurrence of the second event. The conditional *pdf* for X_2, given $X_1 = x_1$, is then the shifted exponential $f_{X_2 \mid X_1}(x_2 \mid x_1) = \lambda e^{-\lambda(x_2 - x_1)}$ for $x_2 > x_1$, as we have seen earlier. Find the probability law for $Y = X_2/X_1$.

12. Assume that $\mathcal{F}$ is an F random variable with degrees of freedom k_1, k_2, and show that

$$X = \frac{k_1 \mathcal{F}}{k_2 + k_1 \mathcal{F}}$$

is a beta random variable with parameters $\alpha = k_1/2$ and $\beta = k_2/2$.

7.3 | Some Limit Theorems

Consider an experiment with sample space S and some event $A \subset S$ whose probability of occurrence is $p = P(A)$ where $0 < p < 1$; that is, A is not certain to occur, and its probability of occurrence is positive. Let us also assume the experiment in question can be repeated independently any number of times with the same value of p for each performance. Three examples of situations we might consider modeling are: (1) the occurrences of defective items coming off a production line (where A refers to the occurrence of a defective item, and defective items are assumed to occur with fixed probability p); (2) the times to failure produced by a given design for a piece of equipment (where $A = \{T \geq t\}$ refers to the failure time being at least some fixed time t, with the same $p = P(A)$ for each device produced); (3) the number of customers to arrive for service at a repair facility in a 1-hour period (where $A = \{N > 0\}$ refers to the number of arrivals being positive, again with $p = P(A)$ fixed for each hour).

A sequence of independent performances of such an experiment can easily be represented by a sequence of independent Bernoulli random variables, where $X_i = 1$ if event A occurs on repetition i of the experiment, and $X_i = 0$ if not, for $i = 1, 2, 3 \ldots, n, n + 1, \ldots$. For any fixed n, then, $T_n = \sum_{i=1}^{n} X_i$ counts the total number of times that event A occurred and $\bar{X}_n = T_n/n$ gives the relative frequency, or *proportion*, of the time that A occurred in the first n repetitions. Granted that the probability of an event is meant to behave like a relative frequency, it would seem reasonable to expect the observed value for $\bar{X}_n$ to be "close" to p, the probability of A occurring. In fact, as the number n of repetitions increases, it seems increasingly likely we should find $\bar{X}_n$ "close" to p. There are many ways in which this convergence of $\bar{X}_n$ to p can be quantified, some of which will be presented in this section.

As above, assume $X_1, X_2, \ldots, X_n, \ldots$ is a sequence of independent Bernoulli random variables, each with the same parameter p; define the derived sequence of totals $T_1, T_2, \ldots, T_n, \ldots$ and the sequence of proportions $\bar{X}_1, \bar{X}_2, \ldots, \bar{X}_n, \ldots$. For any fixed n the random variable $T_n = \sum_{i=1}^{n} X_i$ is binomial with parameters n, p, and $\bar{X}_n = T_n/n$ is a constant times this binomial. Thus

$$E[\bar{X}_n] = \frac{1}{n} E[T_n] = \frac{1}{n} np = p,$$

the same value for all n; the distributions for $X_1, X_2, \ldots, X_n, \ldots$ all have the same mean. The variance of $\bar{X}_n$ is

$$\text{Var}[\bar{X}_n] = \frac{1}{n^2} V[T_n] = \frac{1}{n^2} np(1-p) = \frac{p(1-p)}{n}.$$

The variance of the proportion $\bar{X}_n$ thus decreases as n increases, for any value for p, and in fact, $\lim_{n\to\infty} E[(\bar{X}_n - p)^2] = 0$. Recalling that for any fixed n, $E[(\bar{X}_n - p)^2] = E[\bar{X}_n^2] - p^2$, we also know that $\lim_{n\to\infty} E[\bar{X}_n^2] = p^2$. This convergence of the expected value of $\bar{X}_n^2$ to p^2 as $n \to \infty$ is called *convergence in mean square*. This convergence in mean square is sufficient (but not necessary) for the (weak) Law of Large Numbers to hold, as we shall now discuss.

Recall the Chebyshev inequality:

$$P(|Y - \mu_Y| < k\sigma_Y) \geq 1 - \frac{1}{k^2}, \tag{1}$$

for any random variable Y, where $k > 1$ is any constant. Replacing Y by $\bar{X}_n$ gives

$$P\left(|\bar{X}_n - p| < k\sqrt{\frac{p(1-p)}{n}}\right) \geq 1 - \frac{1}{k^2}. \tag{2}$$

By letting

$$k\sqrt{\frac{p(1-p)}{n}} = \epsilon$$

so

$$k = \epsilon\sqrt{\frac{n}{p(1-p)}},$$

we have

$$P\left(|\bar{X}_n - p| < \epsilon\right) \geq 1 - \frac{p(1-p)}{n\epsilon^2}. \tag{3}$$

Since any probability is bounded by 1, we then have

$$\lim_{n\to\infty} P(|\bar{X}_n - p| < \epsilon) \geq \lim_{n\to\infty} 1 - \frac{p(1-p)}{n\epsilon^2} = 1, \tag{4}$$

for any value of $\epsilon > 0$ no matter how small. This sequence of probability statements converges to 1. The probability that $\bar{X}_n$ differs from p by less than $\epsilon = .001$ (or .000001, or any small number) approaches 1 as $n \to \infty$.

This same argument can be made for any sequence of independent random variables $X_1, X_2, \ldots, X_n, \ldots$ with finite positive variances. Let $X_1, X_2, \ldots, X_n, \ldots$ be independent random variables, each with the same mean μ and the same variance $\sigma^2 > 0$, where $n > 0$ is a fixed integer. Then the *average* $\bar{X}_n = \frac{1}{n}\sum_{i=1}^{n} X_i$ has expected value μ and variance

$$\sigma_{\bar{X}_n}^2 = \frac{\sigma^2}{n}.$$

Thus we also have convergence in mean square for this more general situation, $\lim_{n\to\infty} E[\bar{X}_n^2] = \mu^2$. As before, choosing $\epsilon = k/\sigma_{\bar{X}_n} = k\sqrt{n}/\sigma$ and applying the Chebyshev inequality then gives

$$\lim_{n\to\infty} P(|\bar{X}_n - \mu| < \epsilon) \geq \lim_{n\to\infty} 1 - \frac{\sigma^2}{n\epsilon^2} = 1$$

for any $\epsilon > 0$. This result is known as the (weak) Law of Large Numbers and is summarized in the following theorem.

THEOREM 7.2 If $X_1, X_2, \ldots, X_n$, are independent random variables, each with the same mean μ and the same variance σ^2, then

$$\lim_{n\to\infty} P(|\bar{X}_n - \mu| < \epsilon) = 1,$$

for $\epsilon > 0$, where

$$\bar{X}_n = \frac{1}{n}\sum_{i=1}^{n} X_i.$$

It is important to think carefully about what this type of limiting argument says (and does not say); it literally describes the limit of a sequence of probability statements,

$$\ldots, P(|\bar{X}_{n-1} - \mu| < \epsilon),\ P(|\bar{X}_n - \mu| < \epsilon),\ P(|\bar{X}_{n+1} - \mu| < \epsilon),\ \ldots$$

where $\epsilon > 0$ is any fixed quantity, as small as one likes. With the assumptions we have made, the limiting value for this sequence of probabilities is necessarily 1 for any value of ϵ. This does not, however, imply that every possible observed sequence of $\bar{X}_1, \bar{X}_2, \ldots, \bar{X}_n, \ldots$ values must converge to μ, the common mean. As n increases, the number of possible observed sequences increases very rapidly; even though the proportion of sequences with $|\bar{X}_n - \mu| < \epsilon$ (expressed as $P(|\bar{X}_n - \mu| < \epsilon)$) goes to 1, there can be many different observed sequences that do not share this property.

The (weak) Law of Large Numbers is called *convergence in probability*, describing the convergence of a sequence of probability statements. Although beyond the level of this text, it can be shown that a sequence $X_1, X_2, \ldots, X_n, \ldots$ of independent random variables with the same mean μ and variance σ^2

will also satisfy the *(strong) Law of Large Numbers*. This states that

$$P\left(\lim_{n\to\infty}\bar{X}_n = \mu\right) = 1;$$

in other words, the probability of finding the sequence of averages $\bar{X}_n$ actually converging to the constant μ, as $n\to\infty$, is 1. If the strong law holds, so must the weak, but not necessarily vice versa. This strong law says that only those observed sequences that in fact converge to μ are assigned positive probability (and in total they account for the full measure of 1). The probability assigned to the totality of all sample paths that deviate from this limiting behavior is 0.

A different type of convergence for sequences of random variables is called *convergence in distribution*, based on the *Lévy continuity theorem*, stated here without proof.

THEOREM 7.3 Let $X_1, X_2, \ldots, X_n, \ldots$ be a sequence of random variables, where X_n has *cdf* $F_n(t)$ and moment generating function $m_n(t)$, $n = 1, 2, 3, \ldots$

1. If $\lim_{n\to\infty} F_n(t) = F(t)$, for every t at which $F(t)$ is continuous, and $F(t)$ is the *cdf* for some random variable X, then

$$\lim_{n\to\infty} m_n(t) = m(t),$$

where $m(t)$ is the moment generating function of the random variable X.

2. If $\lim_{n\to\infty} m_n(t) = m(t)$, where $m(t)$ is continuous at $t = 0$, then

$$\lim_{n\to\infty} F_n(t) = F(t),$$

where $F(t)$ is the *cdf* for a random variable X with moment generating function $m(t)$.

The second part of this theorem, saying that the convergence of the sequence of moment generating functions is sufficient for the convergence of the *cdf*s (or the actual probability laws), is quite powerful in examining the limiting behavior of probability laws.

Example 7.11

A software company analyst assumes that the numbers of orders her firm will receive for one of their products, per month over the coming months, will generate a sequence of Poisson random variables $X_1, X_2, \ldots, X_n, \ldots$. She also expects that sales of this product will increase geometrically, which she will represent by changes in the mean value for X_n. Specifically she assumes the *expected* sales for the first month to be $E[X_1] = \mu$, for the second month $E[X_2] = \mu + r\,E[X_1] = \mu(1 + r)$, for the third month $E[X_3] = \mu + r\,E[X_2] = \mu(1 + r + r^2)$, and so on, giving

$$E[X_n] = \mu + r\,E[X_{n-1}] = \mu(1 + r + r^2 + \cdots + r^{n-1})$$

$$= \frac{\mu(1 - r^n)}{1 - r}.$$

What would these assumptions imply about the long-term behavior of the number of monthly orders she expects to receive?

The moment generating function for X_n is

$$m_n(t) = \exp\left\{\frac{\mu(1 - r^n)}{1 - r}(e^t - 1)\right\}.$$

As n increases without limit, then

$$\lim_{n\to\infty} m_n(t) = \exp\left\{\frac{\mu}{1 - r}(e^t - 1)\right\}, \qquad \text{if } |r| < 1. \tag{5}$$

Equation (5) gives the moment generating function of a Poisson random variable X with parameter $\mu/(1 - r)$. This is sufficient to show that the sequence of random variables representing the monthly sales will converge to this probability law, with the assumptions made.

The sequence of averages $\bar{X}_1, \bar{X}_2, \ldots, \bar{X}_n, \ldots$ used in discussing the Law of Large Numbers is "centered," in the sense that $E[\bar{X}_n] = \mu$ for all n. However, as noted earlier, the variances get smaller and smaller, converging to 0. It is interesting to ask what would happen if we were to consider the sequence of standard forms. (This sequence is gotten by subtracting the mean and dividing by the standard deviation for each term in the original sequence; the result is a sequence of independent random variables, each with mean 0 and variance 1.)

Theorem 7.3 makes it straightforward to establish a simple version of the Central Limit Theorem, which states that certain sequences of random variables converge in distribution to the standard normal distribution; this fact is useful in using the normal probability law to approximate a wide range of distributions.

Let $Y_1, Y_2, \ldots, Y_n, \ldots$ be a sequence of independent random variables, each with mean 0 and variance 1, and each with the same moment generating function $m_Y(t)$ (so each has the same probability law). Now construct the sequence of partial sums $S_1 = Y_1, S_2 = Y_1 + Y_2, \ldots, S_n = \sum_{i=1}^{n} Y_i, \ldots$. Then the nth term in this sequence has mean $E[S_n] = 0$ and variance $\text{Var}[S_n] = n$; the means remain constant at 0 but the variances of the S_n values get larger and larger without limit, the farther out in the sequence we go. If we divide S_n by $\sqrt{n}$, then its variance becomes 1; that is, the resulting sequence is then "stable" in the sense that each term in $S_1, (1/\sqrt{2})S_2, \ldots, (1/\sqrt{n})S_n, \ldots$ has mean 0 and variance 1.

Consider the sequence of moment generating functions for these terms. Since the Y_i's are independent and

$$\frac{1}{\sqrt{n}}S_n = \frac{1}{\sqrt{n}}\sum_{i=1}^{n} Y_i$$

is a linear function of $Y_1, Y_2, \ldots, Y_n$, the moment generating function for $(1/\sqrt{n})S_n$ is $(m_Y(t/\sqrt{n}))^n$. What is the value for $\lim_{n \to \infty}(m_Y(t/\sqrt{n}))^n$? The simplest way to answer this question is to investigate the behavior of its natural log,

$$n \ln m_Y\left(t/\sqrt{n}\right) = \frac{c_Y\left(t/\sqrt{n}\right)}{n^{-1}}, \tag{6}$$

where $c_Y(t) = \ln m_Y(t)$ is the cumulant generating function for Y. As $n \to \infty$, both the numerator and denominator of this ratio go to 0 (for any moment generating function, $m_Y(0) = 1$ and $\ln 1 = 0$).

The variable n in Eq. (6) is actually discrete, passing through the positive integers; it is useful to consider both the numerator and denominator as continuous functions of n. In so doing we can employ L'Hôpital's rule, taking the derivative (separately) of the numerator and the denominator with respect to n (not t); the limiting behavior must be the same whether n is considered to vary discretely or continuously.

The derivative of the numerator in Eq. (6) is $c'_Y\left(t/\sqrt{n}\right)\left(-t/2n^{3/2}\right)$, while the derivative of the denominator is $-n^{-2}$, and the ratio of the two derivatives is $c'_Y\left(t/\sqrt{n}\right)\frac{1}{2}t/n^{-1/2}$. Again, as $n \to \infty$ both numerator and denominator go to 0. Remember that $c'_Y(0)$ gives μ_Y, which equals 0. Thus we can employ L'Hôpital's rule once more (the final time), taking the two derivatives separately.

The derivative of this numerator is $\frac{1}{2}tc''_Y\left(t/\sqrt{n}\right)\left(-t/2n^{3/2}\right)$, and the derivative of the denominator is $-\frac{1}{2}n^{-3/2}$. The ratio of these two derivatives is then $\frac{1}{2}t^2 c''_Y\left(t/\sqrt{n}\right)/1$. The limit of this ratio is $\frac{1}{2}t^2 c''_Y(0) = \frac{1}{2}t^2$ since $c''_Y(0)$ gives the variance of Y, which is 1. Thus we have

$$\lim_{n \to \infty} \ln m_Y\left(\frac{t}{\sqrt{n}}\right) = \frac{t^2}{2}$$

which implies that

$$\lim_{n \to \infty} m_Y\left(t/\sqrt{n}\right) = e^{t^2/2},$$

which is the moment generating function for a standard normal random variable. (This use of L'Hôpital's rule in establishing the Central Limit Theorem was given by R. M. Tardiff in *The American Statistician*, Vol. 35, No. 1, February 1981.) The fact that this sequence of moment generating functions converges to the standard normal moment generating function then implies that the sequence of *cdf*s converges to the standard normal. This argument establishes one version of the Central Limit Theorem.

THEOREM 7.4 If $X_1, X_2, \ldots, X_n, \ldots$ is a sequence of independent random variables, each with mean μ and variance σ^2 and moment generating function $m_X(t)$, then the *cdf* for

$$\frac{1}{\sqrt{n}} \sum_{i=1}^{n} \frac{X_i - \mu}{\sigma}$$

converges to the standard normal *cdf* as $n \to \infty$.

It is surprising that one probability law, the standard normal, should occur as the limiting distribution for $(1/\sqrt{n}) \sum_{i=1}^{n} Y_i$, regardless of the details of the probability law for the individual $Y_i = (X_i - \mu)/\sigma$ values; indeed, the result holds for both discrete and continuous X_i. In some senses, this result stands at the "center" of probability theory and finds a wide variety of applications.

Example 7.12

The Central Limit Theorem is the basis for using the normal probability law as an approximation for many distributions, both discrete and continuous. To illustrate this reasoning, suppose $X_1, X_2, \ldots, X_n, \ldots$ are independent Bernoulli random variables, each with the same parameter $0 < p < 1$. The standard form for the ith item in this sequence is then $Y_i = (X_i - p)/\sqrt{pq}$, since the mean and variance of the Bernoulli law are p, and pq respectively. Theorem 7.4 then assures us that the sequence of random variables whose nth term is

$$\frac{1}{\sqrt{n}} \sum_{i=1}^{n} Y_i = \frac{1}{\sqrt{n}} \sum_{i=1}^{n} \frac{X_i - p}{\sqrt{pq}}$$

$$= \frac{\sum_{i=1}^{n} X_i - np}{\sqrt{npq}} = \frac{W - \mu_W}{\sigma_W},$$

where $W = \sum_{i=1}^{n} X_i$ is binomial, with parameters $\mu_W = np$ and $\sigma_W^2 = npq$, converges in distribution to the standard normal. This would suggest (and indeed it is true) that for "large" (but finite) n the probability law for the ratio $(W - np)/\sqrt{npq}$, where W is binomial n, p, is approximately standard normal. This in turn implies that for sufficiently large n the probability law for W itself must be approximately normal with $\mu = np$ and $\sigma^2 = npq$, no matter what the value for $0 < p < 1$.

Recall that the binomial probability distribution is symmetric only for $p = \frac{1}{2}$; for any $p \neq \frac{1}{2}$ the distibution is asymmetric but gets more symmetric as n increases (with p fixed). These facts help illustrate that the normal approximation to the binomial probability law is quite good for moderately small n if $p = \frac{1}{2}$; if p differs from $\frac{1}{2}$, a larger n is required to get the same degree of approximation. Figure 7.6 shows the exact binomial distributions for $n = 20, p = \frac{1}{2}, \frac{1}{4}$ and the approximating normal *pdf* s (with means 10, 5

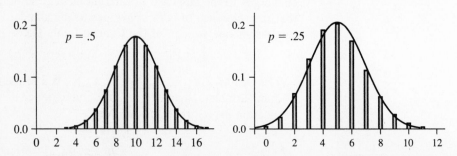

Figure 7.6 Binomial $n = 20, p = .5, .25$ versus normal with distributions

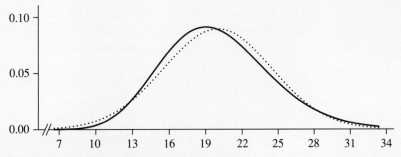

Figure 7.7   Erlang (*solid*), *pdf* with $n = 20, \lambda = 1$, and normal *pdf* (*dotted*)

and variances 5, $\frac{15}{4}$, respectively). Note that the normal *areas* will better approximate the discrete binomial sums for the case where $p = \frac{1}{2}$.

The normal probability law can also provide good approximations for continuous distributions. Assume that $X_1, X_2, \ldots, X_n, \ldots$ is a sequence of independent exponential random variables, each with parameter λ. Since $\mu_X = \sigma_X = 1/\lambda$ for the exponential, the standard form is given by $Y_i = (X_i - 1/\lambda)/\lambda^{-1} = \lambda X_i - 1$, so

$$\frac{1}{\sqrt{n}} \sum_{i=1}^{n} Y_i = \frac{\lambda}{\sqrt{n}} \sum_{i=1}^{n} X_i - \sqrt{n} = \frac{\lambda}{\sqrt{n}} V - \sqrt{n}$$

converges to the standard normal, where V is Erlang with parameters n, λ. Again, if n is "large" we would expect to find

$$\frac{\lambda}{\sqrt{n}} V - \sqrt{n} = \frac{V - \dfrac{n}{\lambda}}{\dfrac{\sqrt{n}}{\lambda}}$$

approximately standard normal. This is equivalent to saying that the Erlang random variable V is approximately normal with mean n/λ and variance n/λ^2. The accuracy of the approximation does not change for differing values of the parameter λ since the random variable involved is $U/\sqrt{n} - \sqrt{n}$ where $U = \lambda V$. It is easy to verify that U is Erlang with parameters $n, \lambda = 1$; λ is called a scale parameter for this reason (as is σ for the normal probability law). Figure 7.7 plots the exact Erlang *pdf*, for $n = 20, \lambda = 1$, with the approximating normal *pdf*. For any finite n the Erlang or gamma probability law is not symmetric, with the "left-hand" tail being shorter than the right, but in the limit this asymmetry disappears. Note as well that the maximum for the Erlang *pdf* is at 19 $(n - 1)$ and the normal is at 20 (n), which difference also disappears as n increases. It is apparent from Fig. 7.7 that some adjustment to the approximating normal will make its areas more accurate approximations to the Erlang areas.

As we have discussed the Central Limit Theorem, each of the individual Y_i values is assumed to have the same probability law with common moment generating function $m_Y(t)$, and then the sequence of values $\frac{1}{\sqrt{n}}\sum_{i=1}^{n} Y_i$ converges in distribution to the standard normal probability law. This same result has been shown to hold in a variety of cases, with weaker assumptions.

If the mean $\bar{X}_n$ is approximately normal, then so is any linear function of $\bar{X}_n$; in particular, the total $\sum_i X_i = n\bar{X}_n$ is approximately normal with mean $n\mu$ and variance $n\sigma^2$.

Example 7.13

Normal random variables can be generated from uniform random variables using the probability integral transform. Another procedure for generating normal random variables is based on the Central Limit Theorem: Suppose $U_1, U_2, \ldots, U_{12}$ are independent uniform random variables on the interval $(0, 1)$. Then each U_i has mean $\frac{1}{2}$ and variance $\frac{1}{12}$, so the sum $V = \sum_{i=1}^{12} U_i$ has mean 6 and variance 1. Assuming $n = 12$ is sufficiently large for the Central Limit Theorem normal approximation to be satisfactory, $V - 6$ is approximately standard normal. That is, if we generate 12 independent uniforms on the interval $(0, 1)$, add them together, and then subtract 6, the resulting random variable should behave as if it had been selected from the standard normal probability law. The number $n = 12$ is used since it is large enough for the approximation to be quite good, and the resulting sum of independent uniform $(0, 1)$ random variables will have the desired variance of 1, without any division being performed. Figure 7.8 plots the normal *pdf* with $\mu = 6$ and $\sigma = 1$, and gives a histogram-type display of the exact *pdf* for the sum of 12 uniform $(0, 1)$ random variables. If the latter *pdf* were also plotted, it overlays the normal distribution within the plotting accuracy available for the graph.

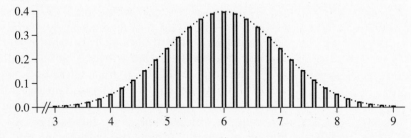

Figure 7.8 Uniform sum (bars), and normal (dotted)

We have seen a number of cases of probability laws that "reproduce" themselves, that is, situations in which sums of independent random variables, all following the same probability law, in turn are described by that same probability law. These cases include the following:

1. The sum of N independent binomial random variables with parameters n_i, p is again binomial with parameters $n = \sum_{i=1}^{N} n_i$ and p.
2. The sum of N independent Poisson random variables with parameters μ_i is again Poisson with parameter $\mu = \sum_{i=1}^{N} \mu_i$.

3. The sum of N independent negative binomial random variables with parameters r_i, p is again negative binomial with parameters $r = \sum_{i=1}^{N} r_i$ and p.

4. The sum of N independent gamma random variables with parameters n_i, λ is again gamma with parameters $n = \sum_{i=1}^{N} n_i$ and λ.

Since sums of independent random variables are well approximated by the normal (for a sufficiently large number of terms in the sum), *each* of these probability laws itself (apart from whether one thinks of the distribution as describing a sum) should then be well approximated by the normal, as long as the appropriate parameter is sufficiently "large." That is, if X is binomial with "large" n or Poisson with "large" μ or negative binomial with "large" r or gamma with "large" n, its probability law should be well approximated by the appropriate normal.

Example 7.14

Assume that the number of telephone calls to an 800 area-code toll-free number, during the 12 hours per day that it is available, is Poisson with parameter $\mu = 900$. Since this is a large value for μ,

$$P(a < X < b) = P\left(\frac{a - 900}{\sqrt{900}} < \frac{X - 900}{\sqrt{900}} < \frac{b - 900}{\sqrt{900}}\right)$$

$$\doteq P\left(\frac{a - 900}{\sqrt{900}} < Z < \frac{b - 900}{\sqrt{900}}\right),$$

where Z is standard normal. Thus

$$P(850 < X < 950) \doteq P(-1\tfrac{2}{3} < Z < 1\tfrac{2}{3}) = .90441930,$$

using special software for evaluation. Exact evaluation of this Poisson value gives .90441108, so the two values are very close indeed, because of the large value for μ.

The discussion in Chapter 4 about using the normal distribution to approximate the binomial and Poisson distributions is based on the Central Limit Theorem, using a "continuity correction," to help adjust for the basic differences in the ranges of the two distributions. If X is a binomial random variable with parameters n, p, where n is large, then the Central Limit Theorem states that the *cdf* for the standardized variable $(X - np)/\sqrt{npq}$ converges to the standard normal *cdf*. It follows, then, that for n sufficiently large,

$$F_X(t) \doteq F_Z\left(\frac{t - np}{\sqrt{npq}}\right),$$

where $t \in R_X$ is any integer between 0 and n. Because R_X is discrete, $P(X \leq t) \equiv P\left(X \leq t + \tfrac{1}{2}\right)$ for $t = 0, 1, 2, \ldots, n$. Adding $\tfrac{1}{2}$-unit like this is called the *continuity correction* and will generally improve the approximation.

It is straightforward to appreciate what this continuity correction is doing, and to see its effect by referring to pictures where the discrete probability distribution is represented by a histogram. The histogram is constructed by drawing bars that are one unit wide, are centered at the integers (the elements of the discrete range), and have heights given by the probability content at that integer (the value of the discrete probability function). If a histogram is constructed in this way, the area of a bar is equal to its height and, even though the random variable is discrete, probabilities of individual values or the probability contents for intervals are visually represented by areas, rather like the continuous case. Figure 7.9 uses this type of histogram to represent the binomial probability law with $n = 10, p = .5$; also shown is the exact value for $P(2 \leq X \leq 4)$, the sum of the areas of the shaded bars centered at the integers 2, 3, and 4.

Although $n = 10$ is hardly large, this case does allow us to see the full probability distribution represented by a histogram. Since we are using $p = .5$, the histogram is symmetric and is fairly well approximated by the appropriate normal density. This approximating normal density function is also pictured on the right ($\mu = 5, \sigma^2 = 2.5$), as is the approximating area under the density if the Central Limit Theorem is directly applied to estimate $P(2 \leq X \leq 4) = P(1 < X \leq 4) = F_X(4) - F_X(1)$. Only half the area of the bar at 4 is included in the normal area, as is half the bar at 1. The approximation is improved by using $F_X(4.5) - F_X(1.5)$ taking the area under the approximating normal for values of x such that $1.5 \leq x \leq 4.5$. Remember that the Central Limit Theorem is a limiting result, literally giving an approximation to a *cdf*; the *cdf* for any discrete random variable whose range is the counting numbers then jumps at integers. A continuous area gives a better approximation by shifting the normal *cdf* $\frac{1}{2}$ unit to the right. The areas of individual bars like those pictured for $n = 10$ actually go to 0 in this limit, so it does not matter whether portions of individual bars are included or not in the limiting process.

The exact binomial value is $P(2 \leq X \leq 4) = .04395 + .11719 + .20508 = .36622$, while the area under the normal *pdf* on the right, between 1 and 4, is about .2578; the latter value is too low because of the difference in probability at $x = 1$ and $x = 4$. Figure 7.10 shows the same binomial histogram

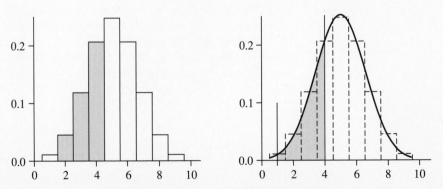

Figure 7.9   Binomial distribution with $n = 10, p = .5$ versus normal *pdf*

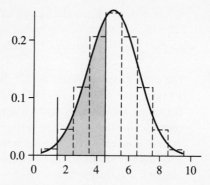

Figure 7.10 Binomial distribution $n = 10, p = .5$, versus normal *pdf* with continuity correction

and normal *pdf*; now, however, the area between 1.5 and 4.5 is shaded in, resulting in an area of about .3625, which is clearly closer in value to the actual sum of the areas of the three bars centered at 2, 3, and 4. even though n is only 10. The approximation (with continuity correction) is good because we have used $p = .5$, which makes the exact histogram symmetric. With more extreme p values (closer to either 0 or 1), a much larger value for n is needed to get the same accuracy.

How large should n be for the normal approximation to the binomial to be "satisfactory"? There are many heuristic rules that can be used to answer this question. Recall that the binomial histogram is not symmetric for $p \neq \frac{1}{2}$; if $p < \frac{1}{2}$, the bulk of the probability is shifted toward 0, and if $p > \frac{1}{2}$, the bulk of the probability is shifted toward n. The approximating normal distribution has mean np and standard deviation $\sqrt{npq}$. With $p < \frac{1}{2}$ we would like the approximating normal *pdf* to have very little probability to the left of 0, since the exact binomial law has no probability to the left of 0. One way to accomplish this is to insist that 0 is at least 3 standard deviations to the left of the mean for the approximating normal *pdf*; that is,

$$np - 3\sqrt{npq} \geq 0,$$

which in turn gives

$$\sqrt{n} \geq 3\sqrt{q/p} \qquad \text{or} \qquad n \geq 9q/p.$$

With $p > \frac{1}{2}$ it is undesirable for the approximating normal *pdf* to have a significant amount of probability to the right of n, the largest possible value for the binomial. Requiring this value (n) to be at least 3 standard deviations to the right of the mean then gives

$$np + 3\sqrt{npq} \leq n;$$

by subtracting np from both sides of this inequality we get

$$3\sqrt{npq} \leq n - np = nq,$$

which then leads to $n \geq 9p/q$ as a requirement. Thus, this line of reasoning leads, for any p, to the requirement

$$n \geq 9\max\left(\frac{q}{p}, \frac{p}{q}\right),$$

the rule given in Section 4.2. With $p = \frac{1}{2}$ this would allow the normal approximation for $n \geq 9$, but if p is as small as .1, it would suggest $n \geq 81$ for using the normal approximation. The value 9 (corresponding to 3 standard deviations on either side of the mean) is somewhat arbitrary but useful. A less conservative approach would use some number smaller than 9 (i.e., less than 3 standard deviations), and a more conservative approach would employ a number larger than 9. The same reasoning is appropriate for the approximation to the Poisson probability distribution.

This normal approximation can be used to approximate the probability that a discrete random variable will equal any particular value in its range, by simply evaluating the height of the approximating normal *pdf* at the integer value of interest. If we think of the histogram representation of the discrete probability distribution, where the height of a given bar represents the probability of observing that element in the range, the area of the bar is thus approximated by the height of the *pdf* (times 1, the width of the bar).

Example 7.15

Suppose 10% of the adults in a given city use consumer product $\mathcal{A}$. An advertising agency wishes to locate 20 of these persons to collect information about their reactions to proposed changes to be made to the product. To find 20 such persons, the agency intends to question adults, essentially at random, until it locates the required number of users. As a simple model for the number of persons the agency will need to contact in order to locate 20 users, we might idealize the situation by assuming there is a constant probability of .1 that any adult selected at random will in fact be found to use the product. If we let X be the number the agency will have to contact, then X is a negative binomial random variable with $r = 20$ and $p = .1$. The expected number of people they will have to contact is $20/.1 = 200$. How likely is it that this will in fact be the observed number required?

The exact value for this probability is

$$P(X = 200) = \binom{199}{19}(.1)^{20}(.9)^{180} = .00936.$$

Since the negative binomial with large r should be approximately normal with mean $\mu = 200$ and variance $\sigma^2 = 2000$, the normal approximating value is

$$P(X = 200) \doteq \frac{1}{\sqrt{4000\pi}} = .00892,$$

with an error of .00044, a little less than 5% of the true value.

Recall that the Poisson probability law reproduces itself; if $X_1, X_2, \ldots, X_n$ are independent Poisson random variables, each with parameter 1, then the total $\sum_i X_i$ is again Poisson with parameter $\mu = n$. If $\mu = n$ is large, the Central Limit Theorem implies that the Poisson distribution is well approximated by the normal.

If X is a Poisson random variable with parameter μ, then

$$P(X = n) = \frac{\mu^n}{n!}e^{-\mu}, \qquad \text{for } n = 0, 1, 2, \dots .$$

X is a discrete random variable, so the two events $\{X = n\}$ and $\{n - \frac{1}{2} < X < n + \frac{1}{2}\}$ are equivalent. When μ is large, the probability of finding $X = n$ should be approximately given by the area under the normal density with mean and variance both equal to μ, over the interval from $n - \frac{1}{2}$ to $n + \frac{1}{2}$, an interval of length 1. This area in turn then should be well approximated simply by the height of the standard normal *pdf* evaluated at n (times 1, the length of the interval). That is, for μ large

$$\frac{\mu^n}{n!}e^{-\mu} \doteq \frac{1}{\sqrt{2\pi\mu}}e^{-(n-\mu)^2/2\mu}. \tag{7}$$

If in fact $n = \mu$ (so now n is large as well), then Eq. (7) gives us

$$\frac{n^n}{n!}e^{-n} \doteq \frac{1}{\sqrt{2\pi n}},$$

or

$$n! \doteq \sqrt{2\pi}n^{n+\frac{1}{2}}e^{-n}. \tag{8}$$

Equation (8) is a classical result known as *Stirling's formula*. With n as small as 10 we have 10!=3,628,800, and the approximation gives 3,598,695.6, for a relative error of less than 1%. The relative error in using this approximation goes to 0 quite quickly with increasing n. This approximation is improved for each n by multiplying by $e^{1/12n}$, that is, by using

$$n! \doteq \sqrt{2\pi}n^{n+\frac{1}{2}}e^{-n+1/12n}. \tag{9}$$

Equation (9) gives 3,628,810 as the approximate value for 10!, considerably closer to the actual value, even for n as small as 10. (Indeed, with $n = 2$ it gives $2! \doteq 2.000652048$.)

Before the easy, widespread access to modern computational power, it was no trivial matter to actually evaluate definite integrals with numerical integration. Since very accurate numerical tables were available for the standard normal *cdf*, and the Central Limit Theorem shows that many probability laws can be approximated by the normal, these numerical results for the standard normal were natural candidates for use in getting approximate values for quantiles of other distributions. The *Cornish–Fisher approximations* ("The Percentile Points of Distributions Having Known Cumulants," R.A. Fisher and E.A. Cornish, *Technometrics* 2 (1960): 209–225) show one way of accomplishing this. Their procedure adjusts the cumulants of a normal distribution to match those of the desired probability law; the quantile from the desired distribution is ultimately expressed as a polynomial function of the corresponding standard normal quantile. Under conditions they describe, the higher-order polynomial terms have less and less effect, allowing the degree of approximation attained to be well controlled.

For example, if W is a χ^2 random variable with ν degrees of freedom (so W is gamma, $n = \nu/2, \lambda = 1/2$), then the first three terms of the Cornish–Fisher expansion of the $100k$th quantile w_k for W are

$$w_k \doteq \nu + \sqrt{\nu}(z_k\sqrt{2}) + \tfrac{2}{3}(z_k^2 - 1), \qquad (10)$$

where z_k is the $100k$th quantile of the standard normal distribution. Note that the first term in Eq. (10) has the multiplier ν, the second has $\nu^{1/2}$, and the third has ν^0. Five more terms are available for this series, continuing to the term with multiplier $\nu^{-5/2}$. As ν gets larger, it is easy to see that these succeeding terms are of less importance (since they involve ν to negative powers). Indeed, if we take only the first two terms of Eq. (10), and solve for z_k, we have

$$z_k = \frac{w_k - \nu}{\sqrt{2\nu}},$$

which then would imply that $(W - \nu)/\sqrt{2\nu}$ is standard normal; recalling that the mean for W is ν and its variance is 2ν, this is exactly the result from the Central Limit Theorem, that for large ν this standard form for W should be well approximated by the standard normal.

Keeping the third term of Eq. (10) as well and solving the quadratic equation, gives the inverse transformation

$$z_k = -\tfrac{3}{4}\sqrt{2\nu} + \sqrt{\tfrac{3}{2}w_k - \tfrac{3}{8}\nu + 1}.$$

This makes it easy to transform the standard normal *pdf* and see how it compares with the exact χ^2 *pdf* with $\nu = 40$ degrees of freedom (and with the original Central Limit Theorem transformation). Figure 7.11 plots the χ^2 *pdf* (dashed line), the Central Limit Theorem approximation $W = \nu + Z\sqrt{2\nu}$ (solid line), and the Cornish–Fisher approximation $W = \nu + Z\sqrt{2\nu} + \tfrac{2}{3}(Z^2 - 1)$ (dotted line). Within the accuracy of the plotting devices available, it is

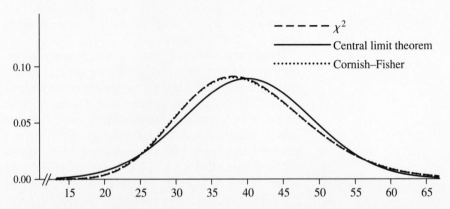

Figure 7.11 χ^2, 40 degrees of freedom (dashes) versus Normal *pdf*s Central Limit Theorem (solid), Cornish–Fisher (dots)

hard to distinguish the χ^2 and Cornish–Fisher *pdf*s. The quantiles gotten from Cornish and Fisher's transformation are clearly much closer to the χ^2 quantiles for any desired k. Their scheme for adjusting standard normal quantiles can give great accuracy in approximating the quantiles for many continuous probability laws.

EXERCISES 7.3

1. Let $X_1, X_2, \ldots, X_n, \ldots$ be independent geometric random variables, each with the same parameter p, and define the sequence of averages whose nth term is

$$\bar{X}_n = \frac{1}{n} \sum_{i=1}^{n} X_i.$$

Will the sequence $\bar{X}_1, \bar{X}_2, \ldots, \bar{X}_n, \ldots$ converge in probability? If so, to what value?

2. Let $U_1, U_2, \ldots, U_n, \ldots$ be independent uniform random variables on the interval (a, b), and define the sequence of averages whose nth term is

$$\bar{U}_n = \frac{1}{n} \sum_{i=1}^{n} U_i.$$

Will the sequence $\bar{U}_1, \bar{U}_2, \ldots, \bar{U}_n, \ldots$ converge in probability? If so, to what value?

3. Let $V_1, V_2, \ldots, V_n, \ldots$ be independent Poisson random variables, each with mean μ, and define the sequence of averages whose nth term is

$$\bar{V}_n = \frac{1}{n} \sum_{i=1}^{n} V_i.$$

Will the sequence $\bar{V}_1, \bar{V}_2, \ldots, \bar{V}_n, \ldots$ converge in probability? If so, to what value?

4. Let $X_1, X_2, \ldots, X_n, \ldots$ be independent negative binomial random variables, each with parameters r, p, and define the sequence of averages whose nth term is

$$\bar{X}_n = \frac{1}{n} \sum_{i=1}^{n} X_i.$$

Will the sequence $\bar{X}_1, \bar{X}_2, \ldots, \bar{X}_n, \ldots$ converge in probability? If so, to what value?

5. A real estate consultant assumes that the arrivals of telephone calls to a suburban office of a large real estate firm during business hours behave like a Poisson process with parameter λ. He counts the number of incoming telephone calls for $n = 10$ working days (each working day being 9 hours long) and observes the following sequence of numbers of calls received on these days: 25, 31, 29, 24, 18, 26, 23, 26, 31, 27. Based on these observed values, what might he guess to be the expected number of calls per business day?

6. *A mathematical exercise*   Let $U_1, U_2, \ldots, U_n, \ldots$ be independent uniform random variables on the interval (a, b), and define the sequence of maximum values whose nth term is $U_{(n)} = \max(U_1, U_2, \ldots, U_n)$. Will the sequence

$$U_{(1)}, U_{(2)}, \ldots, U_{(n)}, \ldots$$

converge in probability? If so, to what value?

7. Let $W_1, W_2, \ldots, W_n, \ldots$ be independent gamma random variables, each with parameters m, λ, and define the sequence of averages whose nth term is

$$\bar{W}_n = \frac{1}{n} \sum_{i=1}^{n} W_i.$$

Will the sequence $\bar{W}_1, \bar{W}_2, \ldots, \bar{W}_n, \ldots$ converge in probability? If so, to what value?

8. Let $Z_1, Z_2, \ldots, Z_n, \ldots$ be independent standard normal random variables, and define the sequence of averages whose nth term is

$$\bar{Z}_n = \frac{1}{n} \sum_{i=1}^{n} Z_i.$$

Will the sequence $\bar{Z}_1, \bar{Z}_2, \ldots, \bar{Z}_n, \ldots$ converge in probability? If so, to what value?

9. Let $Z_1, Z_2, \ldots, Z_n, \ldots$ be independent standard normal random variables, and define the sequence of averages of absolute values whose nth term is

$$|\bar{Z}|_n = \frac{1}{n} \sum_{i=1}^{n} |Z|_i.$$

Will the sequence $|\bar{Z}|_1, |\bar{Z}|_2, \ldots, |\bar{Z}|_n, \ldots$ converge in probability? If so, to what value?

10. Let $Z_1, Z_2, \ldots, Z_n, \ldots$ be independent standard normal random variables, and define the sequence of averages of squares of these values whose nth term is

$$\bar{Z}_n^2 = \frac{1}{n} \sum_{i=1}^{n} Z_i^2.$$

Will the sequence $\bar{Z}_1^2, \bar{Z}_2^2, \ldots, \bar{Z}_n^2, \ldots$ converge in probability? If so, to what value?

11. Let $X_1, X_2, \ldots, X_n, \ldots$ be a sequence of independent exponential random variables, each with parameter λ. For each fixed $n = 1, 2, 3, \ldots$ define $V_n = \min(X_1, X_2, \ldots, X_n)$ to be the smallest of the first n exponential random variables. Does the sequence $V_1, V_2, \ldots, V_n, \ldots$ converge in mean square? If so, to what value?

12. As a model for the spacings between rings in a tree trunk, a biologist assumed that the observed spacings would be well modeled by a sequence

of exponential random variables $X_1, X_2, \ldots, X_n, \ldots$, where the parameter for X_n is $n\lambda/(n + \lambda)$, the reciprocal of the sum of the reciprocals of n and λ, where the rings are counted from the center outward; thus X_1 is the distance from the center of the trunk to the first ring, X_2 is the distance from the first ring to the second, and so on. Does the sequence $X_1, X_2, \ldots, X_n, \ldots$ converge in distribution and, if so, to what probability law does it converge?

13. Let the sequence $X_1, X_2, \ldots, X_n, \ldots$ be defined as described in Example 7.11 but now suppose that $\mathrm{E}[X_n] = \mu - r\,\mathrm{E}[X_{n-1}]$ where $0 < r < 1$; in this case the expected sales decrease in coming years. Show that the sequence

$$X_1, X_2, \ldots, X_n, \ldots$$

converges in distribution, and find the probability law to which it converges.

14. If X is Poisson with $\mu = 900$, find the interval $(900 - c, 900 + c)$ for which $P(900 - c < X < 900 + c) = .99$.

15. Assume that during the rainy season in a certain location the probability of rain occurring each day is .6, and that separate days are well modeled as independent Bernoulli trials. Approximate the probability that it takes at least 100 days for the 50th rainy day to occur after the start of the rainy season.

16. As a gift you receive 100 lottery tickets. Assume that the probability each one wins some prize is $\frac{1}{9}$, and approximate the probability that you will win more than 15 prizes with these tickets.

17. A New Zealand tree farm contains 1 square mile of land that was clear cut and then planted with 750,000 Monterey pine trees. Assume that the increase in trunk diameter (in inches) per year of any one of these trees, as measured 6 inches above the ground, is modeled as an exponential random variable with parameter $\lambda = 1.1$. Increases in trunk diameter to occur from year to year or from tree to tree are assumed to be independent random variables.

    a. What is a good approximate model for the trunk diameter of one of these trees that is 20 years old?

    b. What proportion of 20-year-old trees would you expect to exceed 20 inches in trunk diameter?

18. For the situation described in Example 7.15, where X was assumed to be negative binomial with $r = 20$ and $p = .1$, evaluate the normal approximation to $P(X \leq 190)$ using the continuity correction.

19. Assume that toll-free calls to the 800 number of a major car rental firm arrive at a rate of 400 per hour, like events in a Poisson process, between the hours of 6 AM and 9 PM (Eastern Standard Time). If X represents the number of such calls in a 1-hour period, approximate the value for $P(375 \leq X \leq 425)$, both with and without the continuity correction.

20. A major metropolitan taxi company has 300 taxis on the street during business hours. Assume the probability is .05 that any one of these taxis will have a breakdown of some sort on any given day; also assume that the breakdowns occur independently. Approximate the probability that

the taxi company will suffer at least 10 such breakdowns on any given day.

21. At the height of the season assume that injuries requiring medical assistance occur on a ski slope between 10 AM and 3 PM at the rate of 2 per hour like events in a Poisson process. Let X be the number of such injuries to occur during these hours in a 7-day week. Approximate the value for $P(X \le 60)$.

22. Use the same heuristic reasoning employed in the text for the binomial to derive the rule that the normal approximation to the Poisson probability law should be "satisfactory" if the Poisson parameter μ is at least 9.

23. Suppose that X is binomial with parameters n and $p = \frac{1}{2}$, and that n is very large and even; the probability that X equals its mean value is then

$$P\left(X = \frac{n}{2}\right) = \binom{n}{\frac{n}{2}}\left(\frac{1}{2}\right)^{n/2}.$$

Show that if Stirling's formula is used for the factorials in this expression, the result is the normal approximation for $P(X = n/2)$.

7.4 | Summary

If $Y_1, Y_2, \ldots, Y_n$ are independent continuous random variables with the same *cdf* $F_Y(t)$ and *pdf* $f_Y(t)$, while $Y_{(1)}, Y_{(2)}, \ldots, Y_{(n)}$ represent their ordered values, then the *cdf* for $Y_{(j)}$ is

$$F_{Y_{(j)}}(t) = \sum_{i=j}^{n} \binom{n}{i} (F_Y(t))^i (1 - F_Y(t))^{n-i}, \qquad \text{for } j = 1, 2, \ldots, n.$$

The *pdf* for $Y_{(j)}$ is given by

$$f_{Y_{(j)}}(t) = n f_Y(t) \binom{n-1}{j-1} (F_Y(t))^{j-1} (1 - F_Y(t))^{n-j}.$$

The joint *pdf* for $Y_{(i)}, Y_{(j)}, i \ne j$, is

$$f_{Y_{(i)}, Y_{(j)}}(t_1, t_2) \Delta t_1 \Delta t_2 = n f_Y(t_1) \Delta t_1 (n-1) f_Y(t_2) \Delta t_2$$
$$\times \binom{n-2}{i-1, j-i-1, n-j} (F_Y(t_1))^{i-1}$$
$$\times (F_Y(t_2) - F_Y(t_1))^{j-i-1} (1 - F_Y(t_2))^{n-j}.$$

If $Y = g(X_1, X_2)$ then the conditional *cdf* for Y, given $X_2 = x_2$, is

$$F_{Y|X_2}(t \mid x_2) = \int_{g(x_1, x_2) \le t} f_{X_1 | X_2}(x_1 \mid x_2) \, dx_1.$$

The expected value for this conditional *cdf* is

$$E[F_{Y \mid X_2}(t \mid X_2)] = \int_{-\infty}^{\infty} F_{Y \mid X_2}(t \mid x_2) f_{X_2}(x_2) \, dx_2$$

$$= \iint_{g(x_1,x_2) \leq t} f_{X_1,X_2}(x_1, x_2) \, dx_1 dx_2 = F_Y(t).$$

If $Y = X_1 + X_2$, where X_1, X_2 are independent, then the convolution of $f_{X_1}(x_1), f_{X_2}(x_2)$ is the *pdf* for Y:

$$f_Y(t) = \int_{-\infty}^{\infty} f_{X_1}(t - x_2) f_{X_2}(x_2) \, dx_2.$$

If $Y = X_1/X_2$ is the ratio of two random variables, then its *pdf* is given by

$$f_Y(t) = \int_{-\infty}^{\infty} |x_2| f_{X_1,X_2}(tx_2, x_2) \, dx_2.$$

Student's T distribution is used frequently in problems of statistical inference; it gives the distribution for the ratio of a standard normal random variable divided by the square root of an independent χ^2 random variable over its degrees of freedom. This *pdf*, with k degrees of freedom, is

$$f_T(t) = \frac{\Gamma\left(\dfrac{k+1}{2}\right)}{\Gamma\left(\dfrac{k}{2}\right) \sqrt{k\pi}} \frac{1}{\left(1 + \dfrac{t^2}{k}\right)^{(k+1)/2}}.$$

The F distribution is also widely used in certain problems of statistical inference; it gives the probability law for the ratio of two independent χ^2 random variables, each divided by its degrees of freedom. This *pdf*, with k_1, k_2 degrees of freedom, is

$$f_{\mathcal{F}}(t) = \frac{k_1}{k_2} \frac{\Gamma\left(\dfrac{k_1 + k_2}{2}\right)}{\Gamma\left(\dfrac{k_1}{2}\right) \Gamma\left(\dfrac{k_2}{2}\right)} \frac{\left(\dfrac{k_1}{k_2} t\right)^{(k_1 - 2)/2}}{\left(1 + \dfrac{k_1}{k_2} t\right)^{(k_1 + k_2)/2}}.$$

The (weak) Law of Large Numbers follows from the Chebyshev inequality: If $X_1, X_2, \ldots, X_n, \ldots$ are independent random variables, each with the same mean μ and the same variance σ^2, then

$$\lim_{n \to \infty} P(|\bar{X}_n - \mu| \geq \epsilon) = 0,$$

for any positive ϵ, where

$$\bar{X}_n = \frac{1}{n} \sum_{i=1}^{n} X_i.$$

The Central Limit Theorem states that the probability law for certain sums of independent random variables will converge to the normal. One simple version of this result is the following: If $X_1, X_2, \ldots, X_n, \ldots$ is a sequence of independent random variables, each with mean μ, variance σ^2, and moment generating function $m_X(t)$, than the probability law for

$$\frac{1}{\sqrt{n}} \sum_{i=1}^{n} \frac{X_i - \mu}{\sigma}$$

converges to the standard normal probability law as $n \to \infty$. This provides the basis for using the normal distribution to approximate probability statements for many different distributions. The approximation for the *cdf* of an integer-valued discrete random variable X is generally improved by using the fact that $F_X(t) \equiv F_X\left(t + \frac{1}{2}\right)$, when t is an integer.

EXERCISES 7.4

1. Let X and Y each have density function $f(x) = 2x$, for $0 < x < 1$. Assume X and Y are independent, and find the *pdf* for their ratio $R = X/Y$.
2. Evaluate the $100k$th quantile for $\mathcal{F} = T^2$, where T has Student's T distribution with $k = 1$ degree of freedom. (This is the *cdf* for the F distribution with $k_1 = k_2 = 1$ degrees of freedom.)
3. Evaluate the $100k$th quantile for the F distribution with $k_1 = k_2 = 2$ degrees of freedom.
4. Let $U_1, U_2, \ldots, U_n$ be independent uniform $(0, 1)$ random variables. Show that the probability laws for $U_{(1)}$ and $1 - U_{(n)}$ are identical.
5. Derive a general formula linking the *pdf* for $Y = X_1 - X_2$, the difference between two independent random variables, with the *pdf*s for X_1, X_2.
6. Derive a general formula linking the *pdf* for $Y = X_1 X_2$, the product of two independent random variables, with the *pdf*s for X_1, X_2.
7. Model the lottery tickets you may buy from your state lottery as independent Bernoulli trials with probability of success $p = \frac{1}{9}$ (success meaning the ticket wins some prize). Approximate the probability that you will get 10 winning tickets if you purchase 90 tickets.
8. Show that if X has mean μ and variance σ^2, then

$$P(|X - \mu| \geq \epsilon) \leq \frac{\sigma^2}{\epsilon^2}$$

for any $\epsilon > \sigma$.

9. If X_1, X_2 are independent, each with *pdf* $f(x) = xe^{-x^2/2}$ for $x > 0$, find the probability law for $Y = X_1/X_2$.
10. Derive a "reasonable" rule for applying the normal approximation to the negative binomial probability law. More specifically, if the approximating normal assigns probability 0 (essentially) to intervals more than 3 standard deviations away from the mean, how large should r be for the normal approximation to perform fairly well?
11. Apply the rule you derived in Exercise 9 to the negative binomial probability law with $p = .1$. How large should r be for the normal approximation to be "adequate"? Using the smallest possible value of r from this rule, approximate the value of $P(X > 1 + r/p)$.

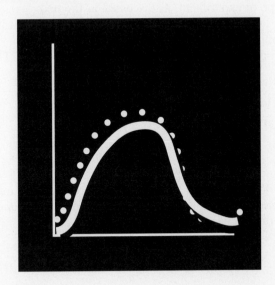

GEOMETRIC AND
RELATED SERIES

By removing the parentheses and carrying out the multiplication, it is easy to see that, if q is any real number,

$$(1 + q + q^2 + \cdots + q^n)(1 - q) = 1 - q^{n+1},$$

since all terms in the product, except the first and last, cancel out. Then, as long as $q \neq 1$ it is true that $(1 - q) \neq 0$, and we can divide both sides of this equation by $(1 - q)$, giving the finite geometric progression

$$1 + q + q^2 + \cdots + q^n = \sum_{i=0}^{n} q^i = \frac{1 - q^{n+1}}{1 - q}.$$

If $|q| < 1$, $\lim_{n \to \infty} q^n = 0$; thus for any $|q| < 1$ we also have

$$\lim_{n \to \infty} \sum_{i=0}^{n} q^i = \lim_{n \to \infty} \frac{1 - q^{n+1}}{1 - q} = \sum_{i=0}^{\infty} q^i = \frac{1}{1 - q}.$$

The series $1 + q + q^2 + q^3 + \cdots$ is called an *infinite geometric progression* or *geometric series.* Since each succeeding term in the series is given by taking q times the preceding term, q is frequently called the "common ratio." This geometric series structure occurs naturally in the probability model of a certain type of experiment, as discussed in the text.

Derivatives of geometric progressions also prove useful in certain probability models. Granted two functions of a real variable are equal to each other, their derivatives then must also be equal. Thus

$$\frac{d}{dq}\left(\sum_{i=0}^{\infty} q^i\right) = \frac{d}{dq}\left(\frac{1}{1-q}\right)$$

that is,

$$\sum_{i=1}^{\infty} iq^{i-1} = \frac{1}{(1-q)^2},$$

as long as $|q| < 1$, giving the value of another infinite series. Equating the derivatives of the two sides of this equation (second derivative of the original geometric series) gives

$$\sum_{i=2}^{\infty} i(i-1)q^{i-2} = \frac{2}{(1-q)^3}.$$

Dividing both sides by 2 we have

$$\sum_{i=2}^{\infty} \frac{i(i-1)}{2}q^{i-2} = \frac{1}{(1-q)^3},$$

or

$$\sum_{i=2}^{\infty} \binom{i}{2}q^{i-2} = \frac{1}{(1-q)^3}.$$

Letting $j = i - 2$ (and recalling that $\binom{j+2}{2} \equiv \binom{j+2}{j}$) this can also be written

$$\sum_{j=0}^{\infty} \binom{j+2}{j}q^j = \frac{1}{(1-q)^3},$$

for $|q| < 1$.

In similar manner, equating the kth derivatives of the original geometric progression gives

$$\sum_{i=0}^{\infty} \binom{k+i}{i}q^i = \sum_{j=k}^{\infty} \binom{j}{k}q^{j-k} = \frac{1}{(1-q)^{k+1}},$$

again for $|q| < 1$. The first summation written is the direct result of taking the derivative k times; the second summation is gotten by replacing the index of summation i by $j = k + i$. Since $\frac{1}{(1-q)^{k+1}} = (1-q)^{-(k+1)}$, this is also called the binomial expansion of $(1-q)^{-(k+1)}$. These series prove useful in studying the negative binomial distribution.

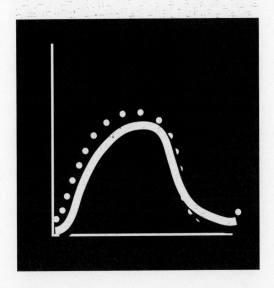

TABLES OF THE BINOMIAL, POISSON, AND STANDARD NORMAL *cdf*s

Table B.1

Binomial cumulative distribution function

$$F_X(t) = \sum_{x=0}^{t} \binom{n}{x} p^x (1-p)^{n-x}$$

n = 5, p =

t	.05	.10	.15	.20	.25	.30	.35	.40	.45	.50
0	.7738	.5905	.4437	.3277	.2373	.1681	.1160	.0778	.0503	.0313
1	.9774	.9185	.8352	.7373	.6328	.5282	.4284	.3370	.2562	.1875
2	.9988	.9914	.9734	.9421	.8965	.8369	.7648	.6826	.5931	.5000
3	1.0000	.9995	.9978	.9933	.9844	.9692	.9460	.9130	.8688	.8125
4	1.0000	1.0000	.9999	.9997	.9990	.9976	.9947	.9898	.9815	.9688

n = 10, p =

t	.05	.10	.15	.20	.25	.30	.35	.40	.45	.50
0	.5987	.3487	.1969	.1074	.0563	.0282	.0135	.0060	.0025	.0010
1	.9139	.7361	.5443	.3758	.2440	.1493	.0860	.0464	.0233	.0107
2	.9885	.9298	.8202	.6778	.5256	.3828	.2616	.1673	.0996	.0547
3	.9990	.9872	.9500	.8791	.7759	.6496	.5138	.3823	.2660	.1719
4	.9999	.9984	.9901	.9672	.9219	.8497	.7515	.6331	.5044	.3770
5	1.0000	.9999	.9986	.9936	.9803	.9527	.9051	.8338	.7384	.6230
6	1.0000	1.0000	.9999	.9991	.9965	.9894	.9740	.9452	.8980	.8281
7	1.0000	1.0000	1.0000	.9999	.9996	.9984	.9952	.9877	.9726	.9453
8	1.0000	1.0000	1.0000	1.0000	1.0000	.9999	.9995	.9983	.9955	.9893
9	1.0000	1.0000	1.0000	1.0000	1.0000	1.0000	1.0000	.9999	.9997	.9990

n = 15, p =

t	.05	.10	.15	.20	.25	.30	.35	.40	.45	.50
0	.4633	.2059	.0874	.0352	.0134	.0047	.0016	.0005	.0001	.0000
1	.8290	.5490	.3186	.1671	.0802	.0353	.0142	.0052	.0017	.0005
2	.9638	.8159	.6042	.3980	.2361	.1268	.0617	.0271	.0107	.0037
3	.9945	.9444	.8227	.6482	.4613	.2969	.1727	.0905	.0424	.0176
4	.9994	.9873	.9383	.8358	.6865	.5155	.3519	.2173	.1204	.0592
5	.9999	.9978	.9832	.9389	.8516	.7216	.5643	.4032	.2608	.1509
6	1.0000	.9997	.9964	.9819	.9434	.8689	.7548	.6098	.4522	.3036
7	1.0000	1.0000	.9994	.9958	.9827	.9500	.8868	.7869	.6535	.5000
8	1.0000	1.0000	.9999	.9992	.9958	.9848	.9578	.9050	.8182	.6964
9	1.0000	1.0000	1.0000	.9999	.9992	.9963	.9876	.9662	.9231	.8491
10	1.0000	1.0000	1.0000	1.0000	.9999	.9993	.9972	.9907	.9745	.9408
11	1.0000	1.0000	1.0000	1.0000	1.0000	.9999	.9995	.9981	.9937	.9824
12	1.0000	1.0000	1.0000	1.0000	1.0000	1.0000	.9999	.9997	.9989	.9963
13	1.0000	1.0000	1.0000	1.0000	1.0000	1.0000	1.0000	1.0000	.9999	.9995

Table B.1

Binomial cumulative distribution function (continued)

					$n = 20,$	$p =$				
t	.05	.10	.15	.20	.25	.30	.35	.40	.45	.50
0	.3585	.1216	.0388	.0115	.0032	.0008	.0002	.0000	.0000	.0000
1	.7358	.3917	.1756	.0692	.0243	.0076	.0021	.0005	.0001	.0000
2	.9245	.6769	.4049	.2061	.0913	.0355	.0121	.0036	.0009	.0002
3	.9841	.8670	.6477	.4114	.2252	.1071	.0444	.0160	.0049	.0013
4	.9974	.9568	.8298	.6296	.4148	.2375	.1182	.0510	.0189	.0059
5	.9997	.9887	.9327	.8042	.6172	.4164	.2454	.1256	.0553	.0207
6	1.0000	.9976	.9781	.9133	.7858	.6080	.4166	.2500	.1299	.0577
7	1.0000	.9996	.9941	.9679	.8982	.7723	.6010	.4159	.2520	.1316
8	1.0000	.9999	.9987	.9900	.9591	.8867	.7624	.5956	.4143	.2517
9	1.0000	1.0000	.9998	.9974	.9861	.9520	.8782	.7553	.5914	.4119
10	1.0000	1.0000	1.0000	.9994	.9961	.9829	.9468	.8725	.7507	.5881
11	1.0000	1.0000	1.0000	.9999	.9991	.9949	.9804	.9435	.8692	.7483
12	1.0000	1.0000	1.0000	1.0000	.9998	.9987	.9940	.9790	.9420	.8684
13	1.0000	1.0000	1.0000	1.0000	1.0000	.9997	.9985	.9935	.9786	.9423
14	1.0000	1.0000	1.0000	1.0000	1.0000	1.0000	.9997	.9984	.9936	.9793
15	1.0000	1.0000	1.0000	1.0000	1.0000	1.0000	1.0000	.9997	.9985	.9941
16	1.0000	1.0000	1.0000	1.0000	1.0000	1.0000	1.0000	1.0000	.9997	.9987
17	1.0000	1.0000	1.0000	1.0000	1.0000	1.0000	1.0000	1.0000	1.0000	.9998

					$n = 25,$	$p =$				
t	.05	.10	.15	.20	.25	.30	.35	.40	.45	.50
0	.2774	.0718	.0172	.0038	.0008	.0001	.0000	.0000	.0000	.0000
1	.6424	.2712	.0931	.0274	.0070	.0016	.0003	.0001	.0000	.0000
2	.8729	.5371	.2537	.0982	.0321	.0090	.0021	.0004	.0001	.0000
3	.9659	.7636	.4711	.2340	.0962	.0332	.0097	.0024	.0005	.0001
4	.9928	.9020	.6821	.4207	.2137	.0905	.0320	.0095	.0023	.0005
5	.9988	.9666	.8385	.6167	.3783	.1935	.0826	.0294	.0086	.0020
6	.9998	.9905	.9305	.7800	.5611	.3407	.1734	.0736	.0258	.0073
7	1.0000	.9977	.9745	.8909	.7265	.5118	.3061	.1536	.0639	.0216
8	1.0000	.9995	.9920	.9532	.8506	.6769	.4668	.2735	.1340	.0539
9	1.0000	.9999	.9979	.9827	.9287	.8106	.6303	.4246	.2424	.1148
10	1.0000	1.0000	.9995	.9944	.9703	.9022	.7712	.5858	.3843	.2122
11	1.0000	1.0000	.9999	.9985	.9893	.9558	.8746	.7323	.5426	.3450
12	1.0000	1.0000	1.0000	.9996	.9966	.9825	.9396	.8462	.6937	.5000
13	1.0000	1.0000	1.0000	.9999	.9991	.9940	.9745	.9222	.8173	.6550
14	1.0000	1.0000	1.0000	1.0000	.9998	.9982	.9907	.9656	.9040	.7878
15	1.0000	1.0000	1.0000	1.0000	1.0000	.9995	.9971	.9868	.9560	.8852
16	1.0000	1.0000	1.0000	1.0000	1.0000	.9999	.9992	.9957	.9826	.9461
17	1.0000	1.0000	1.0000	1.0000	1.0000	1.0000	.9998	.9988	.9942	.9784
18	1.0000	1.0000	1.0000	1.0000	1.0000	1.0000	1.0000	.9997	.9984	.9927
19	1.0000	1.0000	1.0000	1.0000	1.0000	1.0000	1.0000	.9999	.9996	.9980
20	1.0000	1.0000	1.0000	1.0000	1.0000	1.0000	1.0000	1.0000	.9999	.9995
21	1.0000	1.0000	1.0000	1.0000	1.0000	1.0000	1.0000	1.0000	1.0000	.9999

Table B.1

Binomial cumulative distribution function (continued)

t	.05	.10	.15	.20	.25	.30	.35	.40	.45	.50
					$n = 30,$	$p =$				
0	.2146	.0424	.0076	.0012	.0002	.0000	.0000	.0000	.0000	.0000
1	.5535	.1837	.0480	.0105	.0020	.0003	.0000	.0000	.0000	.0000
2	.8122	.4114	.1514	.0442	.0106	.0021	.0003	.0000	.0000	.0000
3	.9392	.6474	.3217	.1227	.0374	.0093	.0019	.0003	.0000	.0000
4	.9844	.8245	.5245	.2552	.0979	.0302	.0075	.0015	.0002	.0000
5	.9967	.9268	.7106	.4275	.2026	.0766	.0233	.0057	.0011	.0002
6	.9994	.9742	.8474	.6070	.3481	.1595	.0586	.0172	.0040	.0007
7	.9999	.9922	.9302	.7608	.5143	.2814	.1238	.0435	.0121	.0026
8	1.0000	.9980	.9722	.8713	.6736	.4315	.2247	.0940	.0312	.0081
9	1.0000	.9995	.9903	.9389	.8034	.5888	.3575	.1763	.0694	.0214
10	1.0000	.9999	.9971	.9744	.8943	.7304	.5078	.2915	.1350	.0494
11	1.0000	1.0000	.9992	.9905	.9493	.8407	.6548	.4311	.2327	.1002
12	1.0000	1.0000	.9998	.9969	.9784	.9155	.7802	.5785	.3592	.1808
13	1.0000	1.0000	1.0000	.9991	.9918	.9599	.8737	.7145	.5025	.2923
14	1.0000	1.0000	1.0000	.9998	.9973	.9831	.9348	.8246	.6448	.4278
15	1.0000	1.0000	1.0000	.9999	.9992	.9936	.9699	.9029	.7691	.5722
16	1.0000	1.0000	1.0000	1.0000	.9998	.9979	.9876	.9519	.8644	.7077
17	1.0000	1.0000	1.0000	1.0000	.9999	.9994	.9955	.9788	.9286	.8192
18	1.0000	1.0000	1.0000	1.0000	1.0000	.9998	.9986	.9917	.9666	.8998
19	1.0000	1.0000	1.0000	1.0000	1.0000	1.0000	.9996	.9971	.9862	.9506
20	1.0000	1.0000	1.0000	1.0000	1.0000	1.0000	.9999	.9991	.9950	.9786
21	1.0000	1.0000	1.0000	1.0000	1.0000	1.0000	1.0000	.9998	.9984	.9919
22	1.0000	1.0000	1.0000	1.0000	1.0000	1.0000	1.0000	1.0000	.9996	.9974
23	1.0000	1.0000	1.0000	1.0000	1.0000	1.0000	1.0000	1.0000	.9999	.9993
24	1.0000	1.0000	1.0000	1.0000	1.0000	1.0000	1.0000	1.0000	1.0000	.9998

Table B.2
Poisson cumulative distribution function

$$F_X(x) = \sum_{j=0}^{x} \frac{\mu^j}{j!} e^{-\mu}$$

x	.1	.2	.3	.4	.5	.6	.7	.8	.9	1.0
					$\mu =$					
0	.9048	.8187	.7408	.6703	.6065	.5488	.4966	.4493	.4066	.3679
1	.9953	.9825	.9631	.9384	.9098	.8781	.8442	.8088	.7725	.7358
2	.9998	.9989	.9964	.9921	.9856	.9769	.9659	.9526	.9371	.9197
3	1.0000	.9999	.9997	.9992	.9982	.9966	.9942	.9909	.9865	.9810
4	1.0000	1.0000	1.0000	.9999	.9998	.9996	.9992	.9986	.9977	.9963
5	1.0000	1.0000	1.0000	1.0000	1.0000	1.0000	.9999	.9998	.9997	.9994
6	1.0000	1.0000	1.0000	1.0000	1.0000	1.0000	1.0000	1.0000	1.0000	.9999

x	1.25	1.50	1.75	2.00	2.25	2.50	2.75	3.00	3.25	3.50
					$\mu =$					
0	.2865	.2231	.1738	.1353	.1054	.0821	.0639	.0498	.0388	.0302
1	.6446	.5578	.4779	.4060	.3425	.2873	.2397	.1991	.1648	.1359
2	.8685	.8088	.7440	.6767	.6093	.5438	.4815	.4232	.3696	.3208
3	.9617	.9344	.8992	.8571	.8094	.7576	.7030	.6472	.5914	.5366
4	.9909	.9814	.9671	.9473	.9220	.8912	.8554	.8153	.7717	.7254
5	.9982	.9955	.9909	.9834	.9726	.9580	.9392	.9161	.8888	.8576
6	.9997	.9991	.9978	.9955	.9916	.9858	.9776	.9665	.9523	.9347
7	1.0000	.9998	.9995	.9989	.9977	.9958	.9927	.9881	.9817	.9733
8	1.0000	1.0000	.9999	.9998	.9994	.9989	.9978	.9962	.9937	.9901
9	1.0000	1.0000	1.0000	1.0000	.9999	.9997	.9994	.9989	.9980	.9967
10	1.0000	1.0000	1.0000	1.0000	1.0000	.9999	.9999	.9997	.9994	.9990
11	1.0000	1.0000	1.0000	1.0000	1.0000	1.0000	1.0000	.9999	.9999	.9997
12	1.0000	1.0000	1.0000	1.0000	1.0000	1.0000	1.0000	1.0000	1.0000	.9999

x	3.75	4.00	4.25	4.50	4.75	5.00	5.25	5.50	5.75	6.00
					$\mu =$					
0	.0235	.0183	.0143	.0111	.0087	.0067	.0052	.0041	.0032	.0025
1	.1117	.0916	.0749	.0611	.0497	.0404	.0328	.0266	.0215	.0174
2	.2771	.2381	.2037	.1736	.1473	.1247	.1051	.0884	.0741	.0620
3	.4838	.4335	.3862	.3423	.3019	.2650	.2317	.2017	.1749	.1512
4	.6775	.6288	.5801	.5321	.4854	.4405	.3978	.3575	.3199	.2851
5	.8229	.7851	.7449	.7029	.6597	.6160	.5722	.5289	.4866	.4457
6	.9137	.8893	.8617	.8311	.7978	.7622	.7248	.6860	.6464	.6063
7	.9624	.9489	.9326	.9134	.8914	.8666	.8392	.8095	.7776	.7440
8	.9852	.9786	.9702	.9597	.9470	.9319	.9144	.8944	.8719	.8472
9	.9947	.9919	.9880	.9829	.9764	.9682	.9582	.9462	.9322	.9161
10	.9983	.9972	.9956	.9933	.9903	.9863	.9812	.9747	.9669	.9574
11	.9995	.9991	.9985	.9976	.9963	.9945	.9922	.9890	.9850	.9799
12	.9999	.9997	.9995	.9992	.9987	.9980	.9970	.9955	.9937	.9912
13	1.0000	.9999	.9999	.9997	.9996	.9993	.9989	.9983	.9975	.9964
14	1.0000	1.0000	1.0000	.9999	.9999	.9998	.9996	.9994	.9991	.9986
15	1.0000	1.0000	1.0000	1.0000	1.0000	.9999	.9999	.9998	.9997	.9995
16	1.0000	1.0000	1.0000	1.0000	1.0000	1.0000	1.0000	.9999	.9999	.9998
17	1.0000	1.0000	1.0000	1.0000	1.0000	1.0000	1.0000	1.0000	1.0000	.9999

x	6.5	7.0	7.5	8.0	8.5	9.0	9.5	10.0	10.5	11.0
0	.0015	.0009	.0006	.0003	.0002	.0001	.0001	.0000	.0000	.0000
1	.0113	.0073	.0047	.0030	.0019	.0012	.0008	.0005	.0003	.0002
2	.0430	.0296	.0203	.0138	.0093	.0062	.0042	.0028	.0018	.0012
3	.1118	.0818	.0591	.0424	.0301	.0212	.0149	.0103	.0071	.0049
4	.2237	.1730	.1321	.0996	.0744	.0550	.0403	.0293	.0211	.0151
5	.3690	.3007	.2414	.1912	.1496	.1157	.0885	.0671	.0504	.0375
6	.5265	.4497	.3782	.3134	.2562	.2068	.1649	.1301	.1016	.0786
7	.6728	.5987	.5246	.4530	.3856	.3239	.2687	.2202	.1785	.1432
8	.7916	.7291	.6620	.5925	.5231	.4557	.3918	.3328	.2794	.2320
9	.8774	.8305	.7764	.7166	.6530	.5874	.5218	.4579	.3971	.3405
10	.9332	.9015	.8622	.8159	.7634	.7060	.6453	.5830	.5207	.4599
11	.9661	.9467	.9208	.8881	.8487	.8030	.7520	.6968	.6387	.5793
12	.9840	.9730	.9573	.9362	.9091	.8758	.8364	.7916	.7420	.6887
13	.9929	.9872	.9784	.9658	.9486	.9261	.8981	.8645	.8253	.7813
14	.9970	.9943	.9897	.9827	.9726	.9585	.9400	.9165	.8879	.8540
15	.9988	.9976	.9954	.9918	.9862	.9780	.9665	.9513	.9317	.9074
16	.9996	.9990	.9980	.9963	.9934	.9889	.9823	.9730	.9604	.9441
17	.9998	.9996	.9992	.9984	.9970	.9947	.9911	.9857	.9781	.9678
18	.9999	.9999	.9997	.9993	.9987	.9976	.9957	.9928	.9885	.9823
19	1.0000	1.0000	.9999	.9997	.9995	.9989	.9980	.9965	.9942	.9907
20	1.0000	1.0000	1.0000	.9999	.9998	.9996	.9991	.9984	.9972	.9953
21	1.0000	1.0000	1.0000	1.0000	.9999	.9998	.9996	.9993	.9987	.9977
22	1.0000	1.0000	1.0000	1.0000	1.0000	.9999	.9999	.9997	.9994	.9990
23	1.0000	1.0000	1.0000	1.0000	1.0000	1.0000	.9999	.9999	.9998	.9995
24	1.0000	1.0000	1.0000	1.0000	1.0000	1.0000	1.0000	1.0000	.9999	.9998
25	1.0000	1.0000	1.0000	1.0000	1.0000	1.0000	1.0000	1.0000	1.0000	.9999

x	$\mu =$ 12	13	14	15	16	17	18	19	20	21
1	.0001	.0000	.0000	.0000	.0000	.0000	.0000	.0000	.0000	.0000
2	.0005	.0002	.0001	.0000	.0000	.0000	.0000	.0000	.0000	.0000
3	.0023	.0011	.0005	.0002	.0001	.0000	.0000	.0000	.0000	.0000
4	.0076	.0037	.0018	.0009	.0004	.0002	.0001	.0000	.0000	.0000
5	.0203	.0107	.0055	.0028	.0014	.0007	.0003	.0002	.0001	.0000
6	.0458	.0259	.0142	.0076	.0040	.0021	.0010	.0005	.0003	.0001
7	.0895	.0540	.0316	.0180	.0100	.0054	.0029	.0015	.0008	.0004
8	.1550	.0998	.0621	.0374	.0220	.0126	.0071	.0039	.0021	.0011
9	.2424	.1658	.1094	.0699	.0433	.0261	.0154	.0089	.0050	.0028
10	.3472	.2517	.1757	.1185	.0774	.0491	.0304	.0183	.0108	.0063
11	.4616	.3532	.2600	.1848	.1270	.0847	.0549	.0347	.0214	.0129
12	.5760	.4631	.3585	.2676	.1931	.1350	.0917	.0606	.0390	.0245
13	.6815	.5730	.4644	.3632	.2745	.2009	.1426	.0984	.0661	.0434
14	.7720	.6751	.5704	.4657	.3675	.2808	.2081	.1497	.1049	.0716
15	.8444	.7636	.6694	.5681	.4667	.3715	.2867	.2148	.1565	.1111
16	.8987	.8355	.7559	.6641	.5660	.4677	.3751	.2920	.2211	.1629
17	.9370	.8905	.8272	.7489	.6593	.5640	.4686	.3784	.2970	.2270
18	.9626	.9302	.8826	.8195	.7423	.6550	.5622	.4695	.3814	.3017
19	.9787	.9573	.9235	.8752	.8122	.7363	.6509	.5606	.4703	.3843
20	.9884	.9750	.9521	.9170	.8682	.8055	.7307	.6472	.5591	.4710
21	.9939	.9859	.9712	.9469	.9108	.8615	.7991	.7255	.6437	.5577
22	.9970	.9924	.9833	.9673	.9418	.9047	.8551	.7931	.7206	.6405
23	.9985	.9960	.9907	.9805	.9633	.9367	.8989	.8490	.7875	.7160
24	.9993	.9980	.9950	.9888	.9777	.9594	.9317	.8933	.8432	.7822
25	.9997	.9990	.9974	.9938	.9869	.9748	.9554	.9269	.8878	.8377
26	.9999	.9995	.9987	.9967	.9925	.9848	.9718	.9514	.9221	.8826
27	.9999	.9998	.9994	.9983	.9959	.9912	.9827	.9687	.9475	.9175
28	1.0000	.9999	.9997	.9991	.9978	.9950	.9897	.9805	.9657	.9436
29	1.0000	1.0000	.9999	.9996	.9989	.9973	.9941	.9882	.9782	.9626
30	1.0000	1.0000	.9999	.9998	.9994	.9986	.9967	.9930	.9865	.9758
31	1.0000	1.0000	1.0000	.9999	.9997	.9993	.9982	.9960	.9919	.9848
32	1.0000	1.0000	1.0000	1.0000	.9999	.9996	.9990	.9978	.9953	.9907
33	1.0000	1.0000	1.0000	1.0000	.9999	.9998	.9995	.9988	.9973	.9945
34	1.0000	1.0000	1.0000	1.0000	1.0000	.9999	.9998	.9994	.9985	.9968
35	1.0000	1.0000	1.0000	1.0000	1.0000	1.0000	.9999	.9997	.9992	.9982
36	1.0000	1.0000	1.0000	1.0000	1.0000	1.0000	.9999	.9998	.9996	.9990
37	1.0000	1.0000	1.0000	1.0000	1.0000	1.0000	1.0000	.9999	.9998	.9995
38	1.0000	1.0000	1.0000	1.0000	1.0000	1.0000	1.0000	1.0000	.9999	.9997
39	1.0000	1.0000	1.0000	1.0000	1.0000	1.0000	1.0000	1.0000	.9999	.9999
40	1.0000	1.0000	1.0000	1.0000	1.0000	1.0000	1.0000	1.0000	1.0000	.9999

Table B.3
Standard normal cumulative distribution function

$$F_Z(z) = \int_{-\infty}^{z} \frac{1}{\sqrt{2\pi}} e^{-t^2/2}\, dt.$$

z	.00	.01	.02	.03	.04	.05	.06	.07	.08	.09
.0	.5000	.5040	.5080	.5120	.5160	.5199	.5239	.5279	.5319	.5359
.1	.5398	.5438	.5478	.5517	.5557	.5596	.5636	.5675	.5714	.5753
.2	.5793	.5832	.5871	.5910	.5948	.5987	.6026	.6064	.6103	.6141
.3	.6179	.6217	.6255	.6293	.6331	.6368	.6406	.6443	.6480	.6517
.4	.6554	.6591	.6628	.6664	.6700	.6736	.6772	.6808	.6844	.6879
.5	.6915	.6950	.6985	.7019	.7054	.7088	.7123	.7157	.7190	.7224
.6	.7257	.7291	.7324	.7357	.7389	.7422	.7454	.7486	.7517	.7549
.7	.7580	.7611	.7642	.7673	.7704	.7734	.7764	.7794	.7823	.7852
.8	.7881	.7910	.7939	.7967	.7995	.8023	.8051	.8078	.8106	.8133
.9	.8159	.8186	.8212	.8238	.8264	.8289	.8315	.8340	.8365	.8389
1.0	.8413	.8438	.8461	.8485	.8508	.8531	.8554	.8577	.8599	.8621
1.1	.8643	.8665	.8686	.8708	.8729	.8749	.8770	.8790	.8810	.8830
1.2	.8849	.8869	.8888	.8907	.8925	.8944	.8962	.8980	.8997	.9015
1.3	.9032	.9049	.9066	.9082	.9099	.9115	.9131	.9147	.9162	.9177
1.4	.9192	.9207	.9222	.9236	.9251	.9265	.9279	.9292	.9306	.9319
1.5	.9332	.9345	.9357	.9370	.9382	.9394	.9406	.9418	.9429	.9441
1.6	.9452	.9463	.9474	.9484	.9495	.9505	.9515	.9525	.9535	.9545
1.7	.9554	.9564	.9573	.9582	.9591	.9599	.9608	.9616	.9625	.9633
1.8	.9641	.9649	.9656	.9664	.9671	.9678	.9686	.9693	.9699	.9706
1.9	.9713	.9719	.9726	.9732	.9738	.9744	.9750	.9756	.9761	.9767
2.0	.9772	.9778	.9783	.9788	.9793	.9798	.9803	.9808	.9812	.9817
2.1	.9821	.9826	.9830	.9834	.9838	.9842	.9846	.9850	.9854	.9857
2.2	.9861	.9864	.9868	.9871	.9875	.9878	.9881	.9884	.9887	.9890
2.3	.9893	.9896	.9898	.9901	.9904	.9906	.9909	.9911	.9913	.9916
2.4	.9918	.9920	.9922	.9925	.9927	.9929	.9931	.9932	.9934	.9936
2.5	.9938	.9940	.9941	.9943	.9945	.9946	.9948	.9949	.9951	.9952
2.6	.9953	.9955	.9956	.9957	.9959	.9960	.9961	.9962	.9963	.9964
2.7	.9965	.9966	.9967	.9968	.9969	.9970	.9971	.9972	.9973	.9974
2.8	.9974	.9975	.9976	.9977	.9977	.9978	.9979	.9979	.9980	.9981
2.9	.9981	.9982	.9982	.9983	.9984	.9984	.9985	.9985	.9986	.9986
3.0	.9987	.9990	.9993	.9995	.9997	.9998	.9998	.9999	.9999	1.0000

APPENDIX C

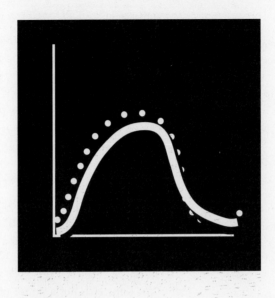

ANSWERS FOR

EVEN-NUMBERED EXERCISES

Chapter 1   Probability

Section 1.1

2. Yes

4. *a*. True   *b*. True   *c*. False   *d*. True   *e*. True
   f. False   *g*. True

6. *a*. $B \cup C = \{x : 0 \le x \le 3\}$
   b. $B \cap C = \{x : 1 \le x \le 2\}$

8. No, at any given time there are many U.S. citizens abroad and many non-U.S. citizens living in the United States.

12. A Venn diagram will do; analytically, if $A \subset B$ then $x \in A \Rightarrow x \in B$, which brings with it the fact that if x does not belong to B it cannot belong to A. Thus $x \notin B \Rightarrow x \in \overline{B} \Rightarrow x \notin A \Rightarrow x \in \overline{A}$.

14. Together with $A \cap B$, these cover S and are all disjoint.

Section 1.2

2. $S = \{(x,y) : x = 1, 2, \ldots, 10, \; y = 1, 2, \ldots, 10, \; x \neq y\}$, where x is the first number drawn and y is the second. $B = \{(1,2),(2,1)\}$

4. Since 18 persons are to be selected, an experimental outcome is simply a listing of the 18 persons from the 45 to be selected.

6. $S = \{1, 2, 3\}$

8. Interchange $P(A \cup B)$ and $P(A \cap B)$ in the result of Theorem 1.4.

10. $S = \{(x_1, x_2, x_3) : x_i = 1, 2, \ldots, 8, \; i = 1, 2, 3, \; x_i \neq x_j\}$
    $A = \{(1, x_2, x_3) : x_i = 2, 3, \ldots, 8, \; i = 2, 3, \; x_2 \neq x_3\}$
    $B = \{(x_1, 1, x_3) : x_i = 2, 3, \ldots, 8, \; i = 1, 3, \; x_1 \neq x_3\}$
    $C = \{(x_1, x_2, x_3) : x_i = 2, 3, \ldots, 8, i = 1, 2, 3, x_i \neq x_j\}$

12. No

14. $P(A \cap B) = 1 - .8 = .2; \; P(A) = .3 + .2;$
    $P(B) = .2 + .2, \; P(A \cup B) = .3 + .2 + .2$

16. $P(S) = 1 = P(A_1) + 2P(A_1) + \cdots + 5P(A_1) = 15P(A_1), \; P(A_1) = \frac{1}{15}.$

Section 1.3

2. $\frac{1}{21}, \frac{2}{21}, \frac{3}{21}, \frac{4}{21}, \frac{5}{21}, \frac{6}{21}$

4. The probability either one wins is $\frac{15}{36}$; the probability of no winner is $\frac{6}{36}$.

6. The 13th falls on a Friday means the 6th, 20th, and 27th do as well. These four dates are the most likely for a Friday.

8. 2^8 persons

10. $\frac{1}{5}, \frac{1}{5}, \frac{1}{25}, \frac{1}{5}$

12. One such assignment is $\frac{1}{6}, \frac{1}{12}, \frac{1}{6}, \frac{1}{12}, \frac{1}{6}, \frac{1}{6}, \frac{1}{12}, \frac{1}{12}$ with the 3-tuples listed in the order given in Example 1.18. The probabilities of winning the first games then are $\frac{7}{12}$ for $\mathcal{A}$ and $\mathcal{D}$, $\frac{5}{12}$ for $\mathcal{B}$ and $\mathcal{C}$.

14. No. Use instead $S_1 = T \times T \times T$, where $T = \{h, t\}$ (standing for heads, tails, respectively) leading to $\frac{1}{8}, \frac{3}{8}, \frac{3}{8}, \frac{1}{8}$ as the single-element event probabilities for the S mentioned in the exercise.

16. The other four splits are $[(1, 5, 9), (2, 6, 7), (3, 4, 8)], [(1, 6, 8), (2, 3, 9), (4, 5, 7)], [(1, 6, 8), (2, 4, 9), (3, 5, 7)],$ and $[(1, 7, 8), (2, 4, 9), (3, 5, 6)]$.

Section 1.4

2. $4!/6! = 1/30.$

4.

4. There are 30 2-tuples for the position numbers of these two individuals, of which 10 have them side by side; $\frac{1}{3}$.

6. $2(4)(3)(4) = 96$

8. $9! = 362{,}880, \quad {}_9P_6 = 60{,}480,$
   $60{,}480 - 720 = 59{,}760, \quad 720/60{,}480 = .0119.$

10. a. $\binom{13}{1}\binom{4}{3}\binom{12}{1}\binom{4}{2} = 3744$

    b. $\binom{4}{1}\binom{13}{5} = 5148.$ (This number includes the 40 straight flushes.)

    c. $\binom{10}{1}4^5 = 10{,}240.$ (This number includes the 40 straight flushes.)

    d. $\binom{13}{1}\binom{4}{3}\binom{12}{2}\binom{4}{1}\binom{4}{1} = 54{,}912$

    e. $\binom{13}{2}\binom{4}{2}\binom{4}{2}\binom{44}{1} = 123{,}552$

    f. $\binom{13}{1}\binom{4}{2}\binom{12}{3}4^3 = 1{,}098{,}240$

12. a. $\binom{9}{5} = 126$ b. $\frac{70}{126} = .556$ c. $\frac{5}{126} = .040$

14. a. $\binom{44}{11}$ b. $\binom{7}{1}/210$ c. $1/210$

16. | 1 | 8 | 28 | 56 | 70 | 56 | 28 | 8 | 1 | | |
    |---|---|----|----|-----|-----|-----|----|----|---|---|
    | 1 | 9 | 36 | 84 | 126 | 126 | 84 | 36 | 9 | 1 | |
    | 1 | 10 | 45 | 120 | 210 | 252 | 210 | 120 | 45 | 10 | 1 |

18. 11 is not a power of 2.

20. 2^{k-1}.

22. $1 - (1 - e^{-1}) = e^{-1} = .36788$

24. $\binom{10}{3\,3\,2\,1\,1} = 50{,}400$
    a. $3360/50{,}400 = .067$ b. $360/50{,}400 = 1/140$

26. $101/\binom{20}{10} = .00055$

28. $4 + 14 + 42 + 102 + 180 + 180 = 522$

32. a. $\binom{100}{10}$ b. $\binom{60}{6}\binom{40}{4}/\binom{100}{10}$ c. $.63855$

Section 1.5

2. $\frac{1}{2}$

4. $\frac{4}{7}$

6. $A \cup B = A \cup (B \cap \overline{A})$

8. a. $\binom{2}{2}\binom{98}{2}/\binom{100}{4}$ b. $\frac{2}{100}$

12. a. $\frac{1}{3}$ b. No

14. a. $(.9)(.8)(.7)(.6) = .3024$ b. $.1, .18$

16. $\frac{5}{15}(\frac{15}{25}) + \frac{10}{15}(\frac{5}{14}) = \frac{46}{105}$

20. $(.99)(.95)/((.99)(.95) + (.05)(.01)) = .9995$

22. a. $\frac{151}{260} = .581$ b. $\frac{71}{144}, \frac{80}{116}$ c. No

Section 1.6

4. a. $1 - (.1)(.2)(.3) = .994$ b. $(.1)(.2)(.7) = .014$
   c. $(.9)(.2)(.3)/.994 = .0543$

6. $(.8)^3 = .512$

8. $(.9)^4 = .6561; 4(.1)(.9)^3 = .2916$

10. *a.* $.4^2 = .16$ *b.* $(.6)(.5) = .3$
   c. $(.4)(.5)(.6) + (.6)(.5)(.4) = .24$

12. $.172$

14. $\frac{67}{256} = .2617$

16. $.3, .260, .252$

Section 1.7

2. $T = \{r, b, g, y\}; S = T \times T \times T;$
   $A = \{(r, r, r), (b, b, b), (g, g, g), (y, y, y)\}$

4. $.7, .26, .03, .01$

6. $\frac{1}{216} \times 1, 3, 6, 10, 15, 21, 25, 27, 27, 25, 21, 15, 10,$
   $6, 3, 1$

8. $11!/(2!)^2 = 9,979,200; 9!/(2!)^2 = 90,720$

10. $10^3 = 1000$

12. $6(5!)/6^5 = .0926$

14. $6!/6^6 = .0154$

16. $1 - (\frac{8}{9})^5, 1 - (\frac{8}{9})^9$

Chapter 2   Random Variables

Section 2.1

2. $p_X(x) = \begin{cases} \frac{1}{12}; & \text{for } x = 6, 7, \ldots, 17 \\ 0, & \text{otherwise} \end{cases}$

4. *a.* $\binom{5}{3}$
   b. $R_W = \{1, 2, 3, 4\}$, and the probabilities of these values occurring are .4, .3, .2, .1, respectively.
   c. $.4$

6. $p_X(x) = (2x - 1)/36$, for $x = 1, 2, \ldots, 6$.

8. *a.* $p_X(x) = \frac{1}{4}$ for $x \in R_X$.
   b. A person chooses one of the four titles at random to purchase; X is the price paid.

10. $R_Y = \{0, 1, 2, 3, 4\}$; $p_Y(y) = $
   $\binom{4}{y}(.1)^y(.9)^{4-y}$, for $y \in R_Y$

12. $R_X = \{0, 1, 2, 3\}$, $p_X(x) = \binom{3}{x}\left(\frac{1}{2}\right)^3$, for $x \in R_X$

14. $p_N(n) = \frac{1}{6}\left(\frac{5}{6}\right)^{n-1}$, for $n = 1, 2, 3, \ldots$;
   $P(N \leq 3) = \frac{91}{216}$; $P(N > 1) = 1 - P(N = 1) = \frac{5}{6}$.

16. *a.* $r = 1$
   b. $\sum_{y=1}^{5} p_Y(y) = .6315$; $\sum_{y=5}^{\infty} p_Y(y) = .0037$

Section 2.2

2. $R_Y = \{0, 1, 2, \ldots, 9\}$ and $Y + 1$ is discrete uniform with parameter 10.

4. $p_X(x) = \begin{cases} \frac{2}{5}, & \text{for } x = 1 \\ \frac{3}{5}, & \text{for } x = 2 \\ 0, & \text{otherwise} \end{cases}$

$F_X(t) = \begin{cases} 0, & \text{for } t < 1 \\ \frac{2}{5}, & \text{for } 1 \leq t < 2 \\ 1, & \text{for } t \geq 2 \end{cases}$

6. $p_T(t) = \begin{cases} .6, & \text{for } t = 0 \\ .3, & \text{for } t = 1 \\ .1, & \text{for } t = 2 \end{cases}$

8. $F_V(t) = p\dfrac{1 - (1-p)^{\lfloor t \rfloor}}{1 - (1-p)} = 1 - (1-p)^{\lfloor t \rfloor}$, for $t > 0$

10. $f_U(u) = \begin{cases} u - 1, & \text{for } 1 < u < 1.5 \\ \frac{3}{4}, & \text{for } 1.5 \leq u \leq 2.5 \\ 3 - u, & \text{for } 2.5 < u < 3 \\ 0, & \text{otherwise} \end{cases}$

$F_U(t) = \begin{cases} 0, & \text{for } t < 1 \\ \dfrac{(t-1)^2}{2}, & \text{for } 1 \leq t \leq 1.5 \\ \dfrac{3t - 4}{4}, & \text{for } 1.5 < t < 2.5 \\ 1 - \dfrac{(3-t)^2}{2}, & \text{for } 2.5 < t \leq 3 \\ 1, & \text{for } t > 3 \end{cases}$

12.

a. $f_X(t) = \begin{cases} \dfrac{t}{3}, & \text{for } 0 \le t \le 2 \\ \dfrac{2(3-t)}{3}, & \text{for } 2 < t \le 3 \\ 0, & \text{otherwise} \end{cases}$

b. $P(X > 1) = 1 - F_X(1) = \frac{5}{6}$

14. $F_U(t) = \dfrac{1}{3} + \dfrac{t}{6} + \dfrac{(t-1)^3}{3}$, for $0 \le t \le 2$

16. Only the fourth of these functions is a *cdf*.

Section 2.3

2. *a.* $p_X(x) = \frac{1}{4}$ for $x = -2, -1, 1, 2$
   b. $E[X] = 0; \sigma_X^2 = 2.5$

4. $E[Y] = .75; \sigma_Y^2 = \frac{35}{48}$.

8. *a.* $R_x = \{1, \frac{1}{2}, \frac{1}{3}, \frac{1}{4}, \frac{1}{5}, \frac{1}{6}, 2, \frac{2}{3}, \frac{2}{5}, 3, \frac{3}{2}, \frac{3}{4}, \frac{3}{5}, 4, \frac{4}{3}, \frac{4}{5},$
   $5, \frac{5}{2}, \frac{5}{3}, \frac{5}{4}, \frac{5}{6}, 6, \frac{6}{5}\}$.
   b. The probabilities are $\frac{1}{36} \times 6, 3, 2, 1, 1, 1, 3, 2, 1,$
   $2, 2, 1, 1, 1, 1, 1, 1, 1, 1, 1, 1, 1, 1$.
   c. 1.429

10. 4.5, .6708

12. $E[U] = (b+a)/2; \sigma_U^2 = (b-a)/\sqrt{12}$

14. The means are both $(n+1)/2$; the variances are
    $(n^2 - 1)/12, (n-1)^2/12$

16. $E[X] = 2b/3; \sigma_X = b/3\sqrt{2}$. Prefer X if $b \le \sqrt{3}$.

18. $E[U] = 1; \sigma_U^2 = \frac{23}{45}$.

20. $\mu_Y = \frac{5}{9}(\mu_X - 32); \sigma_Y = \frac{5}{9}\sigma_X; y_k = \frac{5}{9}(x_k - 32)$

Section 2.4

2. $t_{.5} = b/\sqrt{2}; r_{iq} = b(\sqrt{3} - 1)/2$. Median X is
   smaller for $b < 2\sqrt{2}$.

4. $t_k = a + (b-a)k = a(1-k) + bk$

6. $t_k = \begin{cases} \frac{3}{4} + \sqrt{3k}/4, & \text{for } 0 < k \le \frac{1}{3} \\ \dfrac{7+3k}{8}, & \text{for } \frac{1}{3} < k \le 1 \end{cases}$

8. $Y = \frac{5}{9}(X - 32)$, a linear relationship. Thus
   $y_k = \frac{5}{9}(x_k - 32)$ and $_Y r_{iq} = \frac{5}{9} \, _X r_{iq}$.

10. Both equal 1.

Section 2.5

2. *a.* $(\frac{5}{6})^5(\frac{1}{6}) = 5^5/6^6$.   *b.* $1 - (\frac{5}{6})^5$.   *c.* $\frac{6}{11}$.

4. *a.* $\frac{4}{10}$ *b.* $-.2$, \$.20.

6. Yes, continuous

8. $R_Y = \{1, 3, 6\}$; and $p_Y(y) = \frac{1}{3}, \frac{1}{2}, \frac{1}{6}$, respectively,
   at these values.

10. *a.* $k = \sqrt{3}$ *b.* $\frac{2}{3}$

12. *a.* $t_{.5} = 0, r_{iq} = 1.587$ *b.* $\mu_Y = 0, \sigma_Y = \sqrt{3/5}$

14. $\mu_X = \frac{16}{6}, \sigma_X^2 = \frac{11}{9}$

16. $R_X = \{-3 < t < 1\}; f_X(t) = (3+t)/8;$
    $x_k = -3 + 4\sqrt{k}$

Chapter 3   Discrete Probability Distributions

Section 3.1

2. *a.* X is binomial, with $n = 6$ and $p = \frac{1}{2}$.
   b. $P(X \ge 1) = 1 - \left(\frac{1}{2}\right)^6$ *c.* No

4. $E[Y] = 20(.2) = 4; \sigma_Y^2 = 20(.2)(.8) = 3.2$

6. $q = .5 = p; n = 24; .271$

8. *a.* Binomial, with $n = 5$ and $p = .3$ *b.* 1   *c.* No

10. .5055

12. *a.* Binomial, $n = 15, p = \frac{1}{3}$ *b.* .0882

14. $R_X = \{0, 1, 2, 3\}$; the probabilities of occurrence
    for these values are .315, .485, .185, .015,
    respectively. No.

16. 970, 970, 5.9×10^{-14}. These values all assume
    that the probability of a good loan is .97 for
    each and that the results are independent from
    loan to loan.

18. 2^{-10}

Section 3.2

2. *a.* Geometric, $p = .2$ *b.* .6723

4. 12.5, 3.125

6. *a.* Negative binomial, $r = 3$, $p = .2$
   b. 0.008, 0.0192, 0.03072, 0.04096, 0.049152

8. $\mu_X = q/p$, $\mathrm{Var}[X] = q/p^2$.

10. 7

12. Discrete uniform with $n = 5$

14. $R_X = \{x : x = 1, 2, \ldots, M - r + 1\}$;

$$p_X(x) = \frac{r}{M} \frac{\binom{M - r}{x - 1}}{\binom{M - 1}{x - 1}}.$$

16. Let $N = M + 1$ if no calls occur; then
    $p_N(n) = \frac{1}{3}\left(\frac{2}{3}\right)^{n-1}$, for $n = 1, 2, \ldots M$; and
    $P(N = M + 1) = \left(\frac{2}{3}\right)^M$. As long as $M \geq 4$, the
    answers to parts a and b remain unchanged.

18. .980; the Chebyshev bound is $\frac{8}{9}$.

Section 3.3

2. T is Poisson with $\mu = 12$, so $P(T = 12) = .1144$.

4. *a.* .6703   *b.* .0536   *c.* .8088

6. 2

8. .8347

10. .0563 vs .25

12. .0242

14. .576 hours, or 34.6 minutes

16. Result follows immediately from the series given
    in the hint.

20. *a.* X is Poisson, λt; Y is Poisson, $2\lambda t$.
    b. $P(2X = 2) = \lambda t e^{-\lambda t}$; $P(Y = 2) = 2\lambda^2 t^2 e^{-2\lambda t}$
    c. $\sum_{x=0}^{\infty} (P(X = x))^2$.

Section 3.4

2. $\frac{1}{2}$

4. *a.* .536   *b.* .911

6. $n = 10$

12. .074, .421

14. Binomial, $n = .01m, p = .1$; $\mu_X = .001m$;
    $\sigma_X^2 = .0009m$

Section 3.5

2. .8674, .9992

4. X is geometric, $p = .001$;
   $E[X] = 1000$; $P(X > 1500) = .999^{1500} = .223$.

6. Hypergeometric, $m = 400, r = 15, n = 20$, $\mu_X = \frac{3}{4}$,
   $\sigma_X^2 = .829$

8. Hypergeometric, $m = 10,000, r = 6000, n = 200$,
   120

10. .149

Chapter 4   Continuous Distributions and Transformations

Section 4.1

2. *a.* $E[T_1] = 40$ *b.* $t_5 = 40 \ln 2 = 27.73$
   c. $P(T_1 > 72) = e^{-72/40} = .1653$

6. *a.* $\mu_W = 5/\ln 3 = 4.5514$; $t_5 = 3.1546$ weeks
   b. 16.674 weeks   *c.* .1973

8. *a.* $\mu_{T_2} = 80$ hours; .5578   *b.* $\mu_{T_3} = 120$ hours;
   .8088

12. $n = 6.25, \lambda = 2.5$

14. 1.5 minutes; $\sigma = 1.035$

16. 13

Section 4.2

2. $z_{(1+\gamma)/2}$

4. *a.* .0202   *b.* 4.27 inches   *c.* 1.11 inches

6. *a.* .9332   *b.* 74.71 inches

8. *a.* $(-\infty, 1.282), (-1.282, \infty), (-1.645, 1.645)$
   b. $(-1.645, 1.645)$

10. *a.* .1587   *b.* .0668   *c.* 1.69 gallons

16. $P(X \geq 30) = .1841, P(X < 20) = .1357$.

Section 4.3

2. $\mu_{X_{(2)}} = \frac{2}{3}$, $\sigma^2_{X_{(2)}} = \frac{1}{18}$

4. $\alpha = \frac{21}{25}; \beta = \frac{14}{25}$.

6. Finding the maximum requires
$[(\alpha - 1)(1 - x) - (\beta - 1)x] = 0$, so
$x = (\alpha - 1)/(\alpha + \beta - 2)$ if $\alpha + \beta \neq 2$.

8. The value that maximizes the *pdf* for $X_{(5)}$, the middle ranked value is $\frac{1}{2}$; $\frac{7}{8}$.

10. $\mu = .3$, $\sigma = .0847$.

Section 4.4

2. $P(W = -1) = .5625, P(W = 1) = .4375$;
$E[W] = -.125$

4. $f_W(t) = \lambda e^{-\lambda(t-a)}$, for $t > a$

8. a. $g(X) = a + bX$

b. $\alpha = \dfrac{\mu - a}{b + a - \mu}$;

$\beta = \dfrac{1}{b}\left[\dfrac{(\mu - a)(b + a - \mu)^2}{\sigma^2} - b - a + \mu \right]$

12. $F_W(t) = 1 - e^{-\lambda \ln(1+\beta t)} = 1 - 1/(1 + \beta t)^\lambda$, for $t > 0$

14. χ^2 with one degree of freedom.

16. $w_k = \alpha x_k^{1/\beta} = \alpha[-\ln(1 - k)]^{1/\beta}$.

18. Folded standard normal; $P(X < .9) = .6318$

20. $t_k = z^2_{(k+1)/2}$

22. Geometric, $p = 1 - e^{-\lambda}$

24. $P(Y = j) = e^{-j^3} - e^{-(j+1)^3}$, for $j = 0, 1, 2, \ldots$

Section 4.5

2. $.1333$

4. $.715$

6. $e^{-5/2}$, $.173$, 8:02 AM

8. $F_U(t) = 1 - e^{-\lambda\sqrt{t}}$, for $t > 0$

10. $1 - (1 - 1/n)^n \to 1 - e^{-1}$ as $n \to \infty$

12. $F_W(t) = \begin{cases} \frac{3}{4}t, & \text{for } 0 < t < 1 \\ 1 - \frac{1}{16}(-t + 3)^2, & \text{for } 1 \leq t < 3 \end{cases}$

$f_W(t) = \begin{cases} \frac{3}{4}, & \text{for } 0 < t < 1 \\ \frac{1}{8}(3 - t), & \text{for } 1 \leq t < 3 \end{cases}$

Chapter 5   Jointly Distributed Random Variables

Section 5.1

2. a. $.9^{15} = .2059$ b. $.00143$.
These answers assume the persons form independent multinomial trials, with $p_1, p_2, p_3 = .1, .3, .6$.

4. a. $.0975461$ b. $.21886$ c. $x = 1$

6. $p_{X_1, X_2}(x_1, x_2) = p^2 q^{x_2 - 2}$; X_1 is geometric, parameter p; X_2 is negative binomial with parameters $r = 2, p$.

8. $f_X(x) = 2e^{-2x}$, for $x > 0$; $f_Y(y) = 3e^{-3y}$, for $y > 0$; $P(X > Y) = \frac{3}{5}$

10. Both marginals are triangular;

$f_X(x) = \begin{cases} 1 + x, & \text{for } -1 < x \leq 0 \\ 1 - x, & \text{for } 0 < x < 1 \end{cases}$

12. a. $f_{X,Y}(x, y) = (x + y + 1)/2$, for
$0 < x < 1, 0 < y < 1$
b. $f_{X,Z}(x, z) = (2x + 3)/4$, for
$0 < x < 1, 0 < z < 1$
c. $f_Z(z) = 1$, for $0 < z < 1$

Section 5.2

2. The probability of no heads is $\frac{7}{24}$.

4. a. $p_{X_1, X_2}(x_1, x_2) = \frac{1}{53}P_2$, for $x_i = 1, 2, \ldots, 53$,
where $i = 1, 2$, and $x_1 \neq x_2$
b. $p_{X_2}(x_2 \mid x_1) = \frac{1}{52}$, for $x_1, x_2 = 1, 2, \ldots, 53$, and
$x_1 \neq x_2$.
c. By symmetry, the same as in part b; identical with the answer for part b.

6. a. $(.1)^3$ for each of the 1000 possible triples
$(0, 0, 0), (0, 0, 1), \ldots, (9, 9, 9)$
b. X is discrete with $R_X = \{.000, .001, \ldots, .999\}$
and $p_X(x) = .001$ for $x \in R_X$.
c. $P(X \leq \frac{1}{3}) = .334$

8. X is binomial $\mathcal{N}, p\pi$

10. Y is Poisson, $p\mu = p\lambda t$

12. Binomial, parameters $n - n_1$ and p

14. $P(T_1 \leq 1 \mid T_2 > 2) = e^{-2}/3e^{-2} = \frac{1}{3}$.

16. Uniform on the integers $1, 2, \ldots, x_2 - 1$

Section 5.3

2. *a.* Y_3 is binomial $n = 15, p = \frac{1}{5}$.
   b. Y_2, Y_4 is multinomial $n = 15, p_2 = p_4 = \frac{1}{5}$.
   c. .1622   *d.* .0055

4. Each X_i is hypergeometric, $m = 8000$, $n = 100$, and the appropriate r

6. Multi-hypergeometric with parameters $m = 5400$, $r_2 = 2100, r_3 = 1900, r_4 = 1400$, and $n = 67$; conditional probability $= .0126$

8. X_1, X_2, X_3, X_4 is multinomial with $n = 400$ and p_i values $\frac{30}{125}, \frac{20}{125}, \frac{40}{125}, \frac{35}{125}$

10. $C_1, C_2, \ldots, C_6$ is multinomial with $n = 200$ and p values .0227, .1360, .3413, .3413, .1360, .0227

12. $c = \sigma\sqrt{2\ln 2} = 1.177\sigma$, where c is the desired radius

14. $X_1, X_2, \ldots, X_4$ is multi-hypergeometric, with parameters $m = 500$; $r_1, r_2, r_3, r_4 = 400, 40, 50, 10$; and $n = 20$. The conditional probability function for X_1, X_2, X_3, given $X_4 = 1$, is multi-hypergeometric with $m = 490, r_1 = 400, r_2 = 40, r_3 = 50$, and $n = 19$

16. *b.* No

Section 5.4

2. *a.* $f_X(x) = 6x(1 - x), 0 < x < 1$; $f_Y(y) = 2y, 0 < y < 1$ *b.* Yes

4. $f_X(x) = 2(1 - x)$, $0 < x < 1$

Chapter 6   Expectation, Moments

Section 6.1

2. 7, 12.25, 1.429166667

4. $\mu = \sum_{i=1}^{n} \mu_i = \sigma^2$

6. The means are 0, 10; the variances are 3, 23; the covariance is -5.

10. 42.4

12. $\text{Cov}[T_1, T_2] = 2$; $\rho = .816$

14. *a.* $a_1 = -a_2 = 1/\sqrt{2}$
    b. Any linear function with $b_1 = b_2 = b$;
    $\mu_Z = 6b$, and $\sigma_Z = |b|\sqrt{6}$

Section 6.2

2. $m_1 = (n + 1)/2$; $m_2 = (n + 1)(2n + 1)/6$

4. $m_Y(t) = pe^t/(1 - qe^t)$; $\psi_Y(t) = pt/(1 - qt)$

6. $P(G = -1) = P(G = 1) = \frac{1}{2}$.

8. $c_X(t) = r\ln pe^t - r\ln(1 - qe^t)$

10. $\psi_Y(t) = t^a\psi_X(t^b)$

12. $m_2 = \mu_2 + \mu_X^2$, $m_3 = \mu_3 + 3\mu_X\mu_2 + \mu_X^3$

14. X is degenerate with $P(X = 100) = 1$.

18. $m_X(t) = (pt)^r/(1 - qt)^r$; negative binomial, with parameters r, p

20. $m_V(t) = e^{t\sum_i a_i\mu_i + t^2\sum_i a_i^2\sigma_i^2/2}$, normal, mean $\sum_{i=1}^{n} a_i\mu_i$, variance $\sum_{i=1}^{n} a_i^2\sigma_i^2$

22. All the p_i values are equal.

24. V is χ^2 with $2n$ degrees of freedom.

Section 6.3

2. $\text{E}[Y] = \frac{2}{3}$; $\text{Var}[Y] = \frac{5}{9}$

4. *a.* $\text{E}[Y\,|\,x] = rx$; $\text{E}[XY] = \dfrac{r}{b}$ *b.* $\sqrt{r/(r + 9)}$

8. $\text{E}[X_2\,|\,x_1] = 0$; $\text{E}[X_1\,|\,x_2] = (1 - |x_2|)/2$;
   $\text{Cov}[X_1, X_2] = 0$

12. μx_1, $\sigma^4 + \mu^2\sigma^2$

14. $x_1 - \mu - \rho(x_1 - \mu) = g(x_1)$; $\sigma^2(1 - \rho^2)$

16. 10,000; $\sigma = \sqrt{1,020,000} = 1009.95$

Section 6.4

2. *a.* Erlang with $r = 3$, and $\lambda = \frac{1}{5}$
   b. $P(X_1 < 3) = .4512$, $P(\overline{X} < 3) = .2694$

4. 3.45

6. X is discrete uniform with $n = 6$

8. $\psi_X(t) = p^r/(1 - qt)^r$

10. *a.* $\dfrac{\mu_1^{x_1}\mu_2^{x_2}\mu_3^{x_3}}{x_1!x_2!x_3!}e^{-(\mu_1+\mu_2+\mu_3)t}$
    b. Y is Poisson with parameter $(\mu_1 + \mu_2 + \mu_3)t$.
    c. Multinomial with parameters y, and $\mu_i/(\mu_1 + \mu_2 + \mu_3)$ where $i = 1, 2, 3$.

Chapter 7   Transformations and Limit Theorems

Section 7.1

2. a. $f_{R_{(1)}}(r) = nre^{-nr^2/2}$; $f_{R_{(n)}}(r) = nre^{-r^2/2}(1 - e^{-nr^2/2})^{n-1}$

   b. $m_k = \sqrt{-(2/n)\ln(1-k)}$;
      $M_k = \sqrt{-2\ln(1-k^{1/n})}$

   c. The minimum

4. .4712, .6738

6. .3694

Section 7.2

2. $R_P = \{10p : p = 0, 1, 2, \ldots, 15\}$, each with probability $\frac{1}{20}$, except $P(P = 30j) = \frac{1}{10}$ where $j = 1, 2, 3, 4$.

6. Lognormal

8. $f_X(t) = \begin{cases} \dfrac{t^2}{2} & \text{for } 0 < t < 1 \\ \dfrac{1-(t-1)^2}{2} - (2-t)^2/2 & \text{for } 1 < t < 2 \\ \dfrac{(3-t)^2}{2} & \text{for } 2 < t < 3 \end{cases}$

10. $F_D(t) = 1 - e^{-2\lambda t}$, for $t > 0$

Section 7.3

2. Yes, to $(a+b)/2$

4. Yes, to r/p

6. Yes, to b

8. Yes, to 0

10. Yes, to 1

12. Yes, to the exponential probabilty law with parameter λ

14. $c = 77.4$; $(822.6, 977.4)$

16. .1075

18. .416

20. .9279 from normal, .9302 from Poisson

Section 7.4

2. $t_k = \tan^2 \dfrac{\pi}{2}(1+k)$

6. $f_Y(t) = \displaystyle\int_{-\infty}^{\infty} \dfrac{1}{|x_2|} f_{X_1}\left(\dfrac{t}{x_2}\right) f_{X_2}(x_2)\, dx_2$

8. $f_Y(t) = t/(1+t^2)^2$, for $t > 0$

10. $r = 12$; $P(X > 121) \doteq .482$

INDEX